Cable Modems:
Current Technologies and Applications

Advances in the Information Industry Series

IEEE PRESS

Presented by the
International Engineering Consortium
and the
Institute of Electrical and Electronic Engineers

ISBN 0-7803-5395-1

9 780780 353954 90000

IEEE ORDER NO. PC5829

ISSN 0886-229X
IEEE ISBN: 0-7803-5395-X
IEEE Order Number: PC5829

International Engineering Consortium
549 West Randolph, Suite 600
Chicago, Illinois 60661-2208 USA
+1-312-559-4100 voice • +1-312-559-4111 fax
publications@iec.org • *www.iec.org*

This book and other books may be purchased at a discount when ordered from the publisher in bulk quantities. Contact IEEE Press Marketing, Attention: Special Sales, Piscataway, NJ 08855-1313 or fax +1-732-981-9334.

For more information about IEEE Press products, visit the IEEE Press home page at *http://www.ieee.org/organizations/pubs/press.*

Cable Modems: Current Technologies and Applications

Editors:

John Fijolek
Chief Network Architect
Cable Access Business Unit
3Com

Michelle Kuska
Director of Network Technology
TCI Business Alliance and Technology Co., Inc.

Venkata C. Majeti
President and Chief Executive Officer
Cadant

Kotikalapudi Sriram
Distinguished Member of Technical Staff
Bell Laboratories–Lucent Technologies, Inc.

Executive Editor:

Venkata C. Majeti
President and Chief Executive Officer
Cadant

Managing Editor:

Lisa S. Davidson
Content Development Manager
International Engineering Consortium

International Engineering Consortium

The nonprofit International Engineering Consortium is dedicated to advancing the information industry through the development and dissemination of knowledge. The Consortium sponsors the National and Western Communications Forums as well as ComForums and TecForums focusing on industry-specific issues. The Consortium's vital work can only be accomplished through the cooperation of the industry and university communities. The International Engineering Consortium is affiliated with seventy high-tech universities worldwide.

The annual National Communications Forum and the prior National Electronics Conference have made major contributions to the advancement of the information industry. For more than fifty years, tens of thousands of technical papers have been presented, covering virtually every facet of the electronic sciences. In recent years, emphasis has been placed on the latest developments in telecommunications and computers.

Institute of Electrical and Electronics Engineers

The Institute of Electrical and Electronics Engineers is the world's largest technical professional association with more than 335,000 members in nearly 150 countries. The IEEE promotes the engineering process of creating, developing, integrating, sharing, and applying knowlege about electro- and information technologies and sciences for the benefit of humanity and the profession. Through its members, it is a leading authority in areas ranging from aerospace, computers, and telecommunications to biomedical engineering, electric power, and electronic products including college texts, professional handbooks, monographs on leading-edge developments in technology, career guides, and reprints of collected papers.

University Program

The Consortium provided grants to more than 500 university professors in 1998, enabling their participation in continuing education forums. Professors use both information obtained at forums and contacts with professionals from the industry to develop and improve courses, laboratories, research, and university infrastructures. Corporations valuing improved university educational programs sponsor the University Program. Through sponsorship, corporations have a unique opportunity to influence academic programs and interact with professors and students, thus enriching their industrial experience.

Sustaining Sponsors

Bell Atlantic
BELLSOUTH

GeoPartners Research, Inc.

European Sponsors

Alcatel Network Systems • BT Network Systems • Ericsson Limited • IBM (UK)
Marconi Communications • Nortel (UK) • Proximus-Belgacom Mobile

Consortium Affiliated Universities

The International Engineering Consortium is affiliated with the following leading high-tech universities:

Arizona State University

Auburn University

University of California at Berkeley

Carnegie Mellon University

Case Western Reserve University

University of Colorado at Boulder

Clemson University

Columbia University

Cornell University

Drexel University

Georgia Institute of Technology

Howard University

Illinois Institute of Technology

Iowa State University

Lehigh University

Oklahoma State University

Oregon State University

Pennsylvania State University

Marquette University

Massachusetts Institute of Technology

Michigan State University

Mississippi State University

North Carolina State University at Raleigh

Northwestern University

Ohio State University

Poltechnic University

Purdue University

Renssalaer Polytechnic Institute

Stanford University

Syracuse University

University of Wisconsin at Madison

Texas A&M University

The University of Arizona

University of California at Davis

Univerisity of California at Santa Barbara

University of Illinois

University of Kansas

The University of Kentucky

University of Missouri at Columbia

University of Oklahoma

University of Southern California

The University of Tennessee at Knoxville

University of Florida

University of Illinois at Chicago

University of Maryland at College Park

University of Massachusetts at Amherst

University of Michigan

University of Minnesota

University of Notre Dame

University of Pennsylvania

University of Pittsburgh

University of Texas at Austin

University of Virginia

VA Polytechnic Institute and State University

University of Missouri at Rolla

Worldwide University Affiliates

Ecole Supérieure d'Electricité
Ecole Nationale Supérieure des Télécommunications de Bretagne
Ecole Nationale Supérieure des Télécommunications de Paris
Imperial College of Science, Technology & Medicine
Institut National Polytechnique de Grenoble
Instituto Tecnologico y Estudios Superiores de Monterrey
KAIST
McGill University
Technische Universität München
The Queen's University of Belfast
Universidad Nacional Autónoma de México
Université d'Ottawa
University of Southampton
University of Edinburgh
University of Glasgow
University College London
University of Toronto

Official Media Sponsors

America's Network | CED | Inter@ctive Week

Lightwave | Telephony | tele.com

Wireless Week | X-Change | Wireless Review

Official Media Partners

Business Communications Review
Communications News
Communications Technology
CTI
Global Telephony
Internet Telephony
Outside Plant
Private Cable & Wireless Cable
Satellite Communications
Sounding Board
Telecom Asia
Telecommunications
Via Satellite

Table of Contents

Part I: Overview and Process

Part II: The Physical Layer

Part III: The Medium Access Layer and Performance Issues

Part IV: The Network Routing Layer

Part V: Security

Foreword: The Role of Cable-Modem Technology in Enabling Broadband Data Services

Richard R. Green
President and Chief Executive Officer
CableLabs, Inc.

Cable-modem technology is ready to provide high-speed access to the Internet, World Wide Web, and broadband services. The cable-television industry's transition into data and Internet services is the result of a gradual and systematic effort by CableLabs, cable systems operators, and the vendor community who traditionally supply the telecommunications, data communications, and consumer-electronics markets. The cable industry's efforts in the late 80's and early 90's focused on digital compression and multimedia applications.[1] Research efforts focused on the star/bus topology of the distribution network, digital compression, high-definition television (HDTV), and data services. Innovations in the industry included impulse pay-per-view, games, and cable telephony as new value-added services.

The cable-television industry first attempted to provide data services by offering local-area network (LAN) interconnection over two-way cable networks to business customers. Initially expensive for mass markets, these systems enabled metropolitan-area networking at costs considerably less than those for private-line services. Innovations to support data services over one-way broadcast networks by using the telephone modem network as the return path quickly followed.

At CableLabs, our task is to help the industry reach consensus on service offerings and protocols for various systems deployed in the network. Cable operators have found cable networks to be well suited for expansion into the digital domain at low capital costs. The success of the Internet and the World Wide Web meant that cable data services must also work with systems developed for the data communications industry. This need for data communications–provider compatibility led to the formation of the data over cable service interface specification (DOCSIS) project. DOCSIS defines a reference architecture for the data-services offering and provides requirements for all interfaces within the data-over-cable system.[2] As a part of our commitment to the goal of interoperability, CableLabs has established a forum for interoperability testing among vendor equipment to ensure reliable, time-to-market deployment of broadband services.

The continuous growth of Internet access and the World Wide Web created a critical need for low-cost, high-speed access to data services. Cable operators and CableLabs are taking steps to ensure that this demand can be served to the benefit of residential and business users. In addition, the creation of @Home and other system integrators enable the industry to deploy end-to-end service offerings. It is now appropriate to say that the cable-modem technology has become an industry unto itself, poised to grow and serve an important need in the information society.

Cable Modems: Current Technologies and Applications is a valuable resource for spreading the knowledge of cable modem–technology development and innovation to present and future contributors to the industry. Because many of the authors helped shape the cable data service offering, the book provides a unique perspective based on knowledge and experience.

Cable-modem technology is complicated, and an understanding of the many layers of protocols is critical. *Cable Modems* offers a detailed description of the physical layer, including upstream and downstream communications, the media access control (MAC) layer, routing based on Internet protocol (IP), applications and services, performance, system design, and security of user data. In addition, the book supplies vital information about the network architecture and the evolution of cable data services and provides a perspective on the progress achieved to date. This collection of papers provides insight into all areas of cable-modem technology and thus serves as a valuable reference.

References

1. Richard R. Green, "Broadband Data Service-The Emerging Challenge," presentation delivered to the International Engineering Consortium and published in the Chief Executive Officer Perspectives, Special Edition, June 1997, pp. 75–78.
2. DOCSIS specifications are published at http://www.cablemodem.com. CableLabs is also involved in the testing of vendor equipment to ensure interoperability among multiple suppliers.

Introduction

Venkata C. Majeti
Cadant

John Fijolek
3Com

Michelle Kuska
TCI Business Alliance and Technology Co., Inc.

Kotikalapudi Sriram
Bell Laboratories Lucent Technologies, Inc.

Recent growth of the Internet and World Wide Web has created an awareness of and demand for high-speed access at mass-market prices. The cable data industry is well positioned to serve this market need by the creation of interoperable product standards and through the adaptation of a megabit service offering at $40-50 per month for unlimited Internet access. This book, *Cable Modems: Current Technologies and Applications*, provides insight into cable modems from the perspective of many key contributors to technology and the industry. The book is broadly divided into five subject areas: overview and process, the physical (PHY) layer, the medium access control (MAC) layer and performance issues, the network routing layer, and security.

The overview and process section presents cable data–system architecture, the evolution of the cable-modem industry, gradual industry acceptance of asymmetric systems, the role of data over cable service interface specification (DOCSIS), and legal aspects such as patents, trademarks, and copyrights. Rouzbeh Yassini describes the first experiments with cable-modem technology and the birth of cable-data industry. Venkata Majeti presents early deployment experiences in symmetric and asymmetric cable-data systems and the development of telephone return systems for one-way broadcast cable networks. Michelle Kuska reviews the formation of the Multimedia Cable Network Systems (MCNS) partnership and its objectives leading to DOCSIS requirements, and the process for ongoing maintenance and evolution of requirements. Steve Grzelak et al. present cable data services from the perspective of hybrid fiber/coax (HFC) systems.

Cable modems use a quadrature amplitude modulation, QAM–64 and QAM–256, for the downstream data channel from network to user and quaternary phase shift keying (QPSK) and QAM–16 for the upstream data channel from user(s) to the network. The PHY layer section offers details of downstream and upstream modem technologies. Bruce Currivan reviews the PHY layer design, key tradeoffs that led to major design decisions, and the MCNS DOCSIS and IEEE 802.14 PHY specifications. Victor Hou et al. present a time division multiple access (TDMA) synchronization method for the upstream-data channel to manage the transmission of data from multiple users to the network over a shared medium. This scheme enables independent upstream- and downstream-data channels and scalability of network design. Thomas Kolze presents a complete overview of the upstream-data channel modeling with a discussion of modulation, channel performance, and frequency and time division multiple access (F/TDMA). Krista Jacobson and John Bingham offer insights into the advanced PHY layer in which a synchronized discrete multitone (SDMT) modulation is offered as an alternative to a QPSK/QAM–16 upstream data channel. Jonathan Min and Henry Samueli present an evolution from the current TDMA system by offering an advanced, frequency-agile TDMA system for upstream-data channel.

The MAC layer ensures data integrity in the shared media environment. In a cable-access network, requirements on the MAC layer are most challenging because of the separation of upstream- and downstream-data channels and the large geographic distances of up to 100 miles. Basic requirements on the MAC layer include a method of contention and collision resolution, a means to maintain synchronization, support for fixed and variable-length packets, and the ability to support a wide range of data rates and services. Chris Grobicki and John Ulm present early design experiences in the cable-modem MAC layer. Kotikalapudi Sriram presents a detailed overview of the MAC protocol and its performance and describes the key features required for efficient voice, data, and multimedia services over cable networks. Nada Golmie et al. review contention resolution algorithms presented to the IEEE 802.14 committee and offer a discussion of the MAC performance.

The DOCSIS requirements set precedence to interoperable cable-data systems. Most of the IEEE 802.14 MAC–layer discussions center around the use of ATM cells and support for multiple services such as constant bit rate (CBR), variable bit rate (VBR), unspecified bit rate (UBR), and available bit rate (ABR). The DOCSIS requirements take on a service focus and a ubiquitous IP–based network design for access to the Internet, World Wide Web, and other value-added services. David Hartman and Thomas Quigley present an overview of the MAC layer and implementation and performance considerations for the DOCSIS requirements. Levent Gun and Chester Ruszczyk study the performance of telephone return on one-way–cable networks and conclude that a lower bandwidth–upstream channel does not severely limit the Megabit services offered in the downstream by cable networks.

Access to the Internet and World Wide Web services requires use of transmission control protocol/Internet protocol (TCP/IP) as a communication and networking layer. Special attention is required for IP–address assignment, management, physical and logical network configuration, routing, and security and scalability of network design. Enrique Hernandez-Valencia presents a discussion on cable-network architecture for IP services and offers architecture models for one-way cable with telephone return and two-way cable-data systems. Nurettin Beser discusses the differences between two-way–cable and telephone-return systems from the perspective of dynamic IP–address assignment. Jack Fijolek provides a snapshot of the Internet Engineering Task Force (IETF) efforts relative to cable-data systems and possible future directions for the telephone-return systems. The IP discussions on IP–data services over cable range from dynamic address assignment and management by the use of point-to-point protocol (PPP) and dynamic host configuration protocol (DHCP) to the use of network-address translation, authentication, and quality of service. The technology developed for traditional cable-television networks is also used to provide data services over wireless cable and satellite systems with telephone return (e.g., Hugh's DirectPC services). Hybrid Networks conducted early wireless cable data system trials using wireless cable with a telephone return architecture. Under the wireless licenses, a system operator could use a cable modem for delivering high-speed Internet services on a 6 MHz spectrum using wireless cable networks. Robert Furniss and Keith Ross discuss broadband wireless technology and argue that a service provider may use this technology to complement or compete with cable-television–based technology to provide high-speed data access to small- and medium-sized businesses and high-end customers.

Security is often a main concern with cable-data systems due to the broadcast nature of the distribution network. Cable-data systems address these concerns through measures to ensure the integrity of link-layer data communications across the shared medium, to protect the privacy of user data (from being accessed by others), and to offer service protection from piracy. Chet Birger reviews the need for security in cable-data systems and presents security requirements adopted by the MCNS committee. The MCNS baseline privacy is a low-cost security solution to offer privacy, data integrity, and weak-service protection based on MAC addresses. Robert Rance explores scenarios of security violation in a cable-data system and offers a security architecture based on authentication of the user and encryption of user data.

The perspectives discussed in the book offer a unique insight into cable-data systems. Many of the authors were active participants in shaping cable-modem technologies and the cable-data industry. The discussions of technical and organizational issues described herein benefit from the authors' involvement and first-hand knowledge. Rouzbeh Yassini is the founder and former chief executive officer of LANcity, a company recognized for early contributions to cable-data systems. Michelle Kuska chaired the MCNS working committee that helped create the DOCSIS requirements. Venkata Majeti led Bell Laboratories' early efforts in cable-data systems and the working relationship with Hybrid Networks. Henry Samueli and the Broadcom Corporation are recognized for technical contributions and industry leadership in bringing cable-modem technology to market. Bruce Currivan coordinated the IEEE 802.14 PHY group; his technical knowledge and leadership resulted in expeditious agreements on the PHY standard. Chris Grobicki, John Ulm, Kotikalapudi Sriram and Nada Golmie made significant contributions to the understanding of the MAC layer. Jack Fijolek is an author and a key contributor to DOCSIS cable-modem telephone return–interface requirements. Chet Birger is the author of DOCSIS security requirements.

We hope that the book provides the readers with an overall understanding of cable-modem technology. Future editions of this book will capture ongoing developments and experience with system integration, services,and an end-to-end view of the cable-access networks.

The editors are grateful to Dan Coran in preparation of the book material and Curtis Siller of Lucent Technologies for his review, comments, and assistance in the final stages of publication.

Part I: Overview and Process

History and Evolution of Asymmetric Cable-Modem Systems

Venkata C. Majeti
President and Chief Executive Officer
Cadant

Abstract

The asymmetric cable modem system, particularly for cable data services, was a design concept that originated in 1992. LANcity and Zenith designed and marketed "symmetric LAN over cable" systems that could be deployed in selected areas of cable industry. This required one upstream channel of 6-MHz bandwidth for every downstream channel before data services could be brought to market. The author built a team with the objective of bringing cable data services to mass markets. "Split-Channel Asymmetric Network" (SCAN) is a term coined by the author and Ed Moura (cofounder of Hybrid Networks) to describe a cable data network system and to differentiate from the then–LAN extension systems deployed by LANcity and Zenith. "Split-channel" refers to the separation of upstream and downstream data at the cable head end that increases effective channel bandwidth and offers network control to the service provider. Asymmetric networking is a concept that began with one-way cable with a telephone return, but it proved to be an effective design for two-way cable systems as well. These two concepts, split-channel design and asymmetric networking, led to a point-of-presence architecture that became a natural design choice for later developments. The asymmetric design relieved a requirement to match upstream channel bandwidth for every downstream channel, thus allowing scalability of system for mass markets. The author and his team built early cable data systems and studied various issues from 1993 through 1996. This paper presents many of these early discussions with reference to recent developments as adopted by various standards organizations.

Introduction

SCAN architecture refers to early work by the members of technical staff at Bell Laboratories on cable data networks.[1] In 1992, SCAN was a new system architecture designed to deliver multimedia services to the mass market at affordable prices. A cross-sectional analysis of many access technologies resulted in the selection of cable modems over to other emerging alternatives.

The SCAN project was nurtured under a New Business Opportunities process with an objective to quickly develop a mass market of high-speed service delivery systems. Industry efforts in broadband network design involving hybrid fiber/coax (HFC), switched digital (baseband) video, dual coax, and other full-service networks required an upgrade of the distribution network. Upgrading the distribution network involves significant capital investment, strategic alliances among many parties and, last but not least, digging to put fiber in the ground. The alternative was to focus on a system architecture that would take advantage of already in-ground distribution networks. SCAN used a telephone connection for point-to-point signaling and control and high-speed data delivery over the broadcast cable television network. Today this system architecture is referred to as "Data Over Cable Service Interface Specification with Telephony Return Interface" (DOCSIS-TRI).

This paper will present the architecture and design issues addressed by the SCAN team. Where appropriate, current and standard implementations are reflected with a reference to other source(s) offering detailed discussion.

Cable Data Network Architecture

In 1993 the cable television industry was primarily focused on video-on-demand and near video-on-demand applications with MPEG–encoded digital video transported over ATM networks. Focus on data applications can be broadly characterized into two areas: symmetric LAN over cable systems and asymmetric systems. LANcity and Zenith offered symmetric LAN support over cable networks, while Hybrid Networks and the SCAN team designed asymmetric systems (with telephone return for one-way cable systems and slotted upstream protocol over two-way cable systems).

The cable modem standards were yet to emerge, resulting in a 4-Mbps Zenith modem using vestigial side band (VSB) modulation while LANcity claimed a 10-Mbps capacity using a quadrature phase shift keying (QPSK) system. The Hybrid and SCAN teams used a 4 VSB modem with a 10-Mbps capacity in the downstream (i.e., network to user) direction with a QPSK system in the upstream (i.e., user to network) direction. The rest of cable industry focused on quadrature amplitude modulation (QAM) for video-on-demand applications.

Industry groups like the Digital Audio Video Council (DAVIC), the European Telecommunications Standards Institute (ETSI), the Institute of Electrical and Electronic Engineers (IEEE), the Multimedia Cable Network System (MCNS), the Society of Cable Telecommunications Engineers (SCTE), and CableLabs created standards with the objective of interoperability among vendor equipment. In "Cable Modem Physical Layer Specification and Design," Bruce Currivan presents an overview of the cable modem standards.[2] The cable modem standards adopted 64–QAM and 256–QAM for the downstream (network to user) and QPSK and 16–QAM for the upstream (user to network) communication. It is expected that all future systems will be built to these interoperable standards rather than using proprietary modulation techniques.

Two-Way Cable versus Telco Return

A 1993 survey presented cable industry statistics as about 70% systems up to 450 MHz, 20% supporting 550 MHz, and less than 5% supporting over 550 MHz to deliver cable television.[3] The industry expected to upgrade all systems to 550 MHz or higher to support digital services, including video-on-demand and data applications. Upgrading cable plants to 550 MHz or higher systems and maintaining a competitive edge in broadcast television service (over the emerging satellite television service) was rated higher than conversion to two-way systems. It was expected that only 8% of the cable industry would upgrade their cable plant to two-way before 1998. This was confirmed in many informal discussions with field engineers and managers in the cable industry. It was expected in 1993 that telephony upstream and cable downstream would remain a viable architecture.

The selection of the telephone upstream was the result of a choice between two design alternatives. The first alternative was to use the telephone modem connection as a one-way upstream link with the cable modem as a downstream connection, thus creating two unidirectional communications links. The second alternative was to establish a full two-way communication over the telephone modem and only supplement with cable modems when high-speed downstream was desired. By this time, the SCAN team developed a close working relationship with Hybrid Networks, and another team engaged in one-way cable data service with telephone return. While the hybrid system implemented the first alternative with two unidirectional links, the SCAN team successfully designed and tested both alternatives.

The first alternative, strictly one-way upstream communication over the telephone network with high-speed downstream over a broadcast cable network, was also implemented for satellite television systems. The Hughes Network Systems marketed the system as the DirecPC service offering.[4] This approach was adopted by the MCNS committee as the "DOCSIS Telephone Return Interface." Beser presents an overview of the architectural differences between cable data services offered over a two-way system and one-way cable with telephone return interface.[5]

The second alternative, full two-way communication over a dial-up telephone network with high-speed downstream supplemented by cable networks, was envisioned as a business partnership between an Internet service provider (ISP) and the local cable provider. In many cases, dial-up modem equipment is already in place for access to the Internet and World Wide Web applications. Two alternatives were studied for location of the cable modem equipment: 1) the ISP owns the cable modem equipment and leases a channel from the local cable provider, and 2) the ISP transports downstream data over a private (T1/T3) line to the cable head end, where data is modulated and distributed further down the cable network. The business case analysis showed a positive return for both alternatives. This is due to close proximity of ISPs and local cable providers when the cost of transporting data or a (data modulated into) analog video signal could yield profitable revenues. In our field trials, the SCAN team elected to manage the cable modem equipment and lease a channel from the local cable provider. The choice was made primarily to maintain control of network management, IP address assignment and management, and subscriber problem resolution.

Early Discussions

Within Bell Laboratories, research and development focused on HFC systems and video-on-demand applications. To generate necessary resources and to bring focus to data applications, the SCAN team presented many arguments to distinguish data and online services as viable applications separate from video-on-demand and other television applications. The following is an excerpt from the sometimes heated discussions:

"The SCAN distribution network shares many of the attributes of an end-to-end video delivery system for digital switched or broadcast video application. The emerging multimedia services industry can be broadly categorized into two areas: entertainment services and information services. From a technology and end user equipment perspective, the services can be categorized as TV-centric or PC-centric services. The differences in the PC-centric data services and the TV-centric entertainment services are subtle and distinct. This section provides an overview of the

SCAN architecture with an emphasis on several key requirements that shape the architecture, and highlights the differences between entertainment and information services."

PC–Centric versus TV-Centric Services

Interactive television services such as video-on-demand, near-video-on-demand, enhanced pay-per-view, and home shopping services require point-to-point signaling and control channel. The signaling channel and the program channel forms an asymmetric bandwidth architecture. However, from an end-user perspective, the television is likely to continue as an entertainment medium, while the PC will likely evolve as the interactive medium.

In the case of television, the end-user viewing distance is typically six to eight tube diameters from the screen; i.e., a typical user is sitting faraway from the screen to perform transaction-oriented applications. Also, the latency requirements are very tight. The end user expects the information on the television screen to change after one to two transactions; i.e., the user clicks "12" on the remote and expects a change to channel 12. Presenting textual services on the television requires an interlacing of text and data over the NTSC signal and extracting user stimulus (such as remote control commands on program guides). This makes it difficult for the end user to read and respond to textual services (e.g., e-mail, electronic newspapers and magazines.) Alternately, the signaling and control interface with text and video overlay function is built into the set-top-terminal for basic functions such as video-on-demand. Vito Brugliera at Zenith Electronics Corporation revealed the task to be technically complex and expensive for transaction applications.

For PCs, the end-user viewing distance is typically one to two tube diameters from the screen. This enables a user to read text from the screen and respond to a transaction by typing responses on the keyboard or selecting from available options via mouse clicks. The latency requirements are loose. The end user is willing to accept a lag of up to six to eight transactions, i.e., click on an icon, open a file, escape to the last menu, etc., before the information on the screen changes. The user's expectations are aligned with what will be required in the interactive services.

Viewing distance, latency requirements, and interactivity expectations make the PC a medium of choice for text, graphics, and some multimedia services. Reading news, e-mail, catalog shopping, software downloads, access to bulletin boards, etc. will continue on the PC in the near term. These services may not be easily replaced by televisions or digital set-top-terminals in the near future. However, a common distribution network may be able to provide entertainment services for television, telephony services, and data services for the PC all on one network.

This discussion is important even in the context of today's applications. An industry-wide gap in television as an information appliance led to innovative designs such as WebTV—a low-cost bridge between dial-up modem network access to the Internet and television as a display medium. The personal computer continues to build new value proposition such as Web page creation, access to scanning, image manipulation, audio/video editing tools, animation tools, and enhanced services such as paging. However, browsing, e-mail, chat, and information review services compose the majority of network usage. Thus, a stand-alone cable modem offers significant value to the market over a cable modem network interface card (NIC) that plugs directly into PCs.

Symmetric versus Asymmetric Networks

The Hybrid and SCAN teams initially used 9.6 kbps dialup modems that were prevalent in the mass market at the time and moved to 19.2 kbps modems by end of 1993. This low-speed channel, combined with a cable modem for the downstream, provided an asymmetric system architecture that evolved to two-way cable systems. In contrast, both the LANcity and Zenith systems were designed as "LAN extension" systems.

Channel assignment on a cable television network is based on 6-MHz increments. Each 6-MHz channel can hold one analog NTSC channel, several digital channels, or several digital audio music channels. The user-to-network direction is referred to as the upstream on cable, and the network-to-user direction is referred to as the downstream. A typical cable distribution system employs 5 MHz to 40 MHz for upstream signaling and 50 MHz and up for downstream video distribution (see *Figure 1*). This type of frequency allocation is known as a sub-split system.

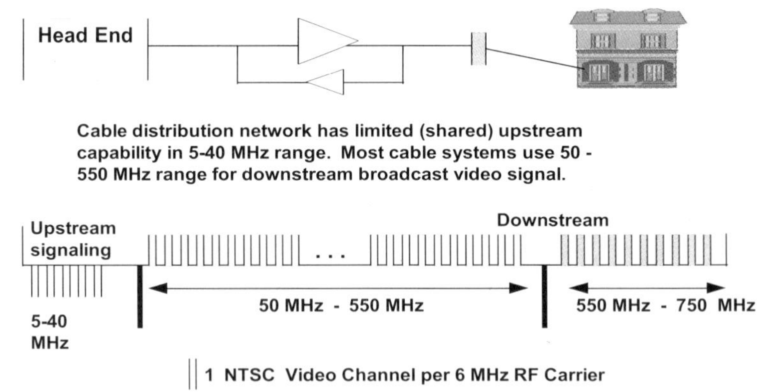

Figure 1: *Bandwidth Use in Sub-Split CATV Distribution*

In a symmetric network, broadband data services can be delivered over cable television networks that support two-way cable by modulating data into 6-MHz NTSC channels in both the upstream and the downstream directions.*Figure 2* shows a conventional cable distribution network providing LAN extension service over a cable network. The available cable spectrum is divided in increments of 6-MHz bands, each carrying one NTSC television programming signal. When a user station (PC) transmits data, the signal is sent out in a 6-MHz upstream channel to the cable head end. At the head end, the signal is modulated onto a separate 6-MHz downstream channel and is broadcast over the cable distribution network. In this arrangement, data transmission over cable does require two 6-MHz cable channels, one for transmitting and the other for receiving. Data transmitted by the end-user PC is sent over a channel in the 5-MHz to 40-MHz area (which was considered below-the-video-usable spectrum), and network data is received on one of the broadcast channels.

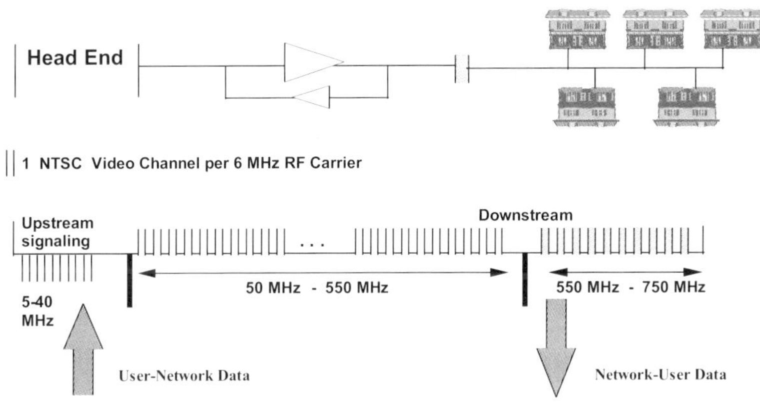

Figure 2: *Symmetric Bandwidth Connection Architecture*

This type of system architecture is referred to as a "LAN extension" system due to the nature of cable network use in this application. Typical cable head end equipment receives data from the sub-split spectrum and remodulates data in one of the downstream channels. In essence, the cable plant is simply used as a wiring cable. One user could transmit data to and receive from another user on the network without external influence or routing. As on other broadcast LANs, any one remote user could overload the bandwidth. The service provider has no control over the LAN bandwidth utilization. An indirect method of discouraging overuse was to monitor packet transmissions on the channel and set up a billing service on usage basis. This feature was considered unacceptable by LAN users and, as a result, was quickly eliminated.

Another key limitation of the symmetric LAN systems was scalability. In order to scale the system for a large number of users, the cable operator had to provide more 6-MHz channels in the sub-split range for LAN service. Noise ingress problems and spectrum use for telephony applications meant that bandwidth available for data applications would be limited.

In addition, the symmetric LAN architecture made inefficient use of the bandwidth. Upstream traffic is retransmitted in a downstream channel to get to the gateway node. The gateway node then redirects LAN traffic back on the upstream channel while sending the outside traffic through a gateway router. The resulting system was inefficient with respect to upstream and downstream spectrum utilization as the effective capacity was reduced to half the capacity of the channel.

The "split-channel" contribution of SCAN team was to introduce the concept of a point of presence at the cable head end and to separate the upstream bandwidth from being retransmitted downstream. The cable head end receives data from all users and retransmits only such data as destined to other users on the cable network. It is thus possible to set limits and control the user-to-user communication and prevent network overload.

The "asymmetric network" contribution of SCAN for two-way cable systems was a natural evolution from telephone return systems. Individual users were assigned 100-kHz RF spectrum (enabling up to 128 kbps using QPSK modulation). A proprietary media access control (MAC) protocol was designed to manipulate reception of multiple upstream links at the cable head end. Further, the 100-kHz spectrum allocated to any given user can be reassigned in the 8-MHz to 18-MHz range to get around noise ingress problems.

Recent developments in cable modem standards embrace the "point of presence" concept and effectively eliminated "LAN extension" systems. An industry standard MAC protocol is defined to generate optimum performance. Hartman and Quigley describe the key features and performance (simulation) results of the MAC layer as adopted by the DOCSIS—a standard set by the MCNS committee.[6] Sriram also presents a detailed discussion of performance analysis of MAC protocols discussed during the IEEE 802.14 committee meetings.[7] The proprietary MAC protocol adopted by the SCAN and Hybrid teams would now be replaced by the standard MAC protocol to enable interoperability among cable modems from multiple vendors.

Addressing and Routing Protocols

Discussions on network addressing and routing protocols ranged from the E.164 standard telephone addressing system to broadband ISDN and exclusive IP addressing systems. Considering that in 1992–1993 the Internet and World Wide Web were yet to gain popularity beyond the technical community, it was a bold decision by the SCAN team to use exclusive IP addressing for data applications. While this decision simplified many issues in connection management, it also complicated service from the user perspective. For example, the then-popular operating system Windows version 3.0 did not support built-in communication for TCP/IP networks. However, TCP/IP communication software was readily available from such vendors as Frontier SuperTCP, NetManage Chameleon, Spry, and SUN Microsystems NFS packages. While many of the lessons learned since 1993 are no longer relevant to future applications, configuration with end-user devices is still considered a major problem in mass-market deployments (see *Figure 3*).

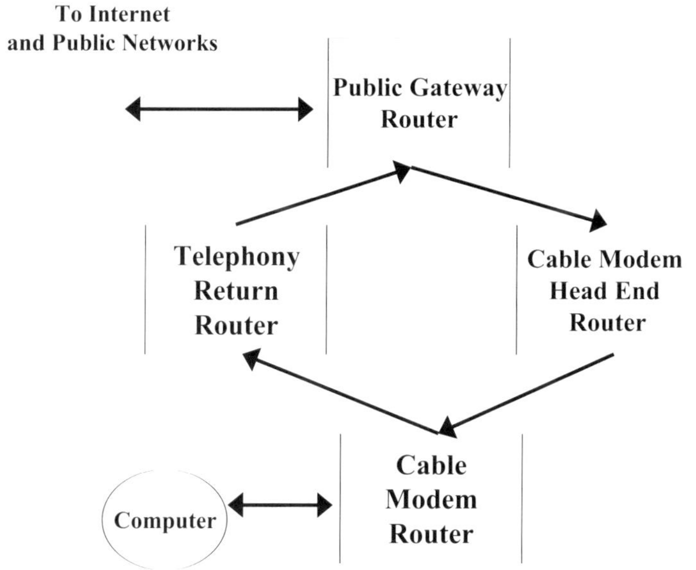

Figure 3: *Routing in One-Way Cable with Telephone Return Systems*

In technical and market trials of the SCAN system, a standard router-to-router communication protocol was established among cable modem, terminal server (telephone upstream aggregator), cable modem head end equipment, and gateway network elements. Proper data flow was arranged by manipulating routing tables in each of these routers so that all data from user to network was routed through telephony return router. Likewise, all data from the network to the user was routed through cable modem head end equipment. The router-based data flow also evolved to two-way systems, leading to the upstream router and downstream router components of a cable head end system. The upstream and downstream router components, however, are modified versions of a true Internet router in the sense that they did not perform the name service, dynamic address assignment, and/or global routing.

The router-to-router data flow also allowed the SCAN team to design a connection management software, called the "director" process, to give users a choice between low-speed telephony downstream or a high-speed cable downstream. The director process would accordingly manipulate the routing tables on the gateway router. This study was an academic exercise to examine the possibility of an ISP with a dial-up modem network to offer cable modem service as a high-speed enhanced service. For strategic reasons, such service was unappealing to service providers.

In the initial configurations, the dial-up modem network and cable modem network were configured as separate IP sub-networks. In addition, each cable channel was configured as a separate IP sub-network in the Hybrid system, thus preventing effective bandwidth management across multiple channels. The SCAN team approach to connection management as a layer above IP routing and the use of the "director" process permitted configuration of all cable downstream channels as single– or multiple–IP sub-networks. Thus, channel bandwidth could also be allocated separately among users of different classes of service. Bandwidth was allocated to a user at the session initiation and released for reallocation to other users at the session termination. Such a scheme also allowed for traffic engineering and administration of cable data networks commensurate with the administration practices of telephone networks.

An important issue often missed by marketing and business enthusiasts at the time was the demand on IP addresses required for each cable modem. The first systems were built with "fixed IP address" routing which meant that each cable modem would require three public IP addresses—one for the PC, one for the telephone modem connection, and one for the cable modem link. Many technical and field trials were done with a serial line interface protocol (SLIP) connection to the telephone modem network, allowing reuse of addresses on the dial-up modem network. Later versions switched to a point-to-point protocol (PPP) connection.

Recent developments in cable data network addressing also incorporated the dynamic host configuration protocol (DHCP), network address translation (NAT), authentication, and accounting systems. Hernandez-Valencia presents a detailed discussion of IP services over cable television networks.[8] Fijolek offers future directions for IP address assignment and management, support for private user addressing on customer premises, and tunneling.[9] As discussed, Beser presents key differences between two-way cable and one-way cable with telephone return systems.[5] Collectively these papers offer the current status and directions for future work in efficient routing of IP services.

Bandwidth Utilization by Applications and Services

Another topic of great interest was how cable modem bandwidth could be used to maximize revenue potential for the service providers. Some members of the SCAN team insisted on true "data" services; others wanted those that offer the full bandwidth on a contention basis at flat monthly rates, while still others preferred a scaled approach to the service offering. The SCAN team eventually built connection management software capable of demonstrating 64 kbps, 128 kbps, and unrestricted bandwidth service offerings. As the business case was discussed in various technical and strategic analysis meetings, the true data service offering of unrestricted bandwidth was considered most appealing to the end user.

In one of the technical and field trials, extensive data collection and analysis was focused on system throughput and the effect of the 9.6 kbps or 19.2 kbps modem link on total system performance. In the end, it was shown that continuous file transfers and some X windows applications that stress downstream bandwidth result in a reduction of system throughput to approximately 80% of the modem capacity. (We conducted trials with a 4VSB modem with a downstream capacity of 10 Mbps and 19.2 kbps upstream. The overall system could only deliver 7.5 Mbps to 8 Mbps resulting in loss of 20% throughput from various networking problems.) However, the system trials offered a validation for split-channel asymmetric network architecture, including one-way cable with telephone return architecture. The asymmetric network architecture was considered reliable and scalable over symmetric LAN extension systems.

The SCAN team discussions focused on bandwidth utilization and system throughput, while today's discussions about quality of service focus on service performance. The quality of service focus is to support real-time and non-real-time services, admission control, and latency requirements. Hartman and Quigly discuss the quality of service analysis for MCNS MAC protocol simulation and find that DOCSIS specification could support voice and data services with adequate performance.[6] In a separate study, Gun and Ruszczyk review the performance implications of telephone return systems and conclude, "while many tradeoffs exist in the decision to deploy a telco return or a two-way system for Internet access, performance as seen by user is not one of them."[10]

Security

Security is often discussed when considering the growth of Internet applications, particularly electronic commerce. Due to the shared bandwidth nature of the cable modem network, perceptions of lack of security are prevalent in the industry. This may not be true, however. In a split-channel system, upstream data from each user is received at the cable head end and forwarded to the appropriate destination. In the downstream direction, however, head end cable modem equipment is the only node permitted to transmit to the users. The data link layer frames are addressed to individual user destinations and can be encrypted. This provides privacy of user data commensurate with many Internet applications. SCAN and Hybrid teams used a single-data encryption standard (DES) scheme to provide privacy over cable data networks.

Security measures such as encryption offer privacy to one user's data from others in a shared medium but offer little or no protection from a MAC clone or spoofing devices used by a hacker for deliberate intentions. This problem, however, is generic to all IP networks and requires expensive digital signature mechanisms to offer security for such services as electronic commerce. Birger and Rance provide a discussion of privacy and security requirements as addressed by cable modems.[11, 12]

Scalability

Cable modem technology is now mature in terms of network architecture, modem technology, standards, and system performance. From a mass market deployment perspective, one has to look at such issues as: cost of goods, port density, automated operations, and maintenance. Also, to support a large and growing market scalability issues can arise with respect to the following:

number of subscribers
number of service providers
number of service types
number of subscribers per service provider
broadcast storm
running out of address space (IPv4 ROAD problem)
increased routing tables
physical port density
service package and price

Some of these issues, particularly broadcast storm, routing, and port density, need to be addressed in order to offer a high-speed service that is profitable as a mass-market deployment. The next-generation, standards-compliant head end systems will have to be designed to support a large number of subscribers while addressing other scalability issues.

Conclusion

In the early stages of cable data networking, LANcity and Zenith brought significant value to the broadband data services market by offering a LAN extension service. The SCAN and Hybrid Networks teams successfully argued for the concept of a point-of-presence and split-channel network architecture. Though many of the variations in technology design and lessons learned from trials are no longer applicable, they serve as a guide to future systems.

Cable modem technology is now mature and offers low-cost, high-speed access to the Internet and World Wide Web. The MCNS standard enables the industry to develop interoperable systems. Kuska presents background information and execution strategy of the MCNS team.[13] The rest of this book presents necessary technical information to understand and develop interoperable cable data networking systems. As executive editor, I believe this book offers a significant value in understanding cable data systems that, in turn, offer significant value to the industry.

Acknowledgments

I would like to thank Mark Rochkind, Rob Richardson, John Shepherd, Dan Carr, Eric Edberg, Norm Schryer, and Tom Killian for their contributions to the SCAN team from 1993 through 1996, and Ed Moura and Jan Gronski, who were members of Hybrid Networks team.

I would like to thank John Fijolek, Michelle Kuska, and Kotikalapudi Sriram, coeditors of this book, for their personal contributions and generous offering of their time during this project.

References

1 Majeti, V. C., Moura, E., and Watson, R. J., "SCAN, A Split-Channel Asymmetric Network Architecture for Delivering Multimedia Services to Mass Market," presented at the National Communications Forum, 1993. Also published in *Annual Review of Communications*, 1995.

2 Currivan, B., "Cable Modem Physical Layer Specifications and Design," included in this compendium on cable modems.

3 "The Cable Television Industry: New Technologies, New Opportunities, and New Competition," Report by The Wasserstein Perella Securities, Inc., March 1993.
4 "DirecPCTM High Speed Information Service," marketing brochure by Hughes Network Systems.
5 Beser, N. B., "The Architectural Differences Between DOCSIS Compliant Bi-directional and Telco Return Systems," included in this compendium on cable modems.
6 Hartman, D. and Quigley, T. J., "Multimedia Cable Network System Media Access Control Protocol Performance Simulation," included in this compendium on cable modems.
7 Sriram, K., "Performance Analysis of Salient Features of MAC Protocols for HFC and Wireless Access Networks," included in this compendium on cable modems.
8 Hernandez-Valencia, E. J., "Architectures for IP Services over CATV Networks," included in this compendium on cable modems.
9 Fijolek, J., "Future Directions for Telephony Return DOCSIS Networks," included in this compendium on cable modems.
10 Gun, L. and Ruszczyk, C. A., "Performance of Telco Return Cable Data Systems," included in this compendium on cable modems.
11 Birger, C., "Data-Over-Cable Security," included in this compendium on cable modems.
12 Rance, R., "Cable Modem Security," included in this compendium on cable modems.
13 Kuska, M., "MCNS/DOCSIS Strategy and Execution," included in this compendium on cable modems.

The Cable Modem's Evolution and Revolution: An Historical View

Rouzbeh Yassini
Founder and Chairman
LANcity

Introduction

The cable modem is profoundly and positively changing society. This technology is enabling the cable communications industry to lead society's transition from its 20th-century emphasis on paper communications to a 21st-century emphasis on electronic media storage and transport. But what is a cable modem, and where did the technology originate? How do cable modems deliver high-speed data over cable services? How did the cable television infrastructure develop, and why is it so well suited for broadband data services? This chapter examines the networking evolution that has positioned cable modem technology at the leading edge of the digital convergence revolution.

Birth of the Cable Television Industry

In 1948 in Astoria, Oregon, the cable television industry was born primarily as a means to provide improved television reception to viewers in remote areas. Regulation of this industry by the Federal Communications Commission began in 1960. In 1972, the first pay-per-view service, Home Box Office (HBO), was launched. Since that time, cable television has become a common source of video entertainment for more than 500 million viewers worldwide and is now found in approximately two thirds of American homes. While cable television became well known as a means for distributing high-quality one-way video services into homes, its marriage to television, a passive receiver of video broadcasts, obscured cable television's ability to provide high-speed, two-way data communications.

I-NETs

In the late 1980s, however, local and state governments, universities, and research institutions began employing the cable television infrastructure for two-way campus data networking. These institutional networks, or I-NETs, served the specific needs of the institutions they supported. Typically, I-Nets provided connectivity between a campus's local area networks (LANs). But the demand for local information on private institutional networks did not provide the impetus for cable modems as the next-generation technology. Rather, it was the demand for access to global information on the World Wide Web that drove the commercial success of the cable modem. Still, I-NETs did serve as the early proving ground for cable modem technology.

LAN Technology

The history of the cable modem is closely linked to the history of internetworking technologies and the explosive growth of the global Internet. By the late 1970s and early '80s, the networking industry had begun developing the LAN. Ethernet, the predominant LAN technology in use today, was first developed in 1973 by the Xerox Corporation. LAN technology, in fact, predated personal computer and engineering workstation development. Originally, LANs were used as an efficient means to link terminals to mainframes and mini-computers.

The LAN industry grew to its second phase with the explosive use of PCs in the commercial marketplace, along with the rapid adoption of engineering workstations for scientific and engineering computing. The need for PCs to share resources (printers, file server) and the growth of network computing in the UNIX environment led to the widespread manufacture and use of network interface cards, bridges, and hubs to interconnect PCs, workstations, and servers across LANs.

At the same time that a vibrant data communications industry was building around LAN technologies, universities, research institutions, and government agencies were experimenting with and developing the building blocks for today's global Internet. These building blocks included the TCP/IP protocol stack; the algorithms and protocols for distributed management of a vast, decentralized network of networks; along with switches—known as routers—for interconnecting these networks.

Internetworking

In the late 1980s and early '90s, LAN and Internet technology merged. Internetworking was then adopted as the principal means to interconnect isolated LANs. Wide-area networks (WANs) evolved to link geographically dispersed LANs, while the TCP/IP family of protocols and routers provided the internetworking glue.

Although the Internet was widely used in research and academia prior to the mid-1990s, its explosive growth did not begin until the introduction of TCP/IP protocol Web browser software onto personal computers. This phenomenal growth continues today and is responsible for catapulting the cable modem to its current position as the premier means to access the Internet and bandwidth-intensive data services.

HFC Network Topology

The cable television infrastructure is ideally suited to provide broadband, two-way data services. In fact, the HFC network topology can provide these broadband services far more efficiently than dedicated wire telephone networks. This efficiency results from the cable television infrastructure design, which typically consists of a hybrid fiber/coaxial (HFC) network laid out in a tree-and-branch topology. A high degree of aggregation of users' traffic is made possible by this tree-and-branch topology, along with the cable network's shared-medium technology. What has resulted is an economy of scale unavailable to traditional telephone networks with their dedicated, twisted-pair local loops connecting each home to a central office.

Cable Modem Beginnings

By the 1980s, early interest in cable modem technology evolved, centered on the IEEE 802.4 token bus over CATV standard and its adoption by the automotive industry as a basis for the manufacturing automation protocol (MAP). Deficiencies in radio frequency (RF) technology at the time made early cable modem products costly to manufacture and maintain, and the resulting broadband data networks expensive to operate. The technical obstacles and lack of financial incentives to solve these problems resulted in a small number of vendors initially interested in the cable data networking market.

Meanwhile, the combined LAN and internetworking industries grew into a multi-billion-dollar industry, providing data networking solutions to corporations, academic and research institutions, and governments. All these solutions employed baseband-networking technology, such as Ethernet, where digital signals are sent without frequency shifting. Although broadband networking's frequency division multiplexing offered the possibility to support multiple communications channels simultaneously, making digital convergence a reality, broadband's technical challenges still proved too daunting for most of the manufacturers determined to overcome cable TV aberrations. One such company, LANcity Corporation of Andover, Mass., not only overcame these technical barriers and hurdles, but also became a commercial success.

LANcity's cable modem technology provided the networking industry with the benefits of many other industry technologies, including networking, software, computers, and content providers. LANcity's cable modem became the critical piece needed by the cable TV industry to extend a data network from within a building to an entire city using the existing, commercial cable TV infrastructure.

LANcity's newly invented cable modem technology was built from the ground up to overcome, at a minimum, the following key cable TV aberrations and requirements (see *Table 1*).

Cable TV Aberrations

Excessive channel group delay
Limited distance due to delay through HFC plant
Upstream Time Division Multiplexing
Microreflections
Varied operating levels
Varying plant losses
Time dependent plant level variations
Adjacent channel operation
Poor Carrier to Noise

Cable TV Requirements
A long propagation delay, which can be greater than packet transmission time
Efficient support of a large number of potential users
Upstream channel noise and ingress
Plug-and-play installation
Interfaces standardized with existing networking solutions offered by the LAN industry
Use of any available frequency up to 750 MHz via software control
Efficient and manageable bandwidth
Significant group delay upstream (800 nanoseconds) and downstream (240 nanoseconds)
A frequency response tilt of up to 5 dB within 6 MHz
Micro reflections
Solid QPSK modulation technology for mass-volume production
A scaleable solution adhering to internetworking principles
SNMP-based management for end-to-end data services
Operation up to a 200-mile round-trip distance
Product installation, manufacturing and deployment occurring in less than one minute
Network privacy over shared media (data and manageability)
A billing interface plus a centralized operations center
Successful implementation of a networking system on top of a two-way cable TV system
Robust-implemented PHY (physical access), and MAC (media access control)
Automatic adjustments for cable TV aberrations

Table 1: *Key Cable TV Aberrations and Requirements*

Cable Modem Mania

The successful launch and deployment of LANcity's cable modem offered the existing commercial cable television industry significant potential for new revenue streams. Early adopters included Continental Cablevision (now MediaOne), Tele-Communications, Inc., Cox Communications, Time Warner, Cablevision, Adelphia, and Jones Intercable. Priced around $400 per unit in 1995, the LANcity cable modem broke the price barrier and helped to create the cable modem industry—and cable modem mania.

Multibillion-dollar industry leaders, including Zenith, Motorola, and Intel, realized the cable modem's potential, then repositioned their networking products for the broadband market. Today, more than twenty-three vendors have entered the cable modem arena, which is now a multibillion dollar industry itself. With the agreement of a DOCSIS standard for the next-generation retail version of the cable modem, mass marketing of this technology is expected by the end of 1998. *Table 2* shows the growth and ranking of cable modem manufacturing in the last eight years.

1990	1993	1995	1998
LANcity Zenith Fairchild Ungermann-Bass Chipcom	LANcity Zenith Intel	LANcity Zenith Motorola Intel Hybrid Com21 Westend	3Com CISCO Panasonic Sony Motorola Intel Bay Networks* Com21 Teryon Hybrid Hayes Thomson Toshiba Libits Daewoo/Cadence NEC Zenith GI Samsung
Estimated Cost 2K to 20K	Estimated Cost 1K to 7K	Estimated Cost $500 to $995	Estimated Cost $150 ~ $250

**Bay Networks purchased LANcity in 1996*

***Table 2:** Growth and Ranking of Cable Modem Manufacturing*

Cable System Installation

To deliver data over cable television services, two categories of equipment are required: head end equipment at the service provider's head end facility, and a cable modem at each subscriber's location. Head end equipment interconnects the broadband CATV network to the service provider's own networking facilities. These facilities are home to various servers employed in the delivery of locally originated data services provided to subscribers, in addition to switches that provide access to the public Internet. Cable modems at subscriber sites are the means by which subscriber equipment, such as PCs, gain access to the broadband CATV network and the data services accessible through this network. A cable modem termination system (CMTS) at the head end controls access to the broadband network, in addition to ultimate access to the data service provided across that network.

DOCSIS

A set of data over cable service interface specifications (DOCSIS) has recently been completed by the Multimedia Cable Network System (MCNS) consortium. Wide acceptance of DOCSIS has accelerated design and development of interoperable head end and cable modem equipment. As a result, more than twenty vendors are now manufacturing DOCSIS-compliant cable modems, in addition to cable modem termination systems and network management systems. By the close of 1998, DOCSIS cable modems will be available to the market through retail outlets. Additionally, it is expected that all major MSOs will then offer DOCSIS-based broadband data services in their respective service areas.

With the addition of software industry giants to the broadband arena, transparent application- and topology-independent networks capable of voice, data, video, and IP-packet-based services will soon be enabled over the cable television infrastructure.

Wall Street and Cable Modems

Since 1996, Wall Street has recognized cable modem technology for its potential to tap the gold mine of revenue contained in the existing, underused cable television infrastructure laid in the ground fifty years ago. By the close of 1998, more than $15 billion will be invested in cable modem technology for deployment of broadband services through the following acquisitions: Continental Cablevision by U.S. WEST; AT&T's merger with TCI; the @Home Network public offering; the purchase of LANcity by Bay Networks; Microsoft's investment in Comcast; and the creation of

Road Runner and Highway 1 and their joint venture. These acquisitions serve as evidence of industry and market trust in the potential of cable modem technology.

Cable Modems: "The Missing Link"

Cable modem technology is having a profound effect on society. By providing last-mile, high-speed connectivity, every household can now be transformed into a powerful information source and office environment. The cable modem revolution will assist in environmental conservation by encouraging home-office productivity over corporate commuting. As a facilitator for education, the cable modem will enable our children (the Internet Generation) to access a wealth of information from home rapidly and efficiently.

Cable modem technology will contribute greatly to the 21st century digital convergence revolution. Because of this invention, the foundation for the information age has been firmly established, the virtual information village has finally been born, and the doorway to the future of computing has been opened.

MCNS/DOCSIS Strategy and Execution

Michelle Kuska
Director of Network Technology
TCI Business Alliance and Technology Co., Inc.

Abstract

In December 1995, high-speed data services had become a very popular offering for cable operators. As a result, cable operators started to evaluate how to grow the business so that high-speed data services could be offered economically on a national basis. It was determined that if cable modems (CMs) could be sold on retail shelves, cable operators would reap the following benefits: service penetration would increase through customer awareness, competition among vendors would drive capital equipment costs down, and cable operators could remove the cost of the CMs from their books. As a result, the cable industry decided to establish the Data Over Cable Service Interface Specification (DOCSIS) project (also known as MCNS). The interface specifications that constitute the core of the project are published at *http://www.cablemodem.com.* This paper described the methodology for producing the MCNS/DOCSIS interface specifications and gives an overview of the specifications.

Introduction

The cable industry established the Data Over Cable Service Interface Specification (DOCSIS) project to facilitate the development of interoperable CMs that could be sold on retail shelves. The project contains three stages: interface specifications, interoperability testing, and certification. The interface specifications detail all the internal and external interfaces for a data over cable system. These documents are cable operator–supported, widely available to vendors, based on working technology, and contain low technical barriers to entry. As the initial interface specification effort wrapped up, cable operators established an environment at CableLabs, the research and development organization for the cable industry, to allow vendors to work out interoperability issues with other vendors. Concurrently wirh the interoperability testing, CableLabs is developing a methodology to certify vendors as compliant to the specification.

Background

In December 1995, cable operators were asking "How will we get interoperable cable modems that can be sold on retail shelves?" Interoperable CMs implied a standard network interface. Waiting for a standard did not appeal to the cable operators because they planned to roll out high-speed data service on a large-scale basis immediately and wanted to have the capital cost of CMs off their books as soon as possible. As a result, the cable industry decided to establish the DOCSIS project (also known as the Multimedia Cable Network System (MCNS) and publish their own network interfaces.

"MCNS" is an abbreviation for MCNS Holdings, L.P. Comcast, Cox, TCI, and Time Warner established MCNS to fund projects of mutual interest. For the first year, MCNS funded DOCSIS. In subsequent years, the project has been funded by CableLabs, a research and development consortium of cable television system operators representing the continents of North America and South America.

Participants

The original participants in DOCSIS were the MCNS partners (Comcast, Cox, TCI, and Time Warner), Continental (now MediaOne), Rogers, and CableLabs. At inception, a management group and a working group were formed. The management group was chaired by Steve Dukes of TCI. The members were Mark Coblitz of Comcast, David Fellows of MediaOne, Alex Best of Cox, Nick Hamilton Piercy of Rogers, Jim Chiddix of Time Warner, and Dick Green of CableLabs. The working group was chaired by Michelle Kuska of TCI. The members were Steve Craddock of Comcast, David Fellows of MediaOne, Jay Rolls of Cox, George Hart of Rogers, Jay Vaughan of Time Warner, and Bob Cruickshank of CableLabs. Over time Doug Jones joined the working group to represent MediaOne and Time Warner. The role of these groups was to define objectives, set timelines, determine strategy, and oversee execution.

Objective

The first thing the cable operators did was to establish the project objective. The reference architecture, shown in *Figure 1*, was self evident. However, the scope of the project was not. Cable operators felt it was valuable to take a network systems perspective and define all of the interfaces so that the equipment could interoperate with the operation support system (OSS), security system, backbone wide-area network (WAN), public switched telephone network (PSTN), and subscriber local-area network (LAN). Defining all the external interfaces to the CM and head end equipment ensures that the equipment can become part of a scaleable service offering by the cable operators. Defining the internal interface for the system, the radio frequency

(RF) interface between the head end equipment and the CM, allows CMs from multiple vendors to work with a given vendor's head end equipment. As a result, the project objective became to "Define all interfaces to and within the data over cable service delivery system." The locations of these interfaces are identified on the reference architecture shown in *Figure 1*.

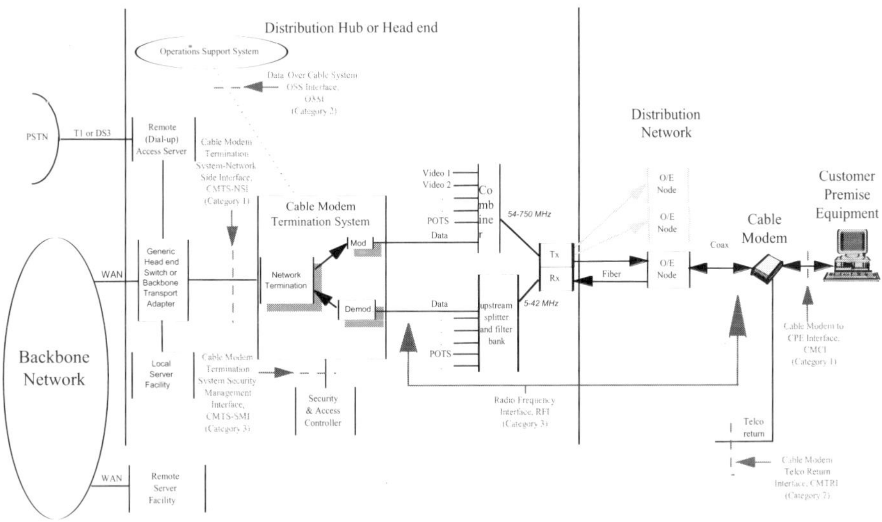

Figure 1: *Data over Cable System Components and Interfaces*

Timeline

The timeline for creating a complete set of interface specifications was set at one year by the management group. The working group started to brainstorm plans for developing the interface specifications. However, at this time it was still unknown how to deliver interface specifications robust enough to develop against within a year.

Service Goal

To meet the timeline, cable operators needed to identify the service goal for the interface specifications. If it were identified too broadly, it would become a rock around the neck of the project. In addition, it had to be set so that documents could be produced without incurring excessive risk or delays by pushing the limits of technology. As a result, the service goal was established as transparent bidirectional transfer of Internet protocol (IP) traffic, between the cable system head end and customer locations, over a hybrid fiber/coax (HFC) cable television system. This is shown in *Figure 2*.

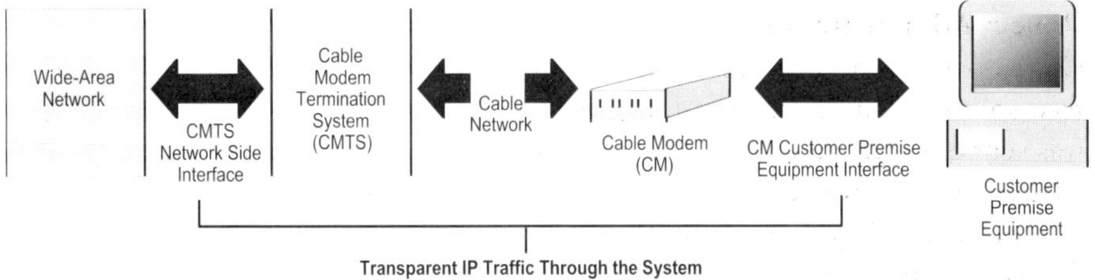

Figure 2: *Transparent IP Traffic Through the Data-Over-Cable System*

In *Figure 2*, the term "cable modem termination system" (CMTS) represents the head end equipment. Each customer location has a CM to interface the customer premise equipment to the network.

Operating Assumptions

As part of the strategy for success, cable operators established project operating assumptions to ensure that the project stayed on track. Knowing how the cable industry viewed the DOCSIS project helped the group avoid tripping itself up while carrying out its mission.

Operating Assumption #1—Promote the commercial deployment of data over cable service in 1996. Cable operators did not want to stall CM deployment with the undertaking of the DOCSIS project. The cable industry needed to continue to deploy CMs as rapidly as possible and not wait for what was admittedly a very risky project.

Operating Assumption #2—Develop interface specifications as soon as possible. The best way to do this was to select existing technology, not developing technology, wherever possible. As a result, the group was driven to incorporate as much pre-existing work as possible.

Operating Assumption #3—Free and open interfaces. Cable operators wanted multiple vendors to develop CMs to the DOCSIS interfaces. These interfaces needed to be available to all vendors as soon as possible.

Operating Assumption #4—Cable operators would be free to purchase or not purchase to the interface specifications. The working group knew that the technology incorporated in the interface specifications had to result in equipment that was cost competitive with proprietary solutions offered by the vendor community.

Project Administration

Cable operators needed a resource to administer the mechanics of creating the interface specifications. A request for proposal (RFP) for a project management group was sent out December 11, 1995. Five responses were received December 22, 1996. Arthur D. Little was chosen as the project manager January 3, 1996. The kick-off meeting was held January 9, 1996. The goal of removing the capital cost of CMs from cable operators' books was set in early December 1995. One month later a plan for doing so was being put into place. The hallmark of the project, speedy execution, was already established at the time of the kick-off meeting. But the question remained, "How would this be done in a year?"

Shortly after selecting a project management group, the mechanics of creating the interface specifications were put into place. Because there were a large number of parties involved in both managing the project and contributing to it, a Web site was established for the dissemination of information. The Web site is*http://www.cable-modem.com.* There are three partitions within the Web site: one for the public, one for vendor participants, and one for the cable operators (MCNS). The public area includes general information and instructions on becoming a vendor participant. The vendor partition allows vendors to access information about the project, download draft documents, and provide input to the process. The MCNS partition provides cable operators with the opportunity to download draft documents, share information, and receive project updates.

The dynamics of the DOCSIS project, with the large number of cable operators and the large number of vendors that are involved, are illustrated in*Figure 3.* Cable operator management of the project creates buy-in by cable operators and reinforces the project direction to vendors in a consistent manner. As a result, the vendors see a united front and are more likely to be responsive since they are not being asked to go in different directions. The project administrator structures vendor participation, and, jointly, the project administrator and vendors execute the project.

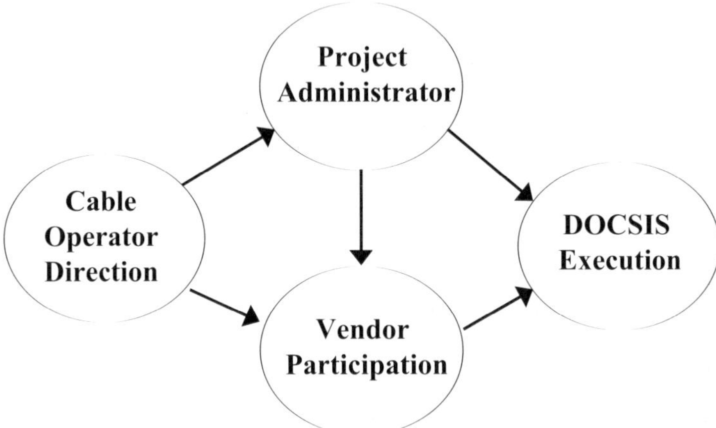

Figure 3: DOCSIS Dynamics

Creating Interface Specifications

In order to start chipping away at the project deliverables, the interface specifications in the reference architecture were separated into three categories. The first category represented interfaces that were straightforward and easily identified. All that was needed was to pick from interfaces currently available. The data interfaces fell into this category. It was fortunate that this category of interfaces existed because it gave the group a chance to work out the mechanics of the interface specification process on relatively simple documents.

The second category represented interfaces that were more complex than those in the first category because no interface existed that could be adopted, but the technical basis for category two interfaces existed. Consequently, work was needed to develop the interfaces specific to the reference architecture. The operation support system interface with management information bases (MIBs) and telephony return interface fell into this category.

The category three interfaces lacked current implementations of the proposed technology. Although the intent was to utilize existing technology as building blocks, developing this group of interfaces was still high risk. The RF interface and the security interfaces fell into this category.

After dividing the interfaces into three categories so they could be tackled as logical groups, the process of creating the documents began with the creation of the category one interfaces. While the category one interface production was going on, the working group continued to discuss strategy for accomplishing the other interfaces.

It was clear what needed to be done: rapid development of interface specifications detailed enough that multiple vendors could develop interoperable product. Taking this into consideration, the working group developed operating principles for document production.

- Select technology based on operating hardware so that the documents produced will represent a system that can be built and will work.
- Have the vendors write the documents.
- Provide architectural direction and make technology choices.
- Facilitate the integration of contributions from multiple vendors.

These operating principles have proven to be sound for producing robust documents in a short time frame. They are currently still being used within the project, and the methodology that resulted is detailed later in this paper.

First, vendors are invited to submit solutions for consideration as an interface specification. The invitation is accompanied by a work in process to help frame the contributions. Contributions are evaluated against the following criteria in descending order of importance:

1. meets basic performance, features, and cost needs for the first three to five years of service
2. offered for use without intellectual property encumbrances
3. has been implemented and tested
4. supports the evolutionary aspects of the architecture
5. provides technology-based performance or feature benefits

Note that these criteria rank technology-based performance or feature benefits last in the decision process to avoid pursuing next-generation technology that would result in missing the one-year deliverable of interface specifications.

Based on the above criteria, a vendor-author is chosen for the specification. Sometimes more than one vendor-author is chosen for a specification based on the technologies involved. The vendor-author(s) identifies the timeline for production of a draft for review by the vendor-authors of the other specifications and the working group. Once these two groups approve the document it goes out as a draft to the wider vendor community on the Web site. The vendor community has one month to provide written comments on the draft. All the comments are reviewed by the vendor-author and the working group for inclusion into the document. The document is then released as an interim specification. Once an interim specification is built and the resulting product is tested for interoperability, the document is given released status. The definitions for the categories of specifications are as follows:

Work in Process: An incomplete document, designed to guide discussion and generate feedback, that may include several alternative requirements for consideration.

Draft: A document in specification format considered largely complete but lacking review by cable operators and vendors. Drafts are susceptible to substantial change during the review process.

Interim: A document that has undergone rigorous cable operator and vendor review, suitable for use by vendors to design to, and suitable for field testing.

Released: A stable document, reviewed, tested, and validated, suitable to enable cross-vendor interoperability.

The Interface Specifications

This section describes the interface specifications by category.

Category One

The CM termination system–network side interface (CMTS–NSI) and the CM to customer premise equipment interface (CMCI) were data interfaces that could be easily specified from those available in the data networking community. All that was needed was consensus on which data interfaces, from the available standards, to identify for each document.

CM Termination System–Network Side Interface (CMTS-NSI): The goal for the CMTS–NSI was to define the minimal communications standards and protocols needed for satisfactory communication between the head end and backbone elements in a data over cable system. To meet this goal, both LAN and WAN interfaces, including 10Base-T, 100Base-T, FDDI & ATM DS3, and OC3-c, were identified. These interfaces were chosen because at least one member of the cable operator community was using them for their data network connection. The CMTS–NSI was beneficial to the cable operators because it focused vendors on providing the network interfaces that cable operators needed.

*CM to Customer Premise Equipment Interface (CMCI):*The CMCI is the interface between the CM and the customer premise equipment. In this instance it was advantageous for the cable operators to identify a single interface. 10Base-T was chosen because it was widely available and low cost. The benefit of identifying a single interface for the CMCI was that it standardized installation procedures across vendors.

Category Two

Category Two interfaces include the operation support system interface (OSSI) and the CM telephony return interface (CMTRI). Although there was significant work to be done to identify the MIB objects, the path forward was clear for the protocols: simple network management protocol (SNMP) for OSSI and point-to-point protocol (PPP) for telephony return.

*Operation Support System Interface (OSSI):*The OSSI was specified to allow cable operators to define, design, and develop the operations support subsystems necessary for the widespread commercial deployment of data-over-cable service. In particular, with a common set of MIB objects supported by all the vendors' equipment, a network management system could be developed that would work with any vendor's equipment.

The OSSI development philosophy was to specify open protocols with a preference for existing, well-known, and well-accepted standards. The scope of the OSSI was to meet the minimal set of requirements for satisfactory network management.

The first document produced for the OSSI was a technical report titled "Operations Support System Framework for Data Over Cable Services."[7] A technical report is not an interface specification but provides a context for understanding or applying an interface specification. The OSS framework technical report outlined business processes and operational scenarios that cable operators might employ when offering data over cable service. This document was designed to provide input to vendors for the production of the OSSI specification, where the specific managed objects and protocols would be identified. The OSS framework technical report was issued by the working group so that the proposal for the MIB objects would be based on plausible service and business models.

The first OSSI solicitation was for the RF MIB. Vendors submitted proposals for the RF MIB based on their current systems. These proposals were evaluated by CableLabs and the vendor-authors of the RF interface specification for inclusion in the OSSI specification. It was determined that having a standards body–endorsed MIB, instead of a proprietary MIB, would be more advantageous to the project because vendors would be more likely to adopt it. As a result, the RF MIB was written in Internet-draft format so that it could be submitted to the Internet Engineering Task Force (IETF) for standardization.

The lead vendor-author on each interface specification is responsible for developing the associated MIB. However, all MIBs are taken through the document development process so that they receive full review and approval before being issued as interim documents.

The benefits to cable operators from publishing the OSSI are as follows:

- single network management system
- single billing support system
- single subscriber management system
- single RF spectrum management system
- uniform set of grade of service parameters

*CM Telephony Return Interface (CMTRI):*The CM telephony return interface (CMTRI) defines the interface requirements for high-speed downstream data over cable and lower speed upstream data over the PSTN.

The goal in developing the CMTRI specification was to specify open protocols, with a preference for existing, well-known, and well-accepted standards. The scope was to identify the minimum set of requirements for satisfactory upstream communication between the CM and the CMTS over the PSTN. A draft CM telephony return interface specification was written and issued to vendors to solicit proposals. It was requested that proposals from vendors meet the following requirements:

- utilize the downstream RF signal as published in the RF interface specification
- support coexistence of cable return and telephony return CMs on the same downstream RF channel
- conform to the architecture published in the draft (specifically, the telephony return modem must be physically attached to or internal to the CM)
- be based on working technology

The resulting specification supports these goals.

The benefit of the CMTRI is that it provides a uniform telephony return interface that allows a cable operator to deploy a channel bank of telephony return dial-up modems used to receive the return path data when an RF return channel is not available. In addition, because CMs may have both RF return and telephony return capability, cable operators can deploy a dual return path CM, using the telephony return initially and switching to the RF return when the cable plant is ready.

Category Three

The category three interfaces include the radio frequency interface (RFI) and the security interfaces.

*Radio Frequency Interface (RFI):*The RFI was developed by two vendors, one responsible for the media access control (MAC) layer and the other responsible for the physical (PHY) layer. Shortly after the first two vendors were chosen as vendor-authors, a silicon vendor was brought in as part of the vendor-author team to provide the group with a sanity check for technology choices. The cable operators needed to know the impact of making one technology choice over another in terms of time to market and cost. All three vendors provided this input in their respective areas of expertise. The MCNS/DOCSIS project owes its success to this team of vendor-authors. Their ability to collaborate on the specifications and make compromises based on the cable operators' initial selection criteria resulted in the separate parties often coming to the same conclusion when a point was under debate.

Once the RFI specification was drafted it was released to the wider vendor community for review. This was done December 4, 1996. This release, combined with the security system release, also in December, met the goal of publishing a complete set of specifications within one year of launching the project. However, over time, documents have been and are still being added to the set to incorporate additional architectures and additional functionality.

Both the cable operators and the vendor-authors were interested in releasing the most robust final specifications possible. The review cycle by the wider vendor community was crucial to make sure the internal team had not overlooked anything or made design errors that would result in a product that did not work or was not interoperable. At this point, there were ninety-five vendors participating in the project through the Web site. Vendors were given one month to provide written comments on the draft specification. Comments were requested that fit the following categories:

- missing items
- items that need clarification
- changes to correct something that is incorrect
- suggested changes that would result in cost reductions, less risk, or faster time to market
- viability of hooks to support evolution and future extensions

These categories were identified to send the message that the project had put a stake in the ground by defining a system. What was sought from the vendor comments were improvements to the document but not proposals for alternate systems.

Midway through the vendor review cycle, DOCSIS held a vendor briefing to allow vendors the opportunity to ask questions about the RFI. The internal project team was interested in getting the highest quality of comments possible, so everything was done to help vendors understand the rationale in creating the document.

At the end of the review cycle, a large number of high-quality comments were received. The vendor-authors processed the comments for inclusion into the document. Many were included. The document was then released as an interim specification in March of 1997.

The following is a summary of the RFI. Within the document the network layer, data link layer, and PHY layer protocols and sublayers are defined.

1. The network layer is IP.
2. The data link layer is comprised of three sublayers:

   - logical link control sublayer
   - link-security sublayer conforming to the security specifications
   - MAC sublayer, suitable for cable system operation, that supports variable-length protocol data units (PDU)

3. The main features of the MAC protocol are:

   - CMTS-controlled mix of contention and reservation transmission opportunities
   - stream of minislots in the upstream
   - bandwidth efficiency through support of variable-length packets
   - extensions provided for future support of asynchronous transfer mode (ATM) or other protocol data units
   - support for multiple grades of service
   - support for a wide range of data rates

4. In the downstream direction, the PHY layer is based on North American digital video specifications (i.e., ITU-T Recommendation J.83 Annex B) and includes these features:

- 64 level and 256 level quadrature amplitude modulation (64-QAM & 256-QAM)
- 6 MHz occupied spectrum coexists with all other signals on the cable plant
- concatenation of Reed-Solomon block code and Trellis code supports operation in a higher percentage of North American cable plants
- variable depth interleaving supports both latency-sensitive and latency-insensitive data
- contiguous serial bit-stream (input and output) with no implied framing provides complete PHY and MAC sublayer decoupling

5. In the upstream direction, the PHY layer features are as follows:

- quadrature phase shift keying (QPSK) and 16 level quadrature amplitude modulation (16-QAM) formats
- multiple symbol rates
- data rates from 320 Kbps to 10 Mbps
- flexible and programmable CM under control of the CMTS
- frequency agility
- time-division multiple access
- support of both fixed-frame and variable-length protocol data units
- programmable Reed-Solomon block coding
- programmable preambles

In addition to specifying the network, data, MAC and PHY protocols, the RFI defines the CM and CMTS initialization and interaction requirements. Specifically, CMs can self-discover the appropriate cable system frequencies for reception and transmission, bit rates, modulation formats, error correction, and power levels. In order to protect service to other users, CMs are not allowed to transmit except under defined conditions. One or more channels in both the upstream and downstream direction can be provisioned to support scaling over a wide range of cable system topologies, service penetration levels, and traffic loading.

In the future, new types of CMs with enhanced capabilities may be introduced that could not be adequately defined within the one year time frame allocated for the creation of the original specifications. These capabilities may include new PHY layer modulation encoding, improvements to, or new options within, the defined PHY layer encoding, and differing traffic flows and classes of service (e.g., synchronous transfer mode telephony). Future-proofing is provided in the protocols to permit new types of CMs to use a standard CMTS to set up communication on an enhanced basis with the head end. In addition, new software can be downloaded to CMs over the cable network.

The benefits of the RFI include utilizing CMs from multiple vendors with a given vendor's head end equipment. This avoids a vendor "owning" a cable system after they have been deployed there. In addition, the consumer may purchase a CM at a retail outlet. As a result, cable operators do not carry the capital cost of the CM on the books.

Security System Specifications: The security specifications are a set of three documents: baseline privacy interface specification, security system interface specification, and removable security module interface specification. The intent of the baseline privacy interface specification is to provide privacy for data transmitted over the cable network. Baseline privacy is required for all CMTSs and CMs. However, baseline privacy is not intended to replace the security system described in the DOCSIS security system specification. The security system is designed to be employed when high-value content or the conditional access system need to be protected. The third document defines a removable security module interface. The removable security module interface allows the subscriber security unit, defined in the security system specification, to be implemented on a removable PCMCIA card.

Baseline Privacy Interface (BPI): Baseline privacy provides CM users with data privacy across the RF network by encrypting traffic flows between the CM and CMTS. A secondary goal of baseline privacy is to provide cable operators with basic protection from theft of service. To provide strong protection against theft of service, the CMTS would need to authenticate client CMs by establishing a unique, verifiable identity for a CM. In addition, the CMTS would need to link the CM identity to a paying subscriber and to the data services that a subscriber is authorized to access. Because data privacy is baseline privacy's principal service goal, baseline privacy's key distribution protocol does not authenticate the CM or CMTS (i.e., it does not employ authentication mechanisms such as passwords or digital signatures). In the absence of authentication, baseline privacy provides basic protection of service by ensuring that a modem, uniquely identified by its 48-bit IEEE MAC address, can only obtain the key material for services it is authorized to access.

The key management protocol for baseline privacy runs between the CM and CMTS. It uses RSA, a public-key encryption algorithm, and the electronic codebook (ECB) mode of DES to secure key exchanges. The cipher block chaining (CBC) mode of the (DES) algorithm is used to encrypt the packet PDU field in both upstream and downstream RF MAC packet data PDUs.

Baseline privacy is beneficial in that it provides a low cost, default, data privacy function. As the project evolves, baseline privacy may be enhanced to provide some security functionality.

Security System Interface (SSI): The security requirements are determined by the value of the information to be protected and the potential threats. Protection is needed for both basic monthly service and high-value premium services. Potential high-value premium services include streaming or broadcast-like services. The intent is not only to protect against the possible loss of revenue from theft of these services but also against the appearance of a breach in the security of the network. The types of attacks that the cable industry may face range from cloned units, modified valid units, cryptographic attacks, insider attacks, attacks on the conditional access scheme, and hybrid attacks of all of the preceding. As a result, the security system was designed to provide the following features:

- Identification and authentication of users—This is the process of recognizing the identity of a user or element of the system and verifying the identity of such user or element.

- Confidentiality of user information—This is the process of delivering information only to the intended recipient, of protecting the user's privacy regarding use of the network, and of protecting transmitted information against eavesdropping.

- Integrity of data and processes—This is physical or logical isolation of separate groups, such as system administrators, subscribers, and HFC network providers, from the resources or services the groups provide or use.

The security system specification defines the protocols, messages, and default encryption algorithms for both keys and content. The basic characteristics of the security system are as follows:

- Encrypted transmission from and to the subscriber over the HFC plant.

- Authentication of units is required. The CM security module must be authenticated to the cable operator, and the cable operator's control system must be authenticated to the CM security module. Digital signatures based on the RSA public key algorithms ensure the integrity of this process.

- Only units that have been authorized to receive content are provided keys to decrypt that content.

- All keys are kept secure. No key used to decrypt content or control messages is outside the boundary of the security module or element in its clear text form.

- The private key that is uniquely a part of a CM security and that is used to communicate in a secure and authenticated fashion must be maintained in a secure area of the security module. This key is never available outside the security module.

- Long-term evolution to a standards-based public key infrastructure, potentially based on the IETF standards.

The benefits of the SSI include long-term protection for the cable operator against theft of service and data. In addition, by specifying the security system, cable operators can use a single-key distribution system for equipment provided by multiple manufacturers.

*Removable Security Module Interface (RSMI):*The removable security module interface (RSMI) allows the subscriber security module in the security system to reside on a card that can be inserted into the CM. This provides an interface that decouples the security system development from the development of the CM.

The RSMI uses a standard PCMCIA physical connector on the motherboard and a PCMCIA card to house the subscriber security module. The electrical and logical interfaces between the RSM and CM are defined in the specification and do not adhere to the provisions of the PCMCIA standard. However, it should be noted that the subscriber security module does not have to be located on a removable security module. It may be located on the motherboard in the CM.

Benefits of the RSM include the ability to incrementally add the protection provided by the security system as the security threat to the data over cable service warrants. In addition, the RSM provides an upgrade path to cable operators if the security system is compromised.

Ongoing Specification Maintenance

Once the initial set of specifications was written and published, the project was faced with ongoing specification maintenance. The primary tasks of ongoing specification maintenance are the following:

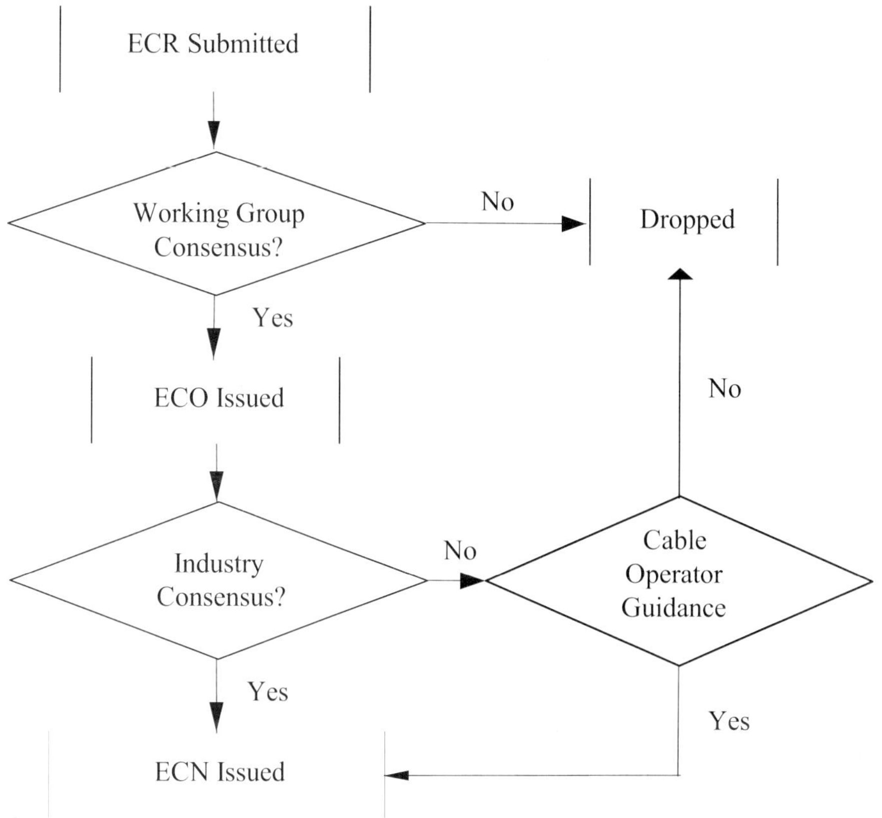

Figure 4: *Change Control Process*

- accept changes to the specifications that correct or clarify
- oversee evolution—Built into the specifications are hooks for more advanced physical layers, provisions for the transport of multiple protocol data unit types, and areas where extensions may be added.

A three-stage change control process was instituted to ensure that any changes to the documents would be reviewed by all interested parties. The change control process is detailed below and shown in *Figure 4*.

The first step involves submitting an engineering change request (ECR). The author of the ECR fills out an ECR form and submits it. Any party can author an ECR. If the ECR is properly filled out, a number and resolution date are assigned. The ECR is posted on the engineering change Web page in the vendor partition. This page includes the ECR number, author's name, a hot link to the ECR, the comment cut-off date, current status, and resolution (e.g., dropped due to insufficient vendor support).

Notification of the ECR is sent out to the appropriate working group. A working group is comprised of interested parties in a particular subject area. Input is accepted from the working group on the support for or the opposition to the ECR up until the resolution date. If the ECR is supported and supports the goals of the project, it is moved to an engineering change order (ECO). An announcement of the ECO is sent out to all working groups and a new resolution date is set. Comments that support or oppose the ECO are taken until the resolution date. A determination is made based on industry (cable operators, vendors, members of any working group) input whether or not to move the ECO to an engineering change notice (ECN). Once an ECO becomes an ECN it is officially part of the applicable interface specification. The ECN is then appended to the appropriate document for inclusion when the next revision is made. The documents are revised when there are enough ECNs to make it prudent.

The ECR, ECO, ECN process is designed to allow interested parties with technical expertise to create an ECR and have it reviewed by their peers. Once approved by a working group, an ECR becomes an ECO and is reviewed by all working groups. This provides full visibility of any potential changes to all parties interested in the DOCSIS documents. ECRs and ECOs are processed in a timely manner to avoid vendors holding up implementation decisions to await resolution of items that have been contested.

Testing

CableLabs has created a facility where vendors can test their equipment with the equipment from other vendors. Test specifications and methods are being developed jointly by CableLabs and the vendor community to verify compliance with the specification and interoperability with other equipment. The testing process began with the PHY layer (i.e., testing the silicon from multiple manufacturers) and is progressing to system testing. The end goal of the testing process is to certify the vendor's equipment as interoperable.

Summary

The process of using business-driven technology choices in the creation of interface specifications, making a test laboratory available for vendors to test interoperability with other vendors, and establishing a certification process is the cable industry's answer to the question, "How will we get interoperable CM that can be sold on retail shelves?"

Acknowledgments

The author wants to thank Roger Hay of Arthur D. Little for contributions to the RF Interface section; Chet Birger, consultant to Bay Networks, for contributions to the Baseline Privacy section; and Bob Wilson of Assured Digital Inc. for contributions to the Security System section and the Removable Security Module section.

The author would also like to thank Stuart Lipoff of Arthur D. Little for his leadership in helping create the specifications and the MCNS working group and management group for their leadership and vision throughout the entire process.

Notes

1 This technical report has since been withdrawn because the information became dated.

Residential Data Services via Hybrid/Fiber Coax Local-Access Networks

Stephen A. Grzelak
Harrison Miles, Jr.
Edward S. Szurkowski
William P. Weber, Jr.
Bell Laboratories–Lucent Technologies

Abstract

The deployment of two-way broadband hybrid fiber-coax (HFC) local-access networks will enable a wide variety of new communication services for consumers and small businesses. Interactive television, video-on-demand, video telephony, and other services focused on TV and video have attracted much attention. Broadband local-access networks also offer compelling data-oriented services that support computers and other devices in homes and small businesses. This paper gives an overview of these new residential data services and applications, the technical challenges involved in their delivery, and architectures and key components of the required networking infrastructure.

Introduction

Residential data services are designed to provide new high-speed data communications capabilities for computers and other devices in the home or small business. Today, a typical home computer might be equipped with a dial-up modem supporting 14.4 or 28.8 kbs over a standard phone line. In contrast, emerging residential data services carried over broadband networks could provide burst speeds up to 30 Mbs—more than 2,000 times faster than an ordinary dial-up modem. Basic rate integrated services digital network (ISDN) service, which is becoming increasingly popular with residential customers, supports only four to five times the bandwidth of common analog modems.

Printed with permission from Bell Labs Technical Journal, vol. 1, no. 1, Summer 1996, pp. 88–99.

Work at Home

High-speed data services for the home enable or enhance several applications. Work at home, for example, is becoming increasingly popular for a variety of social, economic, and governmental reasons. The number of people working from home, full time or part time, is expected to grow from 31 million in 1996 to more than 40 million by 2000.[1, 2]

Currently, most people with home offices use analog dial-up modems to connect their computers to corporate networks and servers. Unfortunately, this arrangement makes home computers much less responsive when accessing corporate resources than machines connected directly to local corporate networks. The difference in connection speed limits the utility of applications that depend on fast access to databases, files, and other resources on remote corporate networks. These include such applications as electronic mail, Intemet or intranet access, file transfers, and PC–based video conferencing.

With high-speed residential data services, computers in home offices could be just as responsive as those in the corporate offices connected to local-area networks (LANs). Trials of these new services indicate that productivity increases significantly compared to analog modem or ISDN arrangements. Driven by this potential for increased productivity, numerous firms with significant numbers of employees who work at home have expressed a willingness to wholly or partially subsidize the cost of providing high-speed residential data services.

Internet and On-Line Services Access

Access to the Internet and on-line services, both national and local, is also a highly promising application for residential data services. From 1989 to 1995, annual sales of computers for home use grew from 3.9 million units to more than 8 million units. During that same period, the number of subscribers to on-line services rose by almost 500%, now exceeding 11 million people. The Internet is projected to grow from 2 million users in 1994 to over 20 million by the year 2000.[3]

Like the work-at-home application, most home computers use standard analog dial-up modems to access the Internet or other services such as America Online* or CompuServe.* With the increasing richness of the available content, particularly on the World Wide Web, the time required to transmit graphics, images, and other multimedia elements over dial-up connections is frustrating to many users.

New Service Opportunity

One of the most attractive aspects of residential data services to current broadband network operators, such as cable television (CATV) multiple system operators (MSOs), is the additional revenue to be realized from their existing broadband plant investment. Early customer market studies have also indicated that this opportunity is truly incremental and will not cannibalize revenues from other CATV services. Subscribers view data services as work at home or educational services, as opposed to premium channels, which they view as entertainment. For example, residential customers signing up for data services from their CATV operator do not appear to be more likely to drop premium TV services such as HBO* or Showtime.*

Many broadband local-access plants, such as the two-way HFC networks operated by CATV MSOs and some telephone companies, are highly asymmetrical, supporting substantially more downstream than upstream bandwidth. This can limit the types of data services that are offered or the number of simultaneous users of a data service. Limitations caused by bandwidth asymmetry can often be overcome by careful access network design, such as restricting the number of subscribers sharing the upstream channel. Reducing the node size of HFC networks would ensure customer access to a range of data services.

Unfortunately, while most broadband local access plants, such as a CATV system, are technically capable of two-way, or bidirectional, communication between a customer's premises and a head end or central office, the embedded base of installations only provides one-way communication. They support forward, or downstream, communication from the head end or central office to the home, but not the reverse, or upstream transmission path (home to central office or head end). Offering data services over such a network requires a separate upstream communication mechanism, but this is not a significant limitation for most data services.

CATV MSOs currently operate the most widely deployed broadband local-access systems. CATV networks pass more than 90% of U.S. households, and approximately 20 % of these plants have been upgraded from all coax to HFC. However, only 5 to 7% of U.S. households are now served by a two-way CATV HFC system.

Given their substantial deployment, CATV HFC networks are becoming the principal broadband access platforms for near-term residential data services. The remainder of this paper gives an overview of the challenges to providing such services on typical CATV plants and describes proposed system architectures and equipment elements that can be used to overcome certain limitations.

System Architecture

Figure 1a shows a high-level architecture for an end-to-end residential data services system operating on a typical two-way CATV plant provided by an MSO. This system, which can also serve small businesses, is designed to support work-at-home applications, Internet, and on-line services access. At a very basic level, the system provides a means for transporting Internet protocol (IP) 6 packets between the computer or other devices in the home or small businesses, and remote networks and servers. Other services involving various forms of multimedia may require transport of additional types of data elements. How these will be supported is described in a later section.

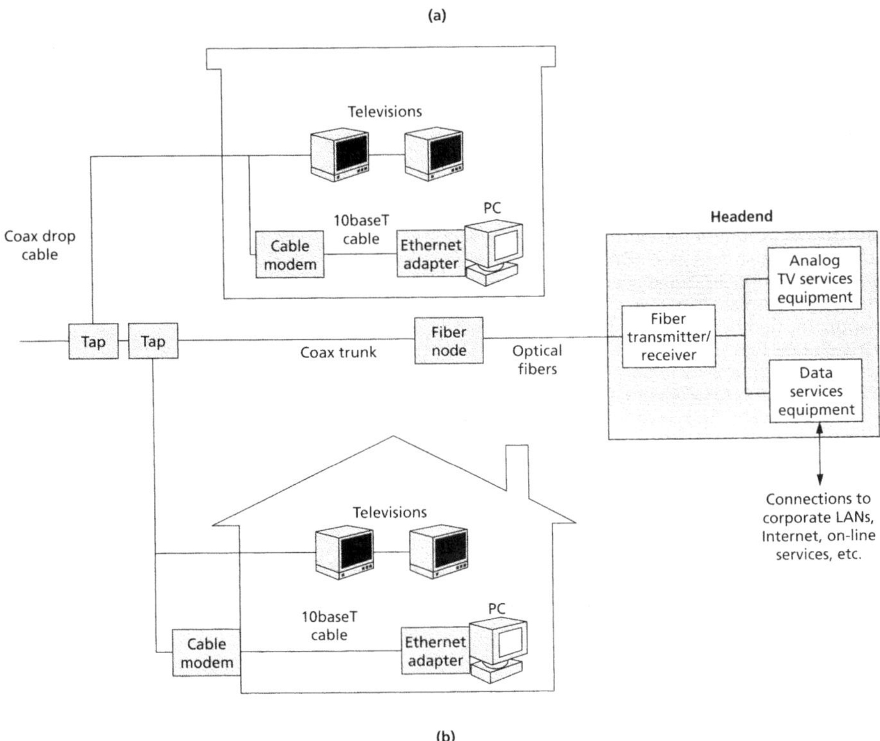

Figure 1: *High-Level Architecture for a Residential Data Services System on CATV Plant with RF Upstream and a) an Interior Cable Modem or b) an Exterior Cable Modem*

Equipment in the Residence or Small Business

In the home or small business, a computer containing an Ethernet interface card is connected via a standard Ethernet lObaseT cable to an external cable modem that emulates a standard LAN connection. Typically, the customer furnishes the computer, with the Ethernet adapter already installed, and the service provider (such as the CATV system operator) supplies and installs the cable modem. The cable modem is connected to the regular CATV radio-frequency (RF) coax drop cable, which also provides video services to televisions in the home or small business.

In this configuration, the cable modem is about the size and form of a standard analog dial-up modem and is a separately powered, stand-alone unit from the personal computer. As cable modems become widely available and standardized, modems may be incorporated directly onto the motherboards of computers designed for home use. The cable modem can function as a simple Ethernet bridge or as a router.

Figure lb shows a cable modem placed on the outside of the house and connected by standard interior 10baseT wiring to the computers and other various devices inside. The cable modem is powered from inside the house, rather than from the network, because it does not support a lifeline service—one that must operate in the event of a general power failure by the public utility. This exterior cable modem arrangement also makes it easier for the network operator to ensure the quality of the data services, because the cable modem is not affected by the widely varying quality of the CATV RF-distribution wiring inside the home.

CATV subscribers often reconfigure the RF cabling inside their homes by moving or adding equipment such as TVs, splitters, and amplifiers. Rewiring by consumers, which may not meet minimum standards for signal levels, reflections, and other parameters, can seriously impede the proper functioning of interior cable modems. Placing the modem on the outside of the house, where it can be connected with high-quality and secure cabling to the network plant, isolates it from these problems. As telephony services are offered on CATV plants, cable modem functionality could also be incorporated directly into the telephony network interface unit (NIU) mounted outside the residence.

CATV HFC Spectrum Plan

In the head end of the CATV system, point-of-presence (POP) equipment connected to the RF distribution plant provides the data services. The POP coexists with equipment that supports traditional analog or digital TV services. In a typical CATV installation, video services equipment and the data POP are fairly independent, sharing only access to the downstream and upstream RF spectrum.

Data Rates. Conventional CATV systems in North America divide their downstream RF spectrum into 6-MHz channels, each of which can carry a single standard National Television Systems Committee (NTSC) TV broadcast signal. Modern CATV systems can offer from 50 to more than 100 channels, depending on the maximum upper frequency that the plant was designed to support (typically 450 to 750 MHz).[8]

The data POP transmits data to the subscribers' cable modems by modulating a digital bit stream onto an RF carrier. The data rate and modulation are selected to ensure that the resulting digital signal occupies a standard 6-MHz downstream channel. Typical data rates range from 10 to 40 Mbs using quadrature phase shift keying (QPSK) or 16-, suited 64-, or 256-level quadrature amplitude modulation (QAM). Given the 6-MHz channel plan, the raw data rate (without forward error correction and various overheads) is determined by the modulation: QPSK supports about 10 Mbs, 16 QAM supports about 20 Mbs, 64 QAM supports about 30 Mbs, and 256 QAM supports about 40 Mbs.

As a practical matter, the modulation scheme chosen, and consequently the data rate supported by a 6-MHz channel, is strongly influenced by various impairments encountered on CATV plants. By requiring more robust modulation techniques to achieve acceptable error rates, impairments such as external and adjacent channel interference, nonlinear active elements that cause beats and other artifacts, and dips or peaks in the frequency response limit the maximum usable downstream transmission speed. Given the variation in channel quality between and within CATV plants, data POPs must support selection of the bit rate to be carried in a 6-MHz channel without expensive upgrades or reconfigurations. Experience in trials has shown that it is often difficult to predict the data rate that can be supported, making it preferable to defer this determination until system installation.

Downstream Data Channels. The characteristics of the downstream data channels generated by the data POP are engineered to allow clean coexistence with other services on the CATV plant, in particular with analog TV distribution. Proper settings of the relative RF levels of the data and analog TV signals allow unconstrained placement within the available downstream channels.

Unfortunately, in many CATV systems, all downstream channels have been allocated to TV and video services—there are no regular channels available to support a residential data offer. In this situation, rather than choosing to displace an existing TV channel, many CATV network operators would like to provide data services using the portions of their plant's downstream spectrum that are not suitable for analog TV signals. Compared with other services, data services can often use less robust spectrum because of the mechanisms typically provided for error correction and recovery.

The frequency response of a CATV plant usually rolls off gradually above the upper frequency that a plant is designed to support. Trial experience has shown that, in this region of a plant's spectrum, one or two additional 6-MHz rolloff channels are still usable for data services if the modulation techniques are appropriate for the observed carrier-to-noise ratio. (A rolloff channel is higher or lower in frequency than the nominal passband of a CATV system. Rolloff channels are not normally suited for transmitting analog TV signals.)

Because no other spectrum is available, many CATV network operators depend on rolloff channels for data services. Using rolloff channels also makes data services a truly incremental business opportunity. Traditional analog services cannot use the rolloff spectrum, so the operator is not forced to choose between an existing revenue-generating analog TV service and new residential data services.

RF Upstream Communication Channels. As this paper noted earlier, standard and rolloff channels can provide downstream transport for data services. Very few CATV plants, however, are equipped to transmit upstream traffic from cable modems to the head end. As *Figure 2* shows, the downstream frequency spectrumof most CATV ranges from about 50 MHz to 450 MHz to 750 MHz. The frequency range from 5 MHz to about 40 MHz has traditionally been reserved for upstream transport, although few services currently use it.

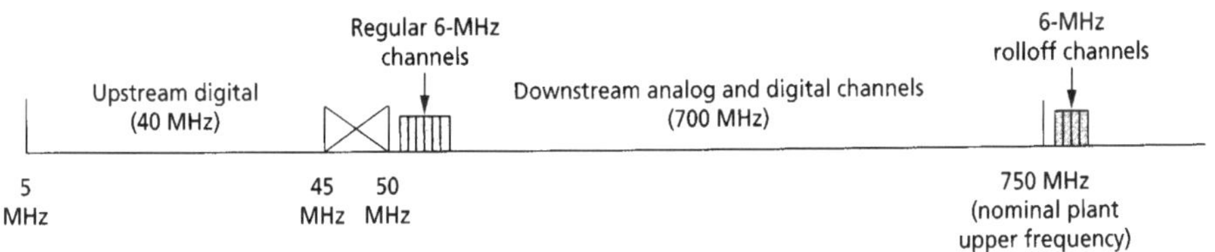

Figure 2: *RF Spectrum of a Typical HFC Access Network (not to scale)*

Consequently, most CATV network operators have not installed the reverse-path amplifiers necessary to use the reserved upstream bandwidth. Operators that have undertaken system upgrades to support two-way RE traffic have found it challenging to achieve adequate quality and reliability.[10] The use and characterization of upstream RF transport on typical CATV plants is an area of significant ongoing work and investigation.

POTS for Upstream Communication. Because of the scarcity of reliable two-way plants, offering residential data services on most CATV systems will require a separate upstream path. This problem is addressed in *Figure 3*, which shows the general

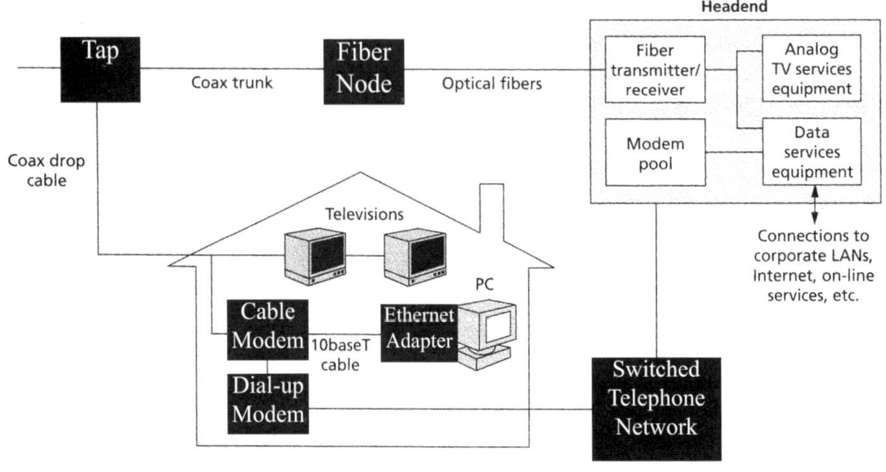

Figure 3: *Residential Data Services System on CATV HFC Plant with POTS-Return and an Interior Cable Modem*

architecture of a residential data services system. It uses a CATV plant to transport the high-speed downstream data and standard analog modems to transport the upstream traffic via the dial-up phone network. In an obvious variation of this architecture (not shown), the cable modem could be placed outside the house, as in *Figure lb.*

In the configuration shown in *Figure 3*, a plain old telephone service (POTS)–return cable modem in the home is connected to both an RF feed from the CATV system and a standard analog phone line. The home computer with an Ethernet adapter is connected to the cable modem, as in the architectures shown in Figures la and lb. The cable modem receives downstream data from the RF feed. When it is activated or has data to transmit, the cable modem dials up a modem pool on the service provider's premises to establish the reverse channel.

The downstream path over the CATV plant will support 10 Mb or more, but the POTS-return upstream path via conventional analog dial-up modems will only support up to 33.6 kbs. Surprisingly, perhaps, trial users report that extreme bandwidth asymmetry does not materially affect the perceived quality of the service for most applications. Compared to using either analog dial-up modems or ISDN for two-way transport, the high-speed downstream channel makes the broadband service appear to be substantially faster. Work-at-home users participating in trials have said that the system usually feels as if it is connected directly to their corporate LAN.

The architecture of the POTS–return broadband residential data services takes advantage of the asymmetry in bandwidth required by most applications. Word processors, spreadsheets, database front-ends, electronic mail readers, Internet browsers, and most other applications of interest to the consumers and work-at-home users tend to generate very little upstream traffic from the home computer to network servers. However, screen updates, file downloads, and other operations would typically require much greater bandwidth from the server to the computer. This has been confirmed by measurements in trials, where the average ratio of downstream to upstream traffic is greater than 10:1, with bursts to 100:1 or more.

On CATV HFC plants that have activated their reverse path, upstream data traffic from the cable modem to the head-end equipment can be transported via RF, although several technical challenges to this approach still exist. The spectrum reserved for reverse traffic is 35 MHz, significantly less than the 400 to 700 MHz allocated for downstream. In addition, the upstream band suffers from significant noise and unstable external interference, caused by problems such as leakage through poorly installed connectors, deteriorating or damaged coax, and missing terminators.

These factors drive an RF–return architecture to provide lower bandwidth for upstream traffic than for downstream, although this is still greater than a POTS–return system. Adaptive carrier-frequency assignment driven by real-time plant characterization, robust RF–modulation techniques, sophisticated contention and reservation protocols, and other technologies might be combined to gain the most efficient utilization of this difficult-to-use spectrum. This is another area of active investigation.

Architectures that provide asymmetrical bandwidth for residential data services systems will become a significant limitation when consumers and small businesses begin to operate servers from their homes or offices. This will require greater bandwidth upstream than downstream, the reverse of the near-term situation in which client computers are expected to be the dominant devices connected to cable modems.

The need for additional upstream bandwidth in HFC networks can be met by reconfiguring the plant architecture to use smaller node sizes, that is, fewer homes served by a fiber feed to and from the head end. This allows a larger fraction of the limited upstream spectrum to be used by each active subscriber. In an extreme situation, an individual small business (or home) could have its own dedicated fiber node, with no other users contending for the available bandwidth.

Low-Level Transport Protocols

One important element of the transport system for the residential data services system—the protocols used to control and format the upstream and downstream data paths—is not explicitly shown in *Figures 1* and *3*. At the lowest physical (PHY) level, this paper has briefly discussed such elements as modulation, channels, and frequencies for the core RF–transport functions. The level of protocol functionality above the PHY level, typically known as the medium access control (MAC) layer, can have a significant impact on the types of services and applications that a residential data services system can support.

The residential cable modem system transports IP packets between computers and other devices on the subscriber's premises and remote servers and networks. Supporting compelling applications involving multimedia, however, is expected to require transport of data types that cannot be effectively encapsulated and transported by an IP–based infrastructure.

Toll-quality voice, for example, depends on consistently low end-to-end transport delay. Many people are familiar with the annoying effects of delay on overseas phone calls carried on satellite circuits. The jitter in typical IP networks necessitates large buffers at the terminals to avoid overruns and underruns. These buffers, combined with compression and decompression processing, can introduce unacceptable delays compared to the switched telephone network.

ADAPt+ Protocol

It is difficult to predict how residential data access services for HFC networks will evolve over time. There are strong trends towards integrating data and nondata networking features within end-user applications. Highly compelling audio+data, voice+data, voice+video+audio+data, and applications for PCs and other devices are all under development. Wide-area network infrastructures are migrating toward asynchronous transfer mode (ATM) as bandwidth and quality-of-service issues become critical for supporting these applications.

As HFC data access systems become popular, they must be gracefully upgradeable to support these new integrated applications. Wholesale replacement of equipment optimized for a restricted class of data-centric applications (for example, those requiring only IP transport) is unacceptable. Although it may be technically feasible to add limited efficiency extensions to primitive transport systems in order to handle additional traffic types, it is vastly preferable to deploy a flexible core transport system from the outset.

Recognizing the need for flexibility in broadband access systems, researchers and developers created the ADAPt MAC protocol and its successor ADAPt+.[11, 12] Created as part of Lucent Technologies' HFC cable data program, ADAPt+ enables dynamic partitioning of a single bit stream. It supports a mix of traffic types, including constant bit rate (CBR) or synchronous transfer mode (STM), ATM, and variable-length packets—such as IP, Internet packet exchange/sequenced packet exchange (IPX/SPX), and Appletalk*—without segmentation into cells. If no traffic of a given type is to be transmitted, no channel bandwidth needs to be allocated. ADAPt+ can be used in both a POTS–return and an RF–return residential data services architecture.

ADAPt+ is a key distinguishing feature of the Lucent cable data system because it enables service providers to update their service capabilities at any time, with minimum changes to their installed infrastructure. For example, to take advantage of the current interest in work-at-home PC support and high-speed Internet access, a service provider could deploy a cable data system that supports only variable-length packet traffic, such as IP. Later, when applications using integrated voice, video, and data become popular, for example, in desktop collaboration systems, network operators can begin to carry CBR traffic on their plants without displacing any data-only cable modems that they have already deployed. New cable modems that support both CBR and data services operating on the same downstream and upstream RF channels can be installed as required by customers.

Data Point of Presence

The data POP is required in the head end or central office to support residential data services. The POP consists of equipment such as the RF plant interfaces, data switches or hubs, various servers, and operations support systems. Connecting the cable modems on the customer's premises with content servers and remote networks, the POP also provides security, authentication, system management, and other capabilities.

Depending on the specific needs or preferences of a service provider, the data POP can be based on ATM or Ethernet (IEEE 802.3) and its variants, such as 100baseT (IEEE 802.3u).[13] *Figure 4* shows a data POP design based on ATM, and *Figure 5* shows a POP architecture using 100baseT. The criteria for choosing between these alternatives include cost, existing data infrastructure, and anticipated service sets, but in either case, the key functions of the POP are essentially identical. The hardware elements may differ, but the software components are mostly the same. In the near term, a 100baseT/10baseT POP implementation is likely to be less expensive than an ATM–based POP for downstream small numbers of subscribers, but it is less scaleable to larger configurations.

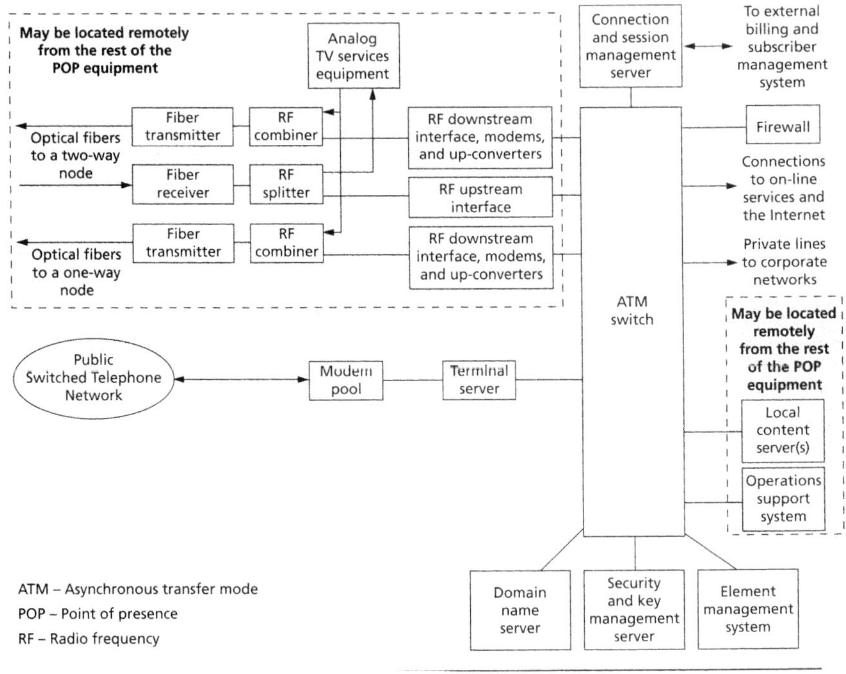

Figure 4: *ATM Data POP Architecture*

In *Figure 4*, an ATM switch forms the backbone of the POP, routing data between the various components. This switch can range in size from 16 ports (for small CATV plants or initial deployments) to several hundred ports, such as the GlobeView®-2000 ATM switch (for large installations serving thousands of subscribers). Private lines to corporate LANs, the Internet, and remote server complexes terminate directly on this switch, perhaps via bridges, to adapt IP networks to the POP's ATM backbone.

Three types of access interfaces connect the different elements of the distribution plant to the POP ATM switch. The downstream access interface accepts data (cells) from the switch for transmission to subscriber cable modems over the downstream RF channel. The access interface encapsulates the ATM cells in the ADAPt+ MAC protocol frame to generate a serial bit stream. An RF modulator modulates this bit stream onto an RF carrier with approximately 45 MHz intermediate frequency (IF). This IF signal is then up-converted to the channel frequency used on the plant and sent to the node's final RF output combiner. Depending on traffic volume, a single downstream access interface, RF modulator, and up-converter combination can drive any number of fiber nodes in the plant.

Figure 5 shows a sample Ethernet (100baseT) POP where data hubs provide the interconnections between system components. This particular architecture has two distinct networks, one on each side of a firewall that protects the servers controlling key POP functions. The firewall is discussed in more detail later in this paper.

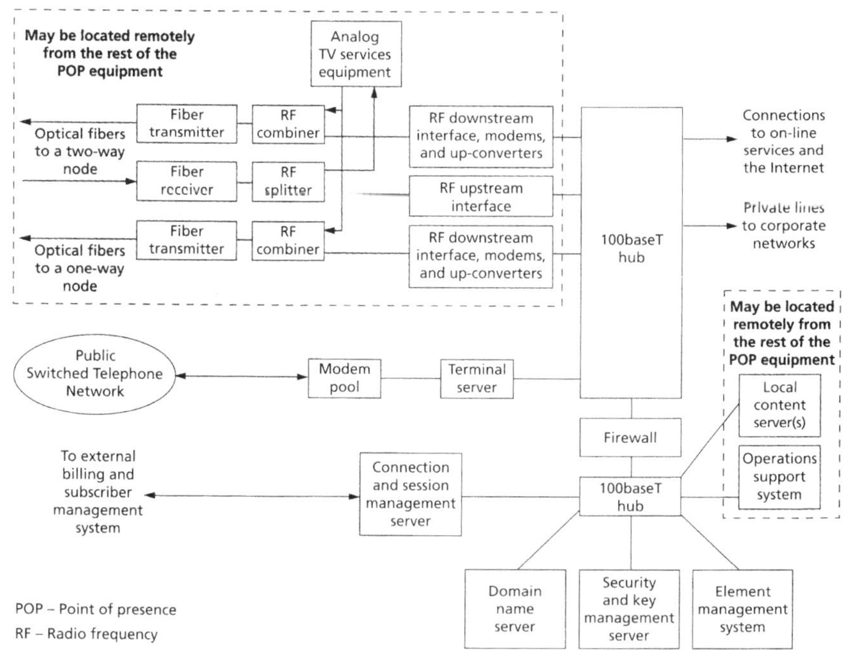

Figure 5: *Ethernet (100baseT) Data POP Architecture*

In an RF–return system, an RF upstream access interface demodulates upstream carriers received on the CATV plant RF–return path, parses the ADAPt MAC protocol frame structure, and passes the ATM cells to the switch for routing as required. As with the downstream, flexible assignments between the fiber nodes in the plant and the access interfaces are supported.

In a POTS–return data services system, a narrowband access interface, with an associated modem pool, accepts dial-up calls from the analog modems attached to the POTS–return cable modems on the subscribers' premises. The Internet standard point-to-point protocol (PPP) is typically run on these links, and the embedded access interface server hardware executes the host side of PPP.[4] The TIP packets from the PPP link are passed to the switch after they are segmented into ATM cells.

In a very large installation, the RF downstream and upstream interfaces can be placed in multiple local head ends and connected via private lines to the remainder of the POP equipment located at a regional center. The speed of these private line connections can range from DS–1 (1.5 Mbs) to OC–12 (622 Mbs) or higher, as required. Partitioning the system in this way takes the minimum amount of space and support resources in the local head ends, with significant economies of scale in the regional center. Similarly, some MSOs are using a regional master head end to provide analog TV services to multiple independent CATV systems clustered in a geographic area.

Security. Security is a serious concern for a residential data-services system, not only to protect system resources from unintentional or malicious harm from subscribers and outsiders, but also to protect the revenues of the network operator. Security measures prevent unauthorized use of the system, snooping on private data transmissions, intentional overloads, and other attacks on the system.

Figures 4 and *5* show the firewalls, which are used in the POP to tightly regulate the type of traffic that can flow between critical components of the system. These firewalls protect the various POP servers, but they do not create bottlenecks for most data flows. Attacks on the system can originate both from subscribers (authorized or not) and from outsiders on corporate LANs or the Internet.

In the ATM configuration in *Figure 4*, the firewall might be physically connected only to the ATM switch, while providing logical protection for all the servers through multiple provisioned virtual circuit (PVC) connections. The PVCs are defined to create two logically separate networks within the same physical switching infrastructure, allowing a single firewall machine to protect multiple servers. The security of this unified firewall arrangement depends on servers being unable to create or change PVCs or SVCs. In the100baseT POP in *Figure 5*, the firewall sits between two physically distinct networks in the POP, one trusted and one untrusted. A second switch, used with an ATM POP, can also create a physical partitioning, though at a somewhat higher cost. The firewall configuration options for the ATM POP are the subject of an ongoing study.

Authentication is the process of certifying that a potential user is authorized to access the system and obtain services. Authentication protocols are executed between client computers and the connection and session management server to verify the identity of the user and to enable the authorized services. On the client side, the authentication would be managed by software furnished by the service provider to each subscriber.

Each home on a fiber node of a CATV HFC plant receives the same downstream signals (in the absence of RE traps). This broadcast of potentially sensitive information in a data services system requires all data transmission channels from the POP to the cable modem to be individually encrypted, with separate keys for each POP–cable modem pair. The assignment and exchange of keys for this encryption are handled by the security server in the POP, with the data encryption standard (DES) or a similar data encryption process handled in the downstream access interface.

Similarly, the upstream signals in an RF–return data services system might also be received by numerous cable modems on a given fiber node. As with downstream signals, these transmissions must also be appropriately encrypted using keys furnished by the security server. A POTS–return system needs less encryption on the upstream channel, because each cable modem uses a private, nonshared dial-up call directly to the POP. The security of this approach is equivalent to that of normal dial-up data communications using standard modems.

The connection and session management server is also responsible for generating billing information used by a service provider. This server is typically interfaced to a subscriber management and billing system to input raw billing records and receive subscriber service update messages.

Other Facilities. Residential data services need operations support to control and monitor the system. The basic configuration for a data POP includes an element management system, which uses simple network management protocol (SNMP) to control and monitor various parameters of the local configuration, such as traffic loading, channel assignments, alarm status, and fault isolation. Element management systems in local POPs can be controlled, in turn, by higher-level operations support systems that monitor the status of a number of POPs in different locations.

The POP may also contain a domain name server (DNS) to map host and resource names into IP addresses. A DNS may not be required if the POP supports only work-at-home services. Instead, the address lookup might be provided by a server on the target corporate network(s). The POP would probably also be included in a DNS if Internet access services were offered.

Local on-line services are an attractive new business opportunity for operators of broadband residential data services systems. Such services might include local chat and bulletin boards, advertising for local businesses, movie schedules, local school event schedules and sports scores, government information, and TV programming schedules. These services would be provided by one or more local servers located in a single head end or in regional facility feeding multiple head ends.

All the servers shown in *Figures 4* and *5* are logical software entities. They can be implemented on independent hardware platforms, or multiple servers might execute on a single CPU. The capability to independently map server software elements to hardware platforms is key for cost-efficient scalability of the POP.

Summary

Offering residential data services on broadband local-access networks is an exciting new business opportunity for network operators. The core technology for providing these services has been developed, but significant work remains to refine the system architectures and create the equipment required to deploy and maintain these services on a large scale. Focused installations of first-generation HFC data services are expected to begin in the second half of 1996. Broader deployment is likely in 1997 and 1998.

Acknowledgments

During the past few years, numerous people have contributed to the residential data services program. Although they are too numerous to name, we acknowledge and recognize the work of our Lucent Technologies colleagues, as well as that of our AT&T colleagues, who made valuable contributions to this program before the company was restructured. We would especially like to thank N. L. Schryer and T. J. Killian of AT&T for sharing their experiences and insights with early trial deployments of residential data services. The comments and suggestions from the reviewers were also very useful.

*Trademarks

Appletalk is a registered trademark of Apple Computer, Inc.
America Online is a registered trademark of Quantum Computer Services, Inc.
CompuServe is a registered trademark of CompuServe Incorporated.
HBO is a registered trademark of Home Box Office, Inc.
Showtime is a registered trademark of Viacom International.

References

1 IDC/Link Resources, *A Comprehensive Review of PC–Based On-Line Services and Forecast of Future Growth and Strategic Directions,* New York, October 1994.
2 Market Strategies, Forecasting the Evolution of the American Information Highway, February 1995.
3 Simba Information, *Online Services: 1994 Review, Trends and Forecasts,* Wilton, Connecticut, August 1994.
4 P. F. Gagen and W. F. Pugh, "Hybrid Fiber-Coax Access Networks," *Bell Labs Technical Journal,* Vol. 1, No. 1, Summer 1996, pp. 28–35.

5 Paul Kagen Associates, *Cable Modems: A Strategic Analysis,* Carmel, California, 1995.
6 J. Postel, ed., *Internet Protocol,* RFC 791, September 1981.
7 ISO/IEC 8802-3:1993 (ANSI/IEEE Std 802.3-199 3), Information technology—Local and metropolitan area networks—Part 3: Carrier sense multiple access with collision detection (CSMA/CD) access method and physical layer specifications, 1993.
8 E. A. Bartlett, *Cable Communications,* McGraw-Hill, New York, 1995.
9 B. Sklar, *Digital Communications: Fundamentals and Applications,* Prentice Hall, Englewood Cliffs, N.J., 1988.
10 "@Home: 'Timing is Key,'" *Multichannel News,* March 11, 1996, p. 6.
11 J. E. Dail et al., "Adaptive Digital Access Protocol: A MAC Protocol for Multiservice Broadband Access Networks," *IEEE Communications Magazine,* Vol. 34, No. 3, March 1996, pp. 104–112.
12 B. T. Doshi et al., "A Broadband Multiple Access Protocol for STM, ATM, and Variable-Length Data Services on HEC Networks," *Bell Labs Technical Journal,* Vol. 1, No. 1, Summer 1996, pp. 36–65.
13 IEEE Std. 802.3u-1995, Supplement to ISO/IEC 8802-3:1993, Local and Metropolitan Area Networks: Media Access Control (MAC) Parameters, Physical Layer, Medium Attachment Computer, Inc. Units, and Repeater for 100 Mbs Operation, Type 100BASE-T (Clauses 21–30), 1993.
14 W. Simpson, ed., *The Point-to-Point Protocol (PPP),* RFC 1661, July 1994.

Cable-Modem Systems and the Law

Gordon E. Nelson
Patent Attorney
Banner and Witcoff, Ltd.

Abstract

This paper is a lay person's description of the areas of the law that are most important for cable modem–related businesses and of how the areas of the law relate to the various aspects of the cable modem business. This paper pays particular attention to the copyright laws, both as they relate to content and to software, and to the patent laws.

Introduction

People build cable modems, set up distribution networks, and provide content to be distributed to make money. You can make money in cable modem–related businesses because there are laws to keep people from copying content, stealing business information, copying inventions, and taking advantage of other people's business reputations. There are also laws against defamation, violation of privacy, and obscenity. Everyone in cable modem–related businesses needs to know both how to use these laws to protect their businesses and how to avoid damaging their businesses by breaking the laws. This paper will first present an overview of the relevant laws and how they relate to the various kinds of cable modem businesses, and will then consider the laws in more detail.

Overview of Intellectual Property Law

The area of law that keeps people from copying content, stealing business information, and so forth is called intellectual property law. It has four chief branches:

> *Copyright* law is what keeps people from simply copying works such as books, radio or TV programs, movies, pictures, music, or software. It thus protects the commercial value of successful works and makes it possible for Stephen King to earn millions writing novels or for Bill Gates to become the richest man in America by selling software.

> *Trademark* law is what keeps people from selling their goods or services under someone else's name. It thus protects the commercial value of a company's reputation.
>
> *Trade secret* law is what keeps people from using secret information belonging to other companies in their businesses, and thus protects the commercial value of a company's know-how.
>
> *Patent* law is what keeps people from using other people's inventions in their products or services, and thus it protects the commercial value of a company's research and development efforts.

A given piece of property may be protected by more than one of the varieties of intellectual property law. The source text for a computer program, for example, may be a trade secret; it is also protected by the copyright law, and it may be an implementation of an invention that is protected by a patent.

The purpose of intellectual property law is to encourage creative people by making creativity profitable. For example, it was not possible to make a living writing books before there was a copyright law. Before the copyright law, it was perfectly legal for a printer to copy any book. As a book became popular, other printers bought copies of the book, set type from it, and began printing. The price of the book quickly dropped to the cost of the paper, typesetting, and printing, leaving no money for the author. After the copyright law, the printer had to have the author's permission to print copies, and that made it possible for authors earn a living writing books.

Using someone else's intellectual property without permission is termed an *infringement.* An infringement may be *direct* (the company or one of its employees does the infringing personally) or *indirect* (the company does something which encourages others to infringe). Penalties for infringement include monetary damages, in which the infringer pays the owner of the intellectual property what the owner lost because of the infringement, and injunctions, which order the infringer to stop infringing. When justice requires it, courts are able to very quickly issue temporary injunctions called temporary restraining orders (TROs). If more than one entity is responsible for an infringement, they are *jointly and severally liable*, that is, if one or more of the infringers cannot pay, the ones who can pay are still responsible for the full amount of the damages. With copyrights, there are also criminal penalties for infringement. There is of course no infringement if the owner of the intellectual property has given permission to use the intellectual property. Such permission is termed a *license.*

Overview of Laws Related to the Publication of Harmful Material

There are a number of laws intended to prevent publication of harmful material. The first group of these laws are intended to prevent publication of material that is unreasonably harmful to individuals. These laws give rise to civil liability: they give the injured individual the right to sue the party that caused the harm for monetary compensation. The second group of laws are regulatory: they permit the government to intervene when people cause social harm. The prime example here are the laws against publication of obscene materials.

Regarding civil liability, the offenses may be classified as follows:

Defamation is publishing information about a person or business that the publisher knows to be false;

Violation of privacy involves publishing private information about someone without the person's permission, for example, publishing a photograph of the person; and

Publicity involves using a person's name or a character developed by the person without the person's permission.

The First Amendment to the U.S. Constitution limits the reach of all these laws. That is particularly true when the injured party is in public life. For example, someone who publishes falsehoods about a politician will be found liable for defamation only if the politician can show not only that the person publishing the falsehoods knew they were false, but also that the person published them with malicious intent. This standard makes it almost impossible for people in public life to win suits for defamation.

Obscenity is not protected by the First Amendment, but anything short of obscenity is. Broadcast media have been a special case in this regard. Because of the scarcity of broadcast channels and the invasive nature of broadcasting, the Supreme Court has consistently held that broadcast media are subject to a greater degree of regulation with regard to objectionable content than bookstores, nightclubs, motion picture houses, and the like. The issue with publication via the Internet was whether it would be treated like broadcasting. The Supreme Court decided the Internet will not be treated like broadcasting, since a user of the Internet must affirmatively search out objectionable material.[1]

The application of these laws to material originating with an Internet content provider is generally straightforward. The Internet content provider is responsible for his or her content just as a newspaper is. Because Internet service providers provide content that they do not originate, such as bulletin boards, the issue becomes more difficult. The analogy then was to bookstores or newsstands, which are liable only if they have actual or imputed knowledge of the offending material. The problem with the bookstore analogy for Internet service providers was that the easiest way to avoid liability was to make no effort to control the content of material provided by third parties. However incompatible with the business needs of certain service providers, this kind of "free for all" approach was to give users some protection against objectionable third-party content. Congress has dealt with this problem with regard to both civil liability and objectionable content by enacting two principles into law:

> that an Internet service provider will not be held liable for content provided by third parties; and
>
> that an Internet service provider will not be held liable on the basis of any attempt by the Internet service provider to limit access to certain services.[2]

The Law and the Different Kinds of Cable Modem Businesses

How the law is relevant to your business depends on which part of the cable modem realm to which it belongs. In general, the relationship between the intellectual property laws and the areas of the cable modem business is as follows:

Content Providers

The copyright law protects the very heart of the content provider's business. Without the copyright law, a provider of content would no more be able to make money than an author could before copyright. Indeed, the ease and precision of digital copying has made the copyright law more important than ever for the content provider. A content provider that wants to remain profitable can neither ignore the copyrights of others nor fail to enforce its own copyrights. The advent of Internet services such as bulletin boards or chat rooms, where the content is provided by Internet users, has made it much harder than before for the content provider to be sure that it is not involved in direct or indirect infringement of someone else's copyrights. It has also made it harder for the content provider to make sure that the content being distributed does not violate the laws against theft of trade secrets, libel, violation of privacy, and obscenity, and that in turn has led to the enactment of the Federal statutes discussed above. Content providers need to protect their own trademarks and trade dress and make sure that the content they are providing does not misuse other people's trademarks and trade dress.

Pure content providers generally don't have to concern themselves much with patents. Content providers should be aware, though, that user interfaces are patentable. Infringement is therefore a concern if the content provider also provides the user interface. If the content provider has a new user interface, it may want to consider protecting it with a patent. For a content provider, the most serious penalty for infringement is generally the injunction, since it will prevent the content provider from continuing to provide the infringing content. A temporary restraining order can be obtained quickly enough to require a content provider to cancel a scheduled broadcast.

Providers of Transmission Media

A business which provides a transmission medium for content may be made a party to a suit for copyright infringement on the basis that it directly infringes by making copies in the course of transmission or that it indirectly infringes by encouraging others to infringe. It may similarly be made a party to a suit for theft of trade secrets or for libel. The real reason for making the transmission business a party is generally that the transmission business is large and therefore has "deep pockets," and will have the money to pay all the damages if it is found to be liable. In general, a transmission business is liable for what it transmits only if it has control over what it transmits. A telephone company, for example, is not liable for slanderous telephone calls made using its network.

It has not yet been completely settled whether the copies of a work made by a transmission company in the ordinary course of transmitting the work constitute copyright infringements. The closest analogy to this kind of copying is the copying performed by a computer system in the course of the normal operation (when a program is copied from the disk drive to the main memory for execution, for example). The copyright statute expressly permits such copying.[3] Copying performed in the course of the normal operation of a network will probably be treated in the same fashion, either expressly by amendment of the statute or by court decision. The same should be the case with the copy that is made when a Web page is cached.

Businesses that provide transmission media do, of course, need to be concerned about patent rights. If the system built by the business is new and the novelties are advantageous, it may be worthwhile for the business to patent it. At the same time, the transmission business may need to be concerned about their system infringing other peoples' patents. The devices that constitute the system may also be covered by patents. In general, though, the purchase of a device gives the purchaser the right to use the device under any patents held by the maker, and a purchaser of a device may contract with the device's maker to require the maker to indemnify the purchaser if the device infringes anyone else's patents.

Cable Modem Makers

Makers of cable modems generally need not be concerned about indirect liability for copyright infringement because their modems have been used to infringe copyrights on content received via the modem. The modems have other legitimate uses, and the modem maker has no control over how the purchaser uses the modem. Most cable modems do, however, have a software component, and the cable modem manufacturer must ensure that everything in the software not licensed from others is original with the cable modem manufacturer, and not a copy of someone else's software. Whatever is not original infringes the copyright in the software and may also represent a theft of a trade secret.

Cable modem makers, like makers of equipment generally, have most to be concerned with patents and trade secrets. They need to protect their own trade secrets and patent their own important innovations, while also ensuring that they do not use trade secrets or patents belonging to others without a license.

Details of Intellectual Property Law

The remainder of this paper discusses the details of intellectual property law and the significance of the various kinds of intellectual property law for cable modem–related businesses.

Details of Copyright Law

What Copyright Protects: Intellectual property law involves a tradeoff between the need to reward authors and inventors and society's need for a free flow of information. In the copyright law, the tradeoff takes the form of prohibiting the copying of the expression contained in an author's work, but not the ideas or information contained in the work. The expression in a work refers to those elements that belong to the author and not to the subject matter. For instance, the copyright in a mathematics text protects the way in which the author presents the mathematics, but does not protect the mathematics itself. It is thus not a copyright infringement to use the author's mathematical ideas in another textbook; it is, however, an infringement if they are presented in the same way. The amount of expression in a work varies with the kind of work. Works of fiction are almost pure expression; the list of numbers in a telephone book, on the other hand, is almost pure factual information. There is nevertheless a copyright in the list of names and telephone numbers. The copyright covers the arrangement of the list on the page, and anyone who simply makes a xerographic copy of the list of numbers in a telephone book without permission infringes that copyright.

Copying in the copyright context covers not only verbatim copies of a given work, but also any other work which uses expression contained in the given work. Works that are not verbatim copies but still use the expression of the given work are termed derivative works. Examples of derivative works are translations into other languages, adaptations, such as movie adaptations of novels or stage plays, and works made by combining the works of others. In the technical realm, translations of source code into other computer languages or into object code are derivative works. The author of a derivative work often adds his or her own expression to the original work. When that is the case, the author of the derivative work has a copyright in the derivative work; however, he or she cannot make the derivative work without the permission of the owner of the copyright in the given work.

Infringement of Copyright: Direct infringement of copyright is a *strict liability* offense; that is, the infringer is liable even if it did not know it was infringing. Since it is not possible to tell from looking at a work whether it is copyrighted, the only safe assumption is that everything is copyrighted. With indirect infringement, there is liability only if the alleged indirect infringer knows of the infringement and is in a position to exercise control.

Copyright law provides a fair use limitation on the author's rights under the copyright law. The limitation provides that use of a portion of a copyrighted work for purposes such as criticism, comment, teaching, news reporting, and the like is not an infringement. What makes a fair use fair is a complex matter. For the present purposes, it is enough to note that a use will likely not be found to be fair if the use is made for commercial purposes and has a substantial detrimental effect on the commercial value of the work.

Controlling Infringement: The material available on the Internet is in digital form; this makes illegal copying and distribution far easier than before, and has increased interest in technical ways of making illegal copying more difficult. There are basically two approaches: making it more difficult for an infringer to gain access to the work, and marking the copy with the source of the work or even the source of the copy. The first approach uses some combination of control of access to the work, restriction of what the user can do with the work, and encryption to protect a work while it is in storage in a server or client or in transit across the network. The second approach uses a technique called *watermarking* to encode information invisibly into the work. The information typically includes an identification of the source of the work and often an identification of the person for whom the copy in question was originally made. The encoded information is resistant to changes in the digital encoding techniques used in the representation, and it cannot be found in the work unless one has the key used to encode the information or a copy of the work that does not have the encoded information. If infringement is suspected, the key can be provided to a watermark reader that is able to read the information in the watermark from the work. It should also be noted that while the Internet has made illegal copying and distribution easier, it has also made detection of at least commercial infringement easier. The reason for this is that anyone who wishes to distribute via the Web can also be found via the Web.

Getting a Copyright: Copyright arises in a work as its author creates it. The copyright is valid throughout the world. It is no longer even necessary to put a copyright notice in a work, though it is still wise to do so. The legal forms of the notice are ©*<owner's name> <year date of creation>*, for example, © *Gordon E. Nelson*, or *Copr. <owner's name><year date of creation>*, or *Copyright <owner's name> <year date of creation>*.

However, before one can sue someone for infringement of your copyright, he or she must register the copyright with the Register of Copyrights in the Library of Congress. The fee for registration is low, and one can do it without assistance. The forms are available from the Copyright Office.[5] In general, if a work is to be widely distributed, it is best to register the copyright before the work is published. Works that are not widely distributed are usually registered only after the copyright is infringed. Unless a work is a work for hire (a work written specifically for another), the original owner of the copyright in a work is the author, and another party can own it only if the author agrees to assign the copyright to the other party. In the case of a work for hire, the copyright belongs to the party for whom the work is written. The most common cases of works for hire are those written by employees. Where the work for hire is written by a non-employee, a business must take care that the non-employee's contract conforms to the requirements set forth in the copyright statute.[6] The term of a copyright in a work that is not a work for hire is the life of the author plus fifty years; the term for a copyright in a work for hire is seventy-five years from the year of first publication.[7]

The Limits of Copyright: As stated above, a copyright protects only the expression in a work, not the ideas or information it contains. If a new work is made using a copyrighted work, but the new work contains none of the expression from the copyrighted work, the new work does not infringe the copyright in the copyrighted work. As a result, a reimplementation of a computer program made without copying the expression of a copyrighted program does not infringe the copyright in the program. This so-called "clean room" technique even makes it possible to use a copy of a copyrighted program obtained legitimately as the starting point for a new program without infringing the copyright in the program. In the clean room technique, a first group of programmers reads the copyrighted program and writes an exact functional description of the program. The functional description is then given to a second group of programmers, which writes a new program from the functional description. Since expression is not function, the functional description by definition cannot contain any of the expression from the copyrighted code and the program written from the functional description cannot infringe the copyrighted code. Of course, the programmers in the second group may not have access to the copyrighted program nor to the programmers in the first group. It should be emphasized here that the copy of the program used by the first group of programmers is legitimate only if it was obtained without infringing anyone's copyright or trade secret rights.

Details of Trade Secret Law

Trade secret law protects a business against the theft of valuable business information that is not generally known. Examples of trade secrets are computer source and object code, customer lists, methods of manufacture, and formulas for materials. In many instances, the writing that embodies a trade secret is also protected by the copyright law. A piece of information is not a trade secret unless its owner does in fact keep it secret. Doing this involves labeling the information as confidential and restricting access to it. A trade secret lasts until it becomes public. The owner of the trade secret may seek damages from the person who made it public without his permission, but there is no trade secret liability for those who distribute it further once it is public. A trade secret may be communicated to others under a non-disclosure agreement, an agreement in which the ones receiving the trade secret promise to keep it in confidence.

Details of Trademark and Trade Dress

The purpose of this area of intellectual property law is to prevent unscrupulous sellers from deceiving buyers about the source of the goods or services by using the trademarks or trade dress belonging to other sellers. Trademarks are the marks that businesses use to identify the products they make. A famous example of such a mark is Coca-Cola®. Trade dress is packaging that has become so closely associated with a product that it identifies the product to the consumer, for example, Coca-Cola's "Coke® bottle–shaped" bottles. A trademark or trade dress is infringed if it is used by someone other than the owner or the owner's licensee to identify products. A trademark is *diluted* if it is famous and is used in such a way that it lessens the capacity of the famous mark to identify and distinguish goods and services.[8]

A constant concern of trademark owners is that the mark will become generic, standing for a class of products rather than the products of a particular manufacturer. If this happens, the mark is no longer a trademark. An example of the problem is the tendency of people to use the Xerox® trademark to indicate any xerographic copy. To prevent a mark from becoming generic, the owner must police it, constantly reminding people to use a true generic term when describing a class of products and to use the trademark only with reference to the owner's products, for instance, "xerographic copies" and "Xerox® brand copiers." Non-registered trademarks are identified by the superscript "TM" and registered marks by the ® symbol. Trademarks may be registered at either the state or national levels. Part of registration is an examination to determine that the trademark is unique enough with regard to the goods or services that it will mark that it will not cause consumer confusion. The U.S. Patent Office is the registry for national trademarks in the United States. If you want to use your trademark in a foreign country, you must register it in that country.

A new problem in the area of trademark protection is ensuring that a business can use its trademark as its Internet domain name. Internet domain names are assigned by Network Solutions, Inc. That organization now permits a trademark owner to intervene when its registered mark is assigned to someone else as a domain name, but this does not solve the underlying problem. Namely, many businesses have similar trademarks, and the techniques used in the trademark realm to distinguish among similar marks are not available for Internet addresses. In addition to protecting their own marks, content providers should as a matter of courtesy make sure that trademarks belonging to others are properly used in the content they provide. The courts may take the appearance of a link to a Web site or of the Web site itself to be a kind of trade dress, and a business that imitates the appearance of another business's links or Web site may be found to have infringed the other business' trade dress. Of course, if the imitation reaches the level of a copying of expression, the imitator will also infringe the other business's copyrights in the appearance of its links and Web site.

Details of Patent Law

What Patents Protect

Patents protect technological concepts implemented in machines, compositions of matter, and methods. This distinguishes patents from copyrights, which protect expression, and trade secrets, which protect information that is not publicly known. The machines, compositions of matter, or methods (the inventions) protected by a patent are defined by the patent's claims Machines or methods implemented using software and a processor may be patented on the same basis as any other machines or methods. During the term of a patent, the patent protects the inventions described by the patent's claims. No one can make, sell, or use a machine, method, or composition of matter which implements the invention in the territory covered by the patent without the permission of the owner of the patent (the patentee.) The patent's term is typically sixteen to eighteen years. Note that a patentee may have a right to exclude others from making, selling, or using the patented invention without itself having the right to make, sell, or use the invention. This happens when the patentee's invention cannot be made without using an invention patented by another. In that case, the patentee must have a license for the other invention in order to make, sell, or use his or her own invention.

When they can be obtained, patents have many advantages over trade secrets or copyrights:

The protection of technical concepts offered by patents is far broader than the protection of expression offered by copyrights. You cannot get around a patent using a clean room technique.

The protection offered by patents does not require secrecy. Once a patent application is filed, the work it discloses can be freely published without affecting the patent rights.

Patents protect not only against copying, but also against independent reinvention of the patented invention. Indeed, if one keeps a concept as a trade secret and someone else patents the same concept, the person who kept the concept as a trade secret cannot practice the trade secret without permission of the patentee.

The main disadvantage with patents is the expense in time and money required to get one. In contrast to trade secrets and copyrights, which cost almost nothing to obtain, obtaining a patent requires paying substantial fees to a patent office and much more substantial fees to the patent attorney who writes the patent application and persuades the patent office to grant you a patent. Even a simple patent application typically costs about $6,000 to write and file, and complex applications are far more expensive. Persuading a patent office to allow a patent generally takes at least two years from when the patent application was filed and often takes much longer. In most countries, once the patent office has allowed a patent, an issue fee must be paid, and once the patent has issued (been published and become effective), the owner must pay maintenance fees to keep the patent in force. A minor disadvantage of patents is that they are published. Thus one cannot simultaneously patent something and keep it as a trade secret. Moreover, in the U.S., the application is required to disclose the best way known to the applicant at the time of filing of making and using the invention.

Legal Requirements for a Patent

For a patent to issue on a patent application, the application's claims must describe an apparatus, method, or composition of matter that is novel and non-obvious when compared with the prior art[9] The prior art is what was publicly known when the invention was made (in the United States) or the first patent application that was filed (everywhere else). "Novel" simply means different in some way from the prior art. "Nonobvious" means the difference is technically significant. Since almost six million U.S. patents have issued, it is clear that the technical significance need not be great. In general, if the inventor is skilled in the technology the invention belongs to and thinks the invention is new and neat, it is in all probability novel and nonobvious. In most cases, the issue is not whether one can get a patent, but rather whether getting a patent makes business sense

Getting a Patent

In the United States, you have to file a patent application within one year of the earliest of the dates of first commercialization or of publication by the inventor of the concept claimed in the application. Publication is anything that makes the concept publicly known, including publishing a paper, giving a public presentation, and putting the concept on a Web page. The requirement that the patent application be filed within one year of the first commercialization keeps an inventor from using an invention commercially as a trade secret for twenty years and then getting a U.S. patent on it for eighteen additional years. In the rest of the world, the inventor must file the patent application in some patent office before the concept claimed in the application is published. The rest of the world does not care about commercialization, as long as the concept remains unpublished. The original owner of any patent is the inventor; it can be owned by another party only if the inventor assigns it to that party.

A patent application consists of a specification, which describes the invention so that those skilled in the technology of the invention can make and use it, and one or more claims. The specification generally includes drawings. The technical content of the specification cannot be changed after it is filed, though the specification may be edited to correct typographical errors and obvious technical errors. The application must therefore be carefully reviewed by the inventor before it is filed. The time to write and review a patent application should be taken into account when coordinating patent filings with publications or public presentations made by the inventors, product announcements, product releases, and the like.

Once a patent application is filed, it must be prosecuted Patent prosecution is the process of reaching agreement with a patent examiner that the invention described by the claims is in fact novel and nonobvious when compared with the prior art. The patent examiner is a Patent Office official who is trained both in the relevant technology and in the patent law. If the examiner knows of prior art that appears to be close to what is claimed, he or she rejects the claims. In the rejection, the examiner cites the prior art upon and explains why it renders the invention unpatentable. The patent attorney then either demonstrates to the examiner that the rejection is wrong and that the claimed invention is novel and nonobvious with regard to the cited prior art, or amends the claims so that the invention they describe is novel and nonobvious when compared with the prior art. The amended claims must of course still describe an invention that is disclosed in the patent application. When the examiner agrees that the invention described by the claims is novel and nonobvious, the examiner allows the case. If the patentee pays the issue fee, the allowed patent issues and its term begins. In most cases, the time from filing to issue is two to three years. The term ends twenty years from the filing date, so the length of the term is twenty years minus the time it took to prosecute the patent application. To keep the patent in force, you have to pay maintenance fees three, five, seven and a half, and eleven and a half years after issue.

Foreign Patents

Patents are national. A U.S. patent, for example, applies only in the United States. To obtain patent protection in another country, the inventor has to file for a patent there. In almost all countries, an inventor can file in the foreign country within a year of the time he or she files in the United States and have the foreign patent application treated as if it were filed on the same date as the U.S. application. The foreign applications have to be prosecuted in the same fashion as the U.S. application. An agent or attorney who has the right to practice before the relevant patent office handles the prosecution. Foreign patent offices have fees similar to those in the U.S. patent office, though actual amounts vary considerably. In addition, if the foreign country is non-English speaking, the application must be translated.

Obtaining foreign patents is an expensive proposition. Where to file foreign patents is an important and difficult business decision. There are two ways of approaching the problem: obtaining patents in countries where the invention will probably be manufactured, and obtaining patents in countries where the main markets for the invention are located. With cable modem equipment, the second approach would seem the best, since there are more countries that can manufacture the equipment than there are countries that have the systems required to use them. A problem in making foreign filing decisions is the length of the patent term: countries that may not be of interest when the patent is filed may be of major importance before the patent's term ends.

Enforcing a Patent

It is up to the patentee to enforce the patent. A patent is enforced by suing the infringer for patent infringement. If the infringer is not willing to settle, the parties end up in the U.S. District Court before a judge and often a jury. Suits for patent infringement are expensive, and their outcomes are uncertain. The issues in patent litigation are usually legally and technically complex, and neither the average judge nor the average jury member is technically trained. Because patent cases are relatively rare, judges tend to be less familiar with the patent law than with other areas of the law. Because the judge and jury have to be taught the technology, the typical patent infringement trial becomes a battle of experts. The experts are expensive, and the outcome may depend more on the jury's perception of the experts than on the technical facts.

What to Patent

Obtaining a patent is easy, though slow and expensive. What is hard is obtaining a good patent. Most patents are quite literally not worth the money it cost to get them. Some patents, on the other hand, are the foundations for profitable businesses, and a few are worth billions of dollars. With patents, then, quality is far more important than quantity. The wisest way for a businessman to approach patenting is to get good patents on the technology at the heart of the business. A good patent is simply one that protects economically important technology. Obtaining good patents is thus a matter of understanding the technology and the marketplace and ensuring that what is being patented makes a difference in the marketplace. An important consideration here is ease of enforcement: it should be easy to tell whether someone is infringing your patent.

Patents on interfaces, human or otherwise, are often good patents. Interface patents have a number of advantages:

They control who can connect to the interface;
Infringement is easy to detect;
Narrow claims are effective, since the infringer must copy the interface exactly.
Broad claims to an interface can be used to help make a particular version of the interface standard: the patent is licensed on the condition that the licensee use only the particular version.

Obtaining a good patent is to some extent made easier by the length of time it takes to prosecute a patent: during prosecution, the patentee's understanding of the technology and the marketplace will increase, and the patent claims can be amended to reflect the improved understanding.

Hedging Your Bets

If there is any chance the inventor will want foreign patents, he or she must file a patent application before publicizing the product; generally, though, one does not know at that point whether the product will succeed, or which of its features will prove most valuable. As a result, it is difficult to figure out either whether the invention is worth patenting at all or which of the product's features ought to be patented. One way of dealing with this problem is a provisional application A provisional application is simply a complete description of the product (schematics, or in the case of software, source code will do) that is filed in the Patent Office. It will typically cost less than $500 to make such a filing. The person who filed the provisional application has one year to change it into a regular application. If a regular application is filed, it receives the provisional application's filing date and is examined in the usual fashion. The inventor can thus file a provisional application on a new product just before announcing the product, and he or she can wait almost a year to determine whether the product merits filing a patent application or which features the application should claim.

Intellectual Property and Your Business

Businesspeople need to concern themselves with two aspects of intellectual property: getting the most out of the intellectual property that belongs to the business, and avoiding liability for the use of other people's intellectual property.

Intellectual Property as an Asset

The first step in making intellectual property an asset in one's business is making sure one owns it. As mentioned earlier, patents are originally owned by their inventors, and copyrights are originally owned by their authors. Consequently, every business needs a clause in its employment contract that obligates its employees to assign the intellectual property they create in the course of their employment to the business. The same is true for contracts with consultants. The employment and consultants' contracts should also contain nondisclosure clauses forbidding employees and consultants from disclosing confidential information belonging to the business.

There are three basic uses for intellectual property: keeping competition out of one's markets, using the intellectual property itself to make money, and using the intellectual property to gain design freedom, that is, the freedom to design products without worrying about intellectual property belonging to the competition. To some extent, these uses are mutually exclusive: for example, if one wants to use intellectual property to keep competition out of one's markets, one cannot license it to others to gain revenue or to gain design freedom.

Keeping the Competition Out

Keeping the competition out of one's markets is the classic use of intellectual property. Copyrights keep others from simply copying and reselling your content and programs; patents keep others from making, using, and selling apparatus and methods that fall within the scope of the claims of your patents; trade secrets keep your competitor from using your advantageous techniques. With deregulation and globalization, intellectual property is now one of the few legal ways of maintaining a monopoly. Of course, the monopoly extends only as far as the intellectual property does. It is illegal to use a patent monopoly, for example, to force a customer to buy a product that is not covered by the patent.

Using Intellectual Property to Make Money

You can use the intellectual property itself to make money in several ways:

> *Raising capital.* In businesses that sell content, businesses that sell software, and businesses that make equipment, the business' most important assets are often its intellectual property. The more intellectual property the business has, the easier it will be to raise capital.
>
> *Using intellectual property for licensing revenue.* If your business cannot get full value from its intellectual property, it makes sense to license it to someone who can. For example, a U.S. content provider will often find it easier to license its copyrights to foreign content providers rather than to enter foreign markets itself The same is the case with a patent or trade secret that is no longer important in your business.
>
> *Using intellectual property to make money when your business fails.* While no one likes to think about business failures, the fact is that the success of a business often has little or nothing to do with the quality of its intellectual property. If a business has good intellectual property, the intellectual property will keep its value even when the business fails.

Design Freedom

Competitors in established industries typically use intellectual property in this manner. Each competitor cross-licenses its technology with its competitors, and thus each competitor can design without fear of infringement suits from other competitors. Of course, a business can participate in cross-licensing only if it has intellectual property that its competitors need.

Liability for Infringement

A business that infringes another's intellectual property is liable for damages and may be subject to an injunction. Damages usually consist of what the owner of the intellectual property lost because of the infringement; however, if the infringement was *willful,* that is, the infringer knew of the existence of the intellectual property and nevertheless went ahead and infringed, the courts may impose punitive damages. With a copyright, these may be up to $100,000; with a patent, this may be up to three times the actual damages. As painful as damages are, they are often less painful than an injunction, which can force a content provider to cease providing the infringing content or force a network operator to cease operating a network that infringes a patent.

Risk Assessment

Because of the increased liability that results from willful infringement, it never makes sense for a business to infringe intellectual property knowingly. However, the amount of effort that a business puts into trying to determine whether a planned product or service infringes depends upon factors such as:

the likelihood that the infringement will be discovered; and
what the owner will do if the infringement is discovered.

With the first factor, patents on software-related inventions can serve as an example. If a patent is addressed to a function performed in the depths of an operating system, there is comparatively little risk that the patent owner will detect an infringement; if the patent is addressed to a user interface, there is a far higher risk that the owner will detect the infringement. There is obviously also a high risk of detection if the patent covers an advertised feature of your product.

With the second factor, if the owner of the intellectual property is not a competing business, an infringement suit can almost always be settled with money. In this situation, the owner of the intellectual property has no interest in putting you out of business; all the owner wants is a share of the money you are making. Further, if the owner is a private person, a monetary amount that is small for the business may be very large for the person. Where the owner of the intellectual property is a competing business, watch out! Their aim may be literally to put you out of business, and if that is the case, settlement will be difficult or impossible. The most effective weapon in this case is often applying one's own intellectual property against the competitor.

Reducing Risk

With copyrights and trade secrets, the best way to reduce risk is to make sure that employees that come from competitors do not bring materials belonging to their former employer with them. It is also important to make sure that employees understand that "freeware" and "shareware" from the Internet are still covered at least by copyright, and are therefore not to be incorporated into products without careful consideration of the legal consequences of doing so.

With patents, there are a number of ways to reduce risk. One is to make indemnity agreements with suppliers. In such agreements, the supplier agrees to make good any losses caused to the purchaser due to patent infringements involving the supplier's products. Of course, an indemnity agreement is worthless if the supplier is unable to pay. If a product is important, it may be worthwhile to have a patent attorney undertake a right-to-use study. In such a study, the attorney compares the design with the relevant patent literature and determines whether there will be any infringement problems. If there appear to be problems, one can "design around" the relevant patents, altering the design so that it no longer infringes the patents, or negotiating licenses for the patents. Right-to-use studies are expensive. Moreover, if one commissions a right-to-use study and then does nothing to forestall any infringement problems found by the patent attorney, the infringement will be willful. The best long-term way of reducing risk in the patent area is to have enough patents in the areas relevant to one's business that one can extract cross-licenses from competitors as required, thereby achieving design freedom.

Conclusion

This paper introduces those areas of the law most relevant to the cable modem business and explains how to avoid liability and use the law to increase the value of the business. The paper pays particular attention to intellectual property law, since copyrights and patents are among the most important assets of many cable modem-related businesses. The paper provides only a general overview and is not intended to replace a consultation with a lawyer.

References

1 *Reno, etal. v. ACLU, et al.,* 117 S. Ct. 2329 (S.Ct. 1997)
2 47 U.S.C. 230(c) and 47 U.S.C. 223(a) and (e)
3 17 U.S.C. 117
4 17 U.S.C. 401(b)
5 Information and Publications Section, LM-455 Copyright Office

Library of Congress, Washington, D.C. 20559

6 17 U.S.C. 101

7 17 U.S.C. 302

8 15 U.S.C. 1125(c), 15 U.S.C. 1127

9 35 U.S.C. 102 and 35 U.S.C. 103

Part II: The Physical Layer

Cable Modem Physical Layer Specifications and Design

Bruce Currivan
Director, Modem Hardware Engineering
Broadcom Corporation

Introduction

To satisfy the strong demand for the Internet and other digital services, high-speed data access is required in the home and business. Of the various media available to provide this wideband data pipe, one of the most promising is the cable television (CATV) cable.[1] With hundreds of MHz of bandwidth and its existing availability in many neighborhoods, it is tempting to think of the cable as a local- or metropolitan-area network (LAN/MAN) waiting to be enabled. To make this vision a reality, industry groups, including the Digital Audio Video Council (DAVIC), the European Telecommunications Standards Institute (ETSI), the Institute of Electrical and Electronic Engineers (IEEE 802.14), Multimedia Cable Network System (MCNS) Holdings, and the Society of Cable Telecommunications Engineers (SCTE), created specifications with the goal of increasing interoperability between equipment.[2] At the same time, the international cable industry began the investment necessary to upgrade cable plants to two-way capability. Behind the scenes, many companies developed cable modems and obtained laboratory and field experience with various modulation approaches. The result of these combined efforts is a set of specifications, a suite of compatible cable modem designs, and a high-performance cable plant infrastructure. The goal is consumer purchasing and hookup of interoperable cable modems.

This paper will review the development of high-speed cable modems from the point of view of the physical (PHY) layer. We will take the following approach:

- review key elements of the MCNS and IEEE 802.14 PHY specifications
- discuss tradeoffs that led to some of the major design decisions
- discuss the cable channel model as it affects modem design
- review implementation considerations in cable modem design

Review of MCNS DOCSIS/IEEE 802.14 Physical Layer Specifications

The MCNS and IEEE 802.14 PHY specifications are nearly identical.[3,4] Historically, the IEEE 802.14 committee developed many key parameters of a physical layer specification through an exhaustive voting process from 1994 to 1997. MCNS, an industry-spearheaded group with strong involvement from cable companies and close ties with the 802.14 group, adopted many of the 802.14 PHY parameters, filled in important missing elements, and issued drafts in March and October 1997. To maximize interoperability, the IEEE 802.14 committee reciprocated by adopting essentially all provisions of the MCNS spec, including some extensions. The main extension is in the forward error correction (FEC); the 802.14 supports both ITU-T J.83 Annexes A and B, though the MCNS specifies only Annex B.

Table 1 and *Table 2* summarize the MCNS specification physical layer parameters upstream and downstream, respectively. A column highlights the key differences between the MCNS and the IEEE 802.14. These summary tables capture parameters that delineate the main design concepts. However, no attempt is made to list exhaustively every parameter. The reader is referred to the specifications themselves for the details.

Interleaver Characteristics

Both Annexes A and B use a convolutional interleaver. *Table 3* lists the interleaver parameters for the Annex B modes, which have variable interleaver depth to permit the user to trade off latency versus burst-error protection. The parameter I is the number of taps in the commutator, and J is the delay increment (in 7-bit symbols). The Annex A interleaver is fixed with I=12 and J=17 (in 8-bit symbols).

Parameter	MCNS Value	IEEE 802.14 Difference	Notes
Modulation format	QPSK and 16-QAM		
Burst symbol rates (ksps)	160, 320, 640, 1280, 2560	5120 added	
Differentially coded modulation	Two bits per symbol denote change in quadrant rather than absolute quadrant. Differential and non-differential modes both available		
Symbol mapping	Gray code or differential		
Pulse shaping	Square-root raised cosine, 25% excess bandwidth ?		
Spurious spectral emissions (summary)	Non-adjacent channels: -53 dBc per 160 kHz Carrier-related, such as harmonics: -47 dBc each, up to 3 terms ACI: -45 to -39 dBc integrated over adjacent channel		See MCNS spec for more detail
Spectral mask, amplitude	Center of passband: 0 to $R_S/4$: ±0.3 dB Edge of passband: $R_S/4$ to $3R_S/8$: -0.5 to 0.3 dB -3 dB point: $R_S/2$: -3.5 to 2.5 dB (1+a) point: $5R_S/8$: < -30 dB		
Integrated phase noise	< -43 dBc from 1 kHz to 1.6 MHz above and below the carrier		
Spectral mask, group delay variation	0 to $5R_S/8$: < 100 ns		
Modulator implementation loss	0.5 dB at SER = 10^{-6} uncoded Cluster SNR > 30 dB		
Frequency range (MHz)	5-42 edge-to-edge	5-65	
Carrier frequency accuracy	±50 ppm over 0 to 40 degrees C for 5 years		
Carrier frequency offset controllability of subscriber by head end	±32 kHz range, 1 Hz resolution, ±10 Hz accuracy		
Symbol rate accuracy	±50 ppm over 0 to 40 degrees C for 5 years		

Table 1: *Upstream Parameter Table*

Parameter	MCNS Value	IEEE 802.14 Differences	Notes
Symbol timing jitter	2% over 2 seconds		
Sideband selection	Modulated waveform = $I \cos \omega t \quad Q \sin \omega t$		
FEC encoding	Reed-Solomon (n,k) Galois field (256) t = 1 to 10, or none $g(x) = (x+\alpha^0)(x+\alpha^1)\ (x+\ \ \alpha^{2t-1})$ $p(x) = x^8+x^4+x^3+x^2+1$		
Codeword length (bytes)	18 255		
Shortened last codeword mode	On/off		Reduces overhead by shortening last codeword if no more data to send
Scrambling	On/off		
Scrambler type	Synchronous, sidestream, non-self-synchronizing. Re-initialized at start of each data burst		
Scrambler polynomial	$p(x) = x^{15}+x^{14}+1$		
Scrambler seed	15 bits, sent to subscriber by head end	May optionally change with time for greater robustness	
Preamble length	QPSK: 0 512 symbols 16-QAM: 0 256 symbols		Preamble selected by head end as substring of 1024-bit preamble superstring
Transmit power level at subscriber (single carrier)	QPSK: +8 to +58 dBmV 16-QAM: +8 to +55 dBmV Step size: 1 dB Accuracy: 2 dB		
Nominal receive power at head end (single carrier)	160 ksps: -16 to +14 dBmV 320 ksps: -13 to +17 dBmV 640 ksps: -10 to +20 dBmV 1280 ksps: -7 to +23 dBmV 2560 ksps: -4 to +26 dBmV		
Maximum receive power at head end (aggregate, 5 MHz to 42 MHz)	+35 dBmV		
Variation in receive power burst-to-burst	±6 dB in addition to nominal range		

***Table 1** (continued)*

TDMA master clock in head end	Frequency: 10.24 MHz Stability: < 5 ppm Drift rate: < 10^{-8} per second Edge jitter: < 10 ns p-p		
TDMA timestamp jitter (jitter in timestamp messages sent downstream from head end to subscribers)	< 500 ns		
TDMA ranging accuracy (error in subscriber transmit burst start time relative to timestamps)	< ‰ symbol + 250 ns		
Inter-burst guard time	5 to 255 symbols		
Maximum burst length	Fixed length bursts: 1 to 255 minislots Variable length bursts: as granted		
Reconfiguration time to change burst parameters	96 symbols plus settling time		Applies to transmit power, modulation type, symbol rate, offset frequency, and ranging offset
Return loss (5 MHz to 42 MHz)	> 6 dB		
Impedance	75 ohms		
Connector	Type F per IPS-SP-401, common input/output		

Table 1 *(continued)*

Parameter	MCNS Value	IEEE 802.14 Differences	Notes
Modulation format	64-QAM and 256-QAM		
Symbol rates (Msps)	64-QAM: 5.056941 256-QAM: 5.360537	64-QAM: 5.056941, 5.064 256-QAM: 5.360537, 5.19	
Nominal channel spacing	6 MHz		
Differential coding	Quadrant coding		
Pulse shaping	64-QAM: square-root raised cosine, 18% excess bandwidth 256-QAM: square-root raised cosine, 12% excess bandwidth		
Spurious spectral emissions (summary)	Inband: < -48 dBc ACI: < -58 dBc in 750 kHz		See MCNS spec for more detail
Integrated phase noise, double-sided	1 to 10 kHz: < -33 dBc 10 to 50 kHz: < -51 dBc 50 kHz to 3 MHz: < -51 dBc		
Center frequency	91-857 MHz	Additional range	
Carrier frequency accuracy	±30 kHz		
Transmit power level at head end (single carrier)	+50 to +61 dBmV		
Nominal receive power at cable modem (single carrier)	-15 to +15 dBmV		
Maximum receive power at cable modem (aggregate, 40-900 MHz)	+30 dBmV		
Interleaving	Annex B variable depth (Table 3)	Annex A fixed depth Annex B variable depth (Table 3)	See ITU-T Recommendation J.83 for more detail

Table 2: *Downstream Parameter*

FEC outer code (summary)	Annex B: Reed-Solomon (128,122) Galois field (128) t = 3 g(x) = $(x+\alpha^1)(x+\alpha^2)...(x+\alpha^5)$ $p(x) = x^7+x^3+1$	Add Annex A as option: Reed-Solomon (204,188) Galois field (256) t = 8 $g(x) = (x+\alpha^0)$ $(x+\alpha^1)...(x+\alpha^{15})$ $p(x) = x^8+x^4+$ x^3+x^2+1	See ITU-T Recommendation J.83 for more detail
Concatenated inner trellis code (summary)	64-QAM: rate 14/15; for every 4 RS symbols (28 bits) entering the trellis-coded modulator, 5 consecutive 64-QAM symbols are output 256-QAM: rate 19/20; for every 38 bits entering the trellis-coded modulator, 5 consecutive 256-QAM symbols are output	Annex A: no inner code Annex B: inner code same as MCNS	See ITU-T Recommendation J.83 for more detail
BER performance, post-FEC (summary)	64-QAM: BER < 10^{-8} at E_S/N_0 > 23.5 dB 256-QAM: BER < 10^{-8} at E_S/N_0 > 30 dB		See MCNS spec for additional requirements, including ACI
Return loss (88-860 MHz)	> 6 dB		
Impedance	75 ohms		
Connector	Type F per IPS-SP-401, common input/output		

***Table 2** (continued)*

Key Trade-offs

This section summarizes the major technical trade-offs that led to the IEEE 802.14 specification.

Upstream

Single vs. multiple carriers One of the basic PHY decisions was to use single-carrier or multicarrier modulation (MCM). Multicarrier modulation uses a large number (tens to hundreds) of narrow carriers stacked next to each other in frequency, sharing the aggregate data throughput.[5, 6] MCM offers the following advantages:

I (# of taps)	J (incre- ment, 7-bit symbols)	Burst protection 64-QAM	Burst protection 256-QAM	Latency, 64-QAM (msec)	Latency, 256-QAM (msec)	Applicability
8	16	5.9	4.1	0.22	0.15	MCNS, IEEE, ITU
16	8	12	8.2	0.48	0.33	MCNS, IEEE, ITU
32	4	24	16	0.98	0.68	MCNS, IEEE, ITU
64	2	47	33	2.0	1.4	MCNS, IEEE, ITU
128	1	95	66	4.0	2.8	MCNS, IEEE, ITU
128	2	190	132	8.0	5.6	IEEE, ITU
128	3	285	198	12	8.4	IEEE, ITU
128	4	379	264	16	11	IEEE, ITU
128	5	474	330	20	14	ITU
128	6	569	396	24	17	ITU
128	7	664	462	28	19	ITU
128	8	759	528	32	22	ITU

Table 3: *Downstream Interleaver Characteristics (ITU J.83 Annex B)*

- It requires little or no adaptive equalization, since each carrier is narrow enough that frequency-selective channel effects can be ignored.
- Narrowband ingress can be avoided simply by turning off one or more carriers.
- Greater throughput can be achieved if the data rate of each narrowband carrier is adjusted to match the signal-to-noise ratio (SNR) seen on the channel at the frequency of that carrier. This concept stems from Shannon's original "water-pouring" solution.
- The long symbol time associated with a narrow carrier gives a natural immunity to low-level impulse noise. A low-level impulse does not last long enough, or have enough energy, to greatly affect a long symbol.
- Additional spectral efficiency (upper bounded by $\alpha = 25\%$ improvement) is achievable, if all upstream transmitters are synchronous to the head end receiver.

On the other side of this trade were the following advantages of single-carrier modulation:

- It was more familiar, and thus it was perceived to have lower risk. Fielded systems existed on cable plants using single-carrier quaternary phase shift keying (QPSK) modulation.

- The prevailing perception was that single-carrier modulation would be more economical in near-term deployments, although this advantage of single-carrier would probably diminish as technology advances.

- Strong impulse noise is handled better by single-carrier modulation, since the time duration of the impulse is limited to (for example) one symbol, and the loss of one symbol results in less data loss for a short single-carrier symbol than for a long multi-carrier symbol. A simple Reed-Solomon code ($t \geq 1$) can correct a single symbol error in single-carrier modulation, but only a more powerful code would correct a multicarrier symbol error.

- Single-carrier modulation has a lower peak-to-root mean square (RMS) ratio and is more robust to minor nonlinearities, allowing higher average transmitted power in the subscriber modulator. This translates into reduced ingress levels relative to the signal in a well-designed and adjusted broadband cable system.

For these reasons, the IEEE 802.14 committee voted to use single-carrier modulation. The proponents of MCM formed an "advanced PHY" group that investigated a future upgrade to other techniques, possibly including multicarrier modulation.

Single vs. Multiple Symbol Rates Although single-carrier, single-data–rate QPSK systems existed in the field, there was a desire to "future-proof" the PHY spec by adding greater capability. A system with multiple symbol rates would have the following advantages:

- Multiple symbol rates permit a higher maximum throughput to be achieved on well-maintained, low-noise cable systems.

- Multiple symbol rates are more scalable, permitting both high- and low-performance channels to coexist on different portions of the same cable system, or on different cable systems.

- Multiple symbol rates provide higher performance, resulting in a longer-lasting standard.

- Multiple symbol rates permit "nailed-up" or dedicated links to be allocated to businesses, using varied amounts of bandwidth such as 64 kbps or 1.544 Mbps.

- A system with narrowband capability could mimic the advantages of a multi-carrier system by deftly placing narrow carriers in between narrowband ingress sources. In addition, one could allocate signal power and modulation type differently to narrowband carriers. This was an important consideration in future-proofing the standard, given the perceived competition between single-carrier and multitone systems.

- Digital filters scale naturally to various symbol rates by changing the sample rate (using interpolation and decimation techniques), providing a conceptually simple (although gate-intensive) implementation approach.

The arguments for a single symbol rate were as follows:

- Since modems existed in the field at a single symbol rate (e.g., 1.28 Msps), and given the industry's desire for rapid deployment, the schedule and the risk favored a single symbol rate.

- Adding multiple symbol rates to a demodulator is one of the most demanding requirements on modem design. Greater dynamic range and more complicated processing are required in the implementation of the transmitter and receiver.

- The system design is more complicated since one now has to consider various combinations of adjacent channel interference cases. For example, the narrowest carrier placed adjacent to the widest carrier may be overwhelmed by a sidelobe of the wide carrier; this is because the wider carrier has proportionately greater power than the narrow carrier, assuming a constant noise floor and equal E_s/N_0 for both carriers (the "constant spectral density" assumption).[7] The solution is either to provide for a guard band in the frequency allocation table for the upstream when placing narrow carriers adjacent to wide carriers, or to tightly specify the sidelobes of the modulator when transmitting wideband carriers. A combination of both techniques is used in practice.

The 802.14 and MCNS standards adopted multiple symbol rates, perhaps largely due to the "future-proofing" sentiment.

QPSK vs. 16-QAM

Another important decision was whether to use only QPSK modulation or to add an option for 16-quadrature amplitude modulation (QAM). Due to the severe noise environment in the upstream, higher constellations such as 64–QAM were not considered for single-carrier upstream signals, although they are part of multicarrier proposals. The argument in favor of only QPSK was as follows:

- QPSK was at the time a *de facto* standard, having been fielded by several companies successfully in upstream modems.
- 16–QAM is more complex to implement than QPSK. 16–QAM requires about 5–6 dB more SNR and is more susceptible to microreflections and distortion (usually requiring an adaptive equalizer in the demod). It requires more complex filtering, better linearity, and greater spectral purity in the transmitter.
- The schedule was of foremost concern: by staying with QPSK, the 802.14 standard could be completed much sooner.

The arguments in favor of adding 16–QAM as an option were the following:

- To "future-proof" the PHY spec by adding greater capability.
- To increase the maximum throughput of the system by nearly a factor of two.
- QAM hardware was already in existence for fielded downstream systems, and could be readily adapted for the upstream application.

A further argument was that the greatest fraction of complexity of 16–QAM is in the receiver, with much less increase in difficulty in the transmitter. After much debate, the 802.14 committee voted to mandate 16–QAM capability at the transmitter but to add it as an option at the receiver. In this way, head ends could add 16–QAM reception as business demanded the increased capacity and technology permitted it to be done economically. There would be no need to upgrade the multitude of installed subscriber transmitters if they were QAM-ready.

Differentially vs. Non differentially Coded Modulation The decision between differentially encoded versus non-differentially encoded modulation involved two considerations: the bit error rate (BER) penalty from differential decoding, and the probability of a cycle slip during a long data burst.[8]

BER penalty: In systems without FEC, the BER penalty for differentially encoded modulation is negligible. Thus differential encoding is often attractive, since it reduces the effect of cycle slips and may enable a receiver to be implemented without a carrier-recovery phase-locked loop. However, with the forward error correction (FEC), the BER penalty associated with differentially encoded modulation is more significant, as large as 1 dB. This is because the demod operaties at a higher BER before the decoder, in a region where the BER-vs.-SNR curve is flatter.

Probability of cycle slip: With long data bursts the probability of a cycle slip in the carrier tracking phase-locked loop may become a factor. With differentially encoded modulation, a cycle slip results in a small number of errors, which the FEC may correct. With non-differentially encoded modulation a cycle slip results in the loss of the entire burst. With the most powerful FEC, operation is enabled at lower signal-to-noise ratios, which tends to increase the incidence of cycle slips in the carrier recovery loop. While the probability of a cycle slip is normally much lower than the loss of a burst due to too many errors for the FEC to correct, it is possible that with exceedingly long bursts and the most powerful FEC, there may be a region of applicability (i.e., long bursts with powerful FEC, such as $t = 10$) where differentially encoded modulation is viable.

It was determined that the added complexity for differentially encoded modulation was small in the transmitter. Hence the decision was made to require the transmitter to support both modes (differentially encoded and non-differentially encoded), with the receiver having the option of implementing either or both.

Trellis Coded Modulation Trellis coding was proposed for use on the upstream.[9] A simple trellis code is very attractive, as it provides approximately 3 dB of coding gain for a very small increase in implementation complexity.[10] This 3 dB is very valuable in the noise-limited upstream channel.

A perception existed, however, that trellis coding would increase the time to market since it was not currently in use on this channel, although telephone modems used it at lower symbol rates. Also, it was noted that the 3-dB gain would not be fully realized in an impulse-noise-limited channel. For these reasons, upstream trellis coding was deferred to the "advanced PHY" subgroup.

TDMA and FDMA

A combination of time division multiple access (TDMA) and frequency division multiple access (FDMA) was selected. TDMA permits each frequency channel to be shared among users. The full (e.g., 5-MHz to 42-MHz) bandwidth of the cable is shared by multiple frequency channels using FDMA. From the point of view of a single-carrier modem, the channel is TDMA.

The alternative to TDMA/FDMA was pure FDMA, which was in field use for upstream telephony over cable. Pure wideband TDMA over the entire 42-MHz upstream band was not considered. TDMA/FDMA was selected over pure FDMA for the following reasons:

- A digital TDMA modem can be easily implemented at the symbol rates of interest (up to 5 Msps) in current ASIC technology.
- A single TDMA modem (2.56 Mbps, for example) can carry a T1 or E1 line (twenty-four or thirty-two voice channels), or an equivalent amount of data, in

a single modem. The same capacity would require 24 or 32 FDMA modems, each, for example, running at a 64-kbps voice rate.

- Similarly, a single TDMA head end demod (2.56 Mbps, for example) can be time-shared among tens to hundreds of users on a single upstream channel. The same capability would require a large number of FDMA demods.

Degree of Synchronism in TDMA The degree of synchronism in TDMA ranging may be divided into two categories: (1) fine or symbol-synchronous, and (2) medium/ coarse or burst-independent. Each is discussed below. In spite of the performance advantages of symbol-synchronous TDMA, burst-independent TDMA was adopted by the committee since it seemed to have lower system complexity, lower technical risk, and greater field experience. In addition, flexibility was designed into the upstream by providing selectable preamble length and guard time, permitting head end implementations to approach the performance of symbol-synchronous TDMA in tightly controlled cable plants.

1) Symbol-synchronous TDMA: In symbol-synchronous TDMA, the system maintains very accurate ranging (on the order of 5% of a symbol period) on all subscribers. Upon arrival at the head end, all subscribers' signals appear to have come from a single subscriber, from the viewpoint of baud synchronization. The head end demodulator does not need to recover baud sync on each burst. Little or no interburst gap is required. As a result, very short bursts are possible without appreciable overhead penalty. This increases throughput and reduces the latency (delay) of the system.

A short preamble may still be inserted on each upstream burst for carrier synchronization, but none is required for baud sync. If differentially coherent QPSK is used (DQPSK), the preamble may be reduced to a single symbol, further decreasing system overhead.

The overhead that is added in order to obtain these advantages includes frequent timing bursts sent from each subscriber to the head end to maintain ranging accuracy. The timing bursts must be sent often enough to track delay variations in the upstream plant, such as those from temperature changes.

For symbol-synchronous TDMA to work properly, the upstream transmitter's symbol-rate oscillator must have very low drift relative to the head end timebase. This may be accomplished in several ways. As one alternative, there may be very tight oscillator specifications (a few parts per million) on the subscriber oscillator; this is to be avoided for reasons of cost. More often, the subscriber transmitter is locked to the head end timebase to track out drift, using, for example, a phaselock loop (PLL) to lock to the received downstream baud clock.

The TDMA ranging bursts may also be used to update equalizer coefficients in the head end transmitter or receiver. In this case, the overhead bursts are required anyhow, so there may be little or no additional overhead to track time offset.

2) Burst-independent TDMA: In burst-independent TDMA, medium/coarse ranging (accuracy of 0.5 to several symbols) is maintained, with no attempt to maintain baud synchronization across upstream bursts from different subscribers. Each burst begins afresh with a preamble that the upstream demod uses to estimate the correct baud sampling epoch. The preamble may also be useful for updating equalizer coefficients. A guard time is required between bursts from different transmitters.

The advantage of burst-independent TDMA is less dependence on the delay stability of the cable plant. The disadvantage is increased overhead; a guard time and preamble are required on each burst.

Interburst Guard Time Interburst guard time was selected to minimize interference between time-adjacent bursts. Since the elemental symbol pulse is tightly bandlimited, significant ringing occurs before and after the pulse center. *Figure 1* shows a raised-cosine pulse with excess bandwidth $\alpha = 0.25$, as observed after the receiver matched filter.[11] The upper trace has a normal linear scale, whereas the bottom trace shows the magnitude in dB at each instant of time. The interference of the tail of one burst on the precursor of the next burst can be estimated by sampling the pulse at a time offset from the symbol center or peak. For QPSK, one can see from the bottom trace of *Figure 1* that a minimum gap of 4 symbol periods (measured from symbol center) gives < -35 dB interference, which is considered sufficient for negligible degradation for QPSK. In addition, there can be an amplitude difference (typically up to 6dB) between adjacent bursts originating from different transmitters that have not been fully adjusted by the ranging process. Furthermore, for 16-QAM, the presence of amplitude modification means that a lowest-amplitude symbol may be surrounded by two largest-amplitude symbols. For these reasons, a more conservative 5-symbol gap was specified.

Preamble Length Preamble length is highly implementation-specific. Hence it was made programmable from 0 bits (no preamble) to 1024 bits (256 16-QAM symbols or 512 QPSK symbols). In practical application, the preamble on each data burst is usually in the range of 8 to 64 symbols. The extremely long preamble lengths are intended for special initialization bursts.

In the 802.14 and MCNS specifications, a 1024-bit preamble "superstring" is downloaded from the head end to the subscribers. Preambles are selected for use on the upstream, under head end control, by selecting substrings of this superstring.

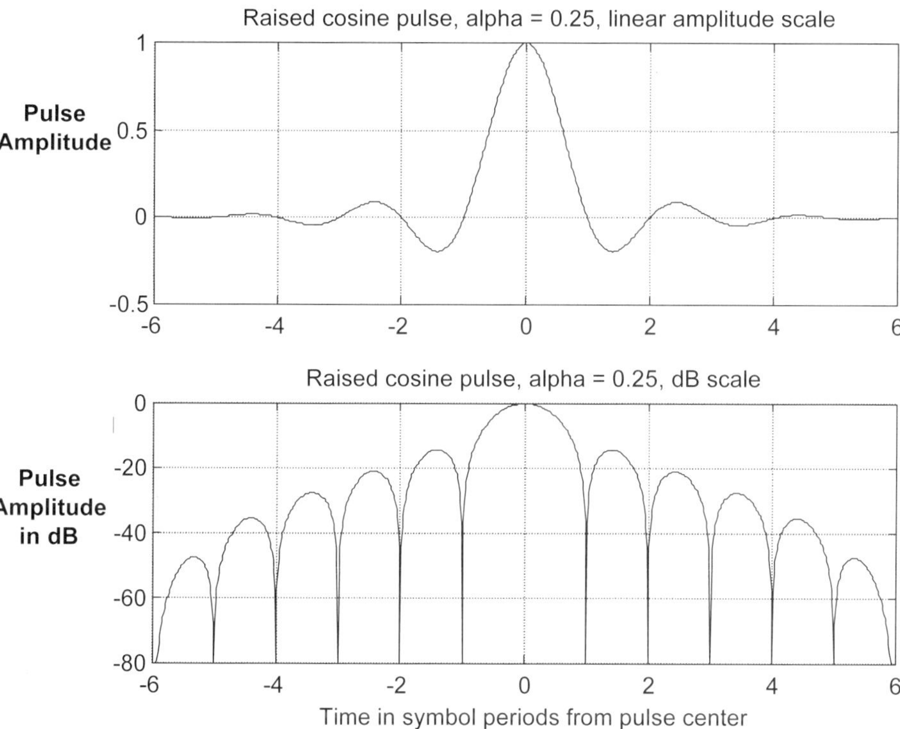

Figure 1: *Raised-Cosine Pulse vs. Time with a=0.25, Linear and Log-Magnitude Scales*

Upstream Symbol Rates The choice of upstream symbol rates was based on the following considerations:

- Ease of implementation of the digital filters led to an early decision that all symbol rates should use the same excess bandwidth. There was also a strong proclivity toward successive rates being related by a simple ratio such as 2.

- The basic T1 bit rate of 1.544 Mbps (which equals 24*64 kbps + 8 kbps) had been adopted by DAVIC as the upstream burst bit rate. The IEEE 802.14 considered following suit, but ultimately decided against it. The reason was that the burst bit rate should be higher than the user rate in order to accommodate PHY overhead (preamble, guard time, and FEC). Hence a burst bit rate greater than 1.544 Mbps would be required to support a T1 user-data link.

- A burst bit rate of 2.56 Mbps was in existence in some fielded systems. This rate was large enough to accommodate a T1 or E1 link plus overhead, and was ultimately selected.

- Another approach was to simplify the clock generation implementation by selecting appropriate rates. This led to one proposal for a highest rate of 533.33 ksps, which resulted from dividing a commonly available clock frequency by an integer.[12]

- Another consideration was to match the signal bandwidth to the upstream noise spectrum. This led to a selection in the MCNS spec of 2.56 Msps as the highest rate, with lower rates obtained by successive divisions by 2. The known noise spectrum commonly observed in the upstream was analyzed, and it was shown that the resulting signal bandwidths could efficiently fit between interference sources.

The rates selected by IEEE 802.14 and MCNS are given in *Table 4.*

Symbol Rate (Msps)	QPSK Burst Bit Rate (Mbps)	16-QAM Burst Bit Rate (Mbps)	Notes
0.160	0.320	0.640	
0.320	0.640	1.28	
0.640	1.28	2.56	
1.28	2.56	5.12	
2.56	5.12	10.24	
5.12	10.24	20.48	IEEE-802.14 only

Table 4: *IEEE 802.14 and MCNS Upstream Symbol Rates*

TDMA Tme Transfer Method In a TDMA system, each user transmits during an assigned time slot. Hence the subscribers must have a common notion of time, to prevent one user from transmitting on top of another and causing interference. The head end is the master clock, and must communicate its time to the subscribers. Various gradations of timing accuracy affect the following functions: carrier frequency and phase, baud frequency and phase, and TDMA burst start time. This section discusses only the latter, transmit burst start time. Two methods were traded off within IEEE 802.14 for transferring the head-end clock time and frequency to the subscribers: locking to the downstream baud clock and sending downstream timestamps.

Baud Clock Method: In this method, the subscriber locks to the downstream baud clock (in the 5- to 7-MHz range), which provides a reference frequency that drives a counter in the subscriber. The phase (time) of the counter is coarsely set by a downstream time-marker message, and then finely controlled by ranging messages sent downstream from the head end. This method takes advantage of the timing information present in the downstream baud clock. Downstream ranging messages are still required, but can be sent very infrequently, as they only need to track the delay drift of the cable plant.

Time Stamp Method: The time stamp method has been described using the following informal analogy.[13] Imagine that I have a new watch, and I also have a friend who works at the U.S. Naval Observatory, an installation where a highly accurate cesium clock is available. My friend offers to call me from time to time and tell me what time it is. Each time he does so, I look at my watch and notice its time error. For example, I might notice that my watch is 40 seconds fast. Rather than just resetting the time, I open up the case of my watch and adjust the small screw that controls the frequency of the quartz oscillator by a slight amount to decrease its frequency. After several calls from my friend, the crystal in my watch will converge to a good replica of the frequency of the master cesium standard at the Observatory. The time on my watch will be an accurate replica of the time at the master station, with an offset due to the delay in the telephone system over which his voice traveled. Notice that my friend does not have to call me at any exact interval; he can call every few days or when it is convenient, and if I miss an update or get an extra one, it is not significant.

In a similar manner, the head end of a cable system broadcasts timestamp messages in the downstream to all subscribers at some nominal rate such as ten times per second. At each message, the subscriber compares the time stamp to the contents of a local time counter. It filters the error and controls a local oscillator. After a few tens of messages, the subscriber has a good replica of the frequency and time of the head end timebase, with a time offset due to the delay or distance (range) from the head end. Round-trip ranging messages are used to measure and track this delay. From then on, the subscriber's clock is accurate and may be used to determine when to send each upstream TDMA burst. Timestamp messages must be sent often enough to track changes in cable plant delay.

The timestamp method was adopted by the committee for time transfer after much discussion.[14] It has the following advantages:

- It does not depend on physically locking to the downstream baud clock. Hence timing is maintained when the downstream receiver changes channels.
- It is largely implementation-independent, and can be used even if the modulation technique is changed.

Carrier Frequency Control The carrier frequency control problem is analogous to the TDMA time transfer. The carrier frequency difference between the head end and subscriber oscillators must be adjusted to within about 100 ppm or better to permit the head end to readily acquire each upstream burst transmission. Three methods for carrier control were weighed: locking subscribers to a downstream baud clock, long-loop control of subscriber fine frequency (the adopted method), and wide-range head end carrier search. In the first two methods, the subscriber is slaved to the head end, whereas in the third method, the subscriber operates an open loop and the head end tracks the subscriber. Each method is described below.

Locking Subscriber to Downstream Baud Clock: In this method, the subscriber locks to the downstream baud clock and uses it as a reference for upstream carrier frequency adjustment. There are two sub-methods of doing this: phase lock and frequency lock.

Phase-Lock Method: In this method, the subscriber locks a conventional PLL to the downstream baud clock. The same PLL used for baud clock regeneration may be used. There are two disadvantages to this method. First, phase noise on the downstream baud recovery loop is multiplied in the PLL and can cause a problem on the upstream signal, especially if a 16-QAM modulation is being used. Second, the PLL must re-acquire whenever the downstream receiver changes channels.

Frequency-Lock Method: In this novel method, the subscriber measures the downstream baud clock rate using the local subscriber clock as a reference. An up-down counter or other frequency estimator could be used as a frequency discriminator. The frequency error is averaged and applied to control the subscriber's upstream transmitter frequency. The upstream transmitter may use a digitally controlled synthesizer.

The burst nature of the upstream TDMA channel is taken advantage of in the following way: the upstream frequency is only adjusted when the transmitter is muted between bursts; during transmissions, the frequency is frozen. This way, no phase noise is coupled from the baud tracking loop into the upstream carrier signal.

Long Loop Fine Control of Subscriber Frequency In this method, the head end measures the frequency of the upstream bursts received from the subscriber, and sends correction messages downstream to control the fine frequency offset of the subscriber to within 50 Hz or better.

This method shares the implementation burden between the head end and subscriber. The head end may perform the frequency measurement using a frequency estimator, or by locking a second-order PLL to the upstream signal and reading the PLL integrator contents. The subscriber may implement the fine upstream frequency control (approximately 5-Hz steps) using direct digital synthesis (DDS) with a numerically controlled oscillator (NCO). This complexity was deemed acceptable in currently available technology, and this method was adopted after much debate.

Wide Range Head End Carrier Search In this method, the head end assumes all responsibility for carrier tracking. The advantage is that the subscriber's frequency synthesizer is simplified, as its step size may be 160 kHz instead of 5 Hz. The head end searches over the frequency uncertainty range (up to several kHz) on each burst. The disadvantage is that more resources (longer preamble or less robustness) must be dedicated to frequency acquisition, at the expense of other parameters in the system.

The head end may implement a lookup table with a frequency offset entry for each subscriber. This would obviate the need for a wide frequency search on each burst, at the expense of more memory and greater complexity in the head end controller.

Downstream The downstream physical layer was more mature than the upstream when the IEEE 802.14 began its deliberations. Hence there were fewer major trade-offs on the downstream.

QAM vs. VSB

An early decision was to adopt QAM as the downstream modulation format. The other major contender was vestigial sideband (VSB) modulation. Both techniques had been field-tested and gave nearly equivalent performance.[15] QAM was adopted because it was more readily adapted to high-speed application specific integrated circuit (ASIC) implementation (and hence was considered more future-proof than VSB), and it had greater support in the industry.

Annex-A vs. Annex-B FEC

A particularly difficult decision for the 802.14 committee was which type of FEC coding to adopt. Two options existed in the field: Annexes A and B of ITU-T J.83.[16, 17] In addition, there was the option of developing a new coding specification. Other trellis codes are known to exist with approximately 1 dB better performance than Annex B and comparable complexity.[18] This option was considered too risky, though, given the schedule for releasing a standard. After much debate, the IEEE 802.14 selected Annex A. MCNS selected Annex B. The IEEE 802.14 then adopted both Annexes A and B as options.

The main differences between Annexes A and B are summarized in *Table 5.* The most telling difference is that Annex B adds an inner trellis code, obtaining 1.7 dB additional effective coding gain.[18] Raw coding gain does not take into account the redundancy of the waveform, while effective coding gain does.

Parameter	Annex A	Annex B
Inner code	None	Trellis
Outer code	Reed-Solomon, t=8, GF(256)	Reed-Solomon, t=3, GF(128)
Primary initial usage area	Europe	North America
Raw/effective coding gain	5.0/3.6 dB	7.3/5.3 dB

Table 5: *Summary of FEC Options: Annex A vs. Annex B*

Channel Model

One of the primary tasks of the IEEE 802.14 committee was the development of a channel model, [7, 19–25] especially for the upstream channel. The channel model was required for the selection of the upstream modulation techniques and parameters. The following summary discusses the key elements of the channel model from the point of view of its impact on implementation.

Upstream

Figure 2 summarizes the impairments in the upstream CATV channel. The three most prominent effects are microreflections, ingress, and impulse/burst noise. Each impairment and its effect on implementation are discussed below.

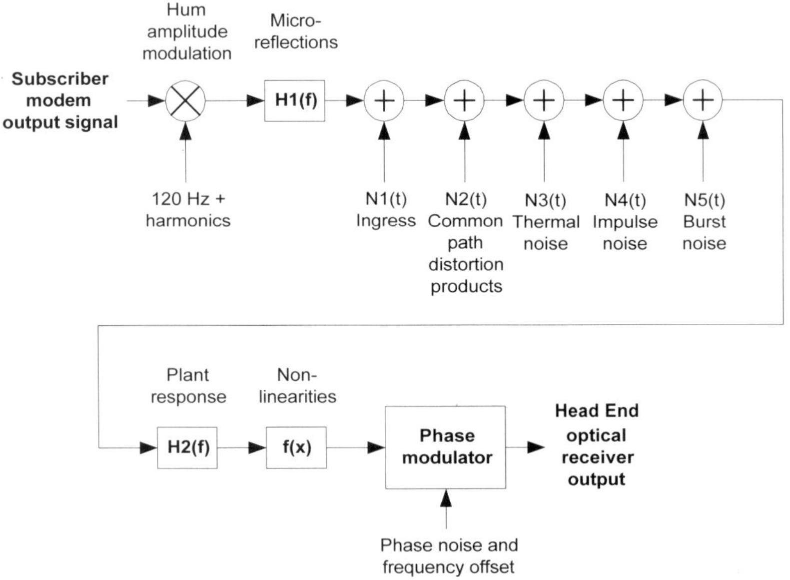

Figure 2: *Upstream Cable Channel Model*

Hum Modulation

Hum modulation is specified in the MCNS spec as not greater than –23 dBc (7%). It is slow enough (120 Hz plus harmonics) that it can be tracked out by a properly designed automatic gain control (AGC) loop or adaptive equalizer. Some demodulator implementations include a digital AGC loop to track out hum.

Microreflections

Individual microreflections are defined in the MCNS spec as –10 dBc up to 0.5 μs, –20 dBc up to 1.0 μs, and –30 dBc beyond 1.0 μs. These echoes may be great enough to distort a 16-QAM upstream signal beyond recognition. Hence, an adaptive equalizer is generally required in the upstream burst demod. Four to eight T/2-spaced feedforward taps and two to four T-spaced feedback taps are typically used in MCNS head end demods. In addition, predistortion or Tomlinson-Harashima precoding may be used in the upstream transmitter.

Ingress

Ingress may be typified by narrowband interference such as AM radio stations that leak into the cable. Active nulling systems have been considered for use at the head end to cancel ingress, but such systems can actually add noise. This is because a clean replica of the noise may not be available at the head end to use in the cancellation process, as the cable system is geographically widely dispersed. Spread spectrum techniques have also been proposed to combat ingress; such systems make some improvement against moderate ingress levels, but expend precious upstream bandwidth and tend to spread the narrowband ingress on top of the clean parts of the spectrum. Avoidance of the ingress by frequency agility and placing of carriers away from ingress sources is seemingly the most powerful technique. Since ingress is worst at lower frequencies (5 MHz to 15 MHz) due to the presence of man-made interference sources like HF radios and international shortwave stations, these low portions of the band may be avoided or allocated to more robust-application traffic, such as available-bandwidth Ethernet. The use of a robust modulation such as QPSK can also permit the signal to overpower low levels of ingress. Multitone modulation techniques can help avoid ingress on an even finer scale. On a broader system level, cable plant maintenance may be increased to reduce ingress levels; this can be expensive if other techniques are not also employed.

Common Path Distortion Products

Common path distortion products are the result of intermodulation of the downstream channels (on 6 MHz centers in North America) in passive, diode-like nonlinearities that may occur in corroded cable-plant connectors. The result is frequency-selective noise, which varies with time and weather. Common path distortion products are considered to be an addition to the noise floor. Mitigation techniques are to avoid carrier frequencies that are multiples of 6 MHz when the problem is occurring, allocate carrier power and modulation type accordingly, and to repair the cable plant as soon as possible.

Thermal Noise

Thermal noise is generally not the dominating factor in the upstream cable plant. Usually ingress, common path distortion products, and impulse or burst noise predominate. However, in a high-performance multitone system, portions of the upstream band may be identified where the modulation level may be increased up to 64–QAM or higher, limited by the thermal noise floor.

Impulse and Burst Noise

Impulse and burst noise may be thought of as short bursts of band-limited Gaussian noise. Impulse noise may be arbitrarily defined as having a duration of less than 1 μs, and burst noise as having a duration from one to hundreds of microseconds. The MCNS spec for burst noise is not longer than 10 μs at a 1-kHz average rate for most cases. The upper limit of duration may be placed at 100 to 300 μs.[26] The following mitigation techniques have been considered.

Coding: Reed-Solomon codes have an inherent burst-noise correction capability since each RS symbol is (typically) 7 to 8 bits long, and multiple RS symbols can be corrected by the code. For example, with $t = 10$, a burst of noise up to 40 QPSK symbols long or 20 16-QAM symbols long, can be corrected by the code. At the lowest symbol rate of 160 ksps, this corresponds to 250 μs of burst protection capability for QPSK and 125 μs for 16-QAM, enough to combat almost all noise bursts. At the highest symbol rate of 5.12 Msps, this corresponds to 7.8 μs of burst protection capability for QPSK and 3.9 μs for 16-QAM, which provides protection against shorter bursts.

Interleaving: If long upstream bursts are used, and if the data is not delay-sensitive, then interleaving may be used in the upstream. Interleaving breaks up bursts of noise into random-looking pieces that can be corrected by the FEC decoder. The MCNS and 802.14 specs do not include upstream interleaving, although it has been considered for "advanced modems."[27]

Blanking: If the receiver front end has extra dynamic range, some of it may be used for blanking. If the received signal exceeds a threshold, it is replaced with zero. This technique excises the noise burst before it can cause narrowband filters to ring. Limiting (clipping) may also be used instead of blanking.

Low Symbol Rates: Low symbol rates imply long symbol durations. Short noise impulses (if limited or blanked at the receiver front end) may then only corrupt a small portion of the symbol, resulting in the correct demodulation of that symbol.

Multitone modems take advantage of this property with extremely long symbol durations.

Retransmission: If the noise burst is too long, the upstream data burst is lost and must be retransmitted. PHY- and MAC-level retransmission are not included in the MCNS and 802.14 specs for reasons of complexity, although they would be very effective. Application-layer retransmission is used for Internet data traffic.

Interpolation: In the case of voice traffic, missing samples due to impulses of noise may be filled in by interpolation, often without noticeable loss of voice quality.

Plant Response

The diplexers that separate upstream and downstream frequencies exhibit amplitude and group-delay distortion,[25] especially at the band edges (near 5 MHz and 42 MHz). Mitigation techniques include avoidance of the band edges, the use of lower order modulation (QPSK), the use of lower symbol rates near the band edge, multitone modulation (with tones that are very narrow compared to the rate of change of the distortion versus frequency), and adaptive equalization.

Nonlinearities

The hybrid fiber/coax plant exhibits two major types of nonlinearities: laser and amplifier.

Laser Nonlinearity: The laser nonlinearity occurs in the fiber transmitter at the fiber node. It is an abrupt clipping that occurs above a certain aggregate power level. If the laser clips, it affects the entire upstream spectrum sharing the laser, not just the channel that caused the overload. As the upstream is the sum of many uncorrelated channels and ingress sources, the amplitude of the aggregate signal seen by the laser resembles a Rayleigh distribution. The envelope has a finite probability of clipping at any given instant.

The aggregate signal into the laser is backed off according to a design margin, but clipping still occurs on a regular basis when the amplitudes and phases of many or all upstream transmitters happen to line up to a sufficient degree. When clipping occurs, all channels see the difference between the clipped and unclipped signal as an impulse of noise. Thus the mitigation techniques for impulse noise apply.

If a subscriber or ingress source exceeds its average allotted power, it may cause longer periods of laser clipping. This event is disruptive for the whole upstream spectrum. The mitigation techniques for burst noise apply, together with enforcement of power allocations by the cable operator.

Amplifier Nonlinearity: Amplifier nonlinearities are less severe (more gradual) than laser clipping. The result is a filling in of the noise floor with intermodulation products. The mitigation techniques for thermal noise apply. Careful coordination of upstream channel frequencies can also be helpful.

Phase Noise Phase noise results mainly from the modem equipment itself. Mitigation techniques include the choice of robust modulation techniques, specification of good-quality oscillators, and proper design of carrier tracking loops.

Frequency Offset Frequency offset may result from the modem equipment itself. It may also come from frequency-stacking upconverters (see *Figure 3*) employed in the upstream cable plant to reuse the upstream spectrum. If these "stackers" are not locked to a reference signal from the head end, a frequency offset may occur. Mitigation techniques include the carrier tracking methods described earlier.

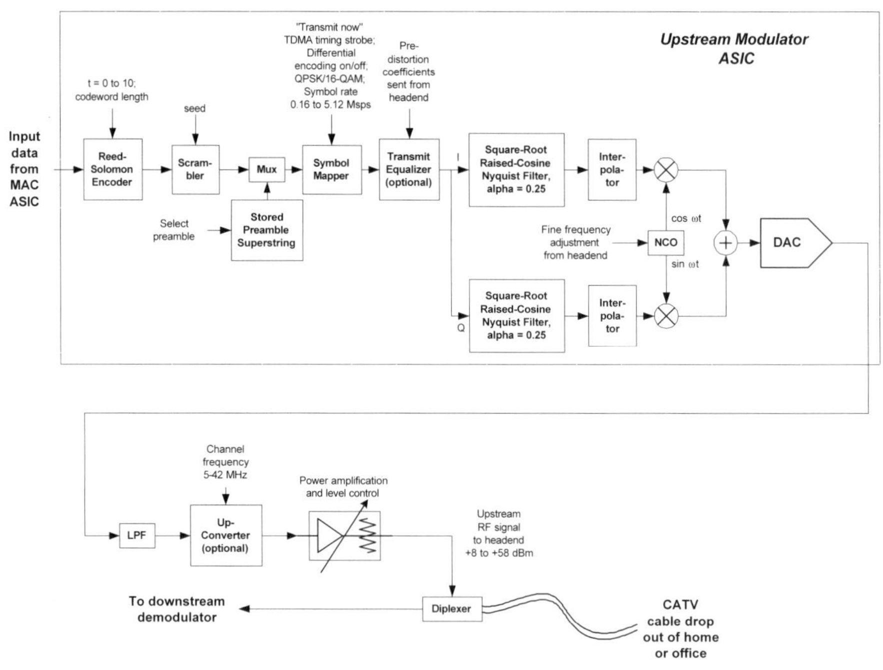

***Figure 3:** Upstream Burst Modulator Block Diagram (example)*

Downstream

The downstream channel model is similar to the upstream model, but more benign.[28] There is far less man-made noise at the downstream frequencies (50 MHz to 800 MHz). As the plant was designed for and carries analog TV transmission, it is maintained "tight" for that purpose, and ingress is less of a problem. Since the downstream symbol rates are generally higher (5 Msps to 7 Msps), and higher modulations are used (64 QAM and 256 QAM), adaptive equalization is a necessity. Normally eight to sixteen T-spaced feedforward taps and zero to sixteen T-spaced feedback taps are used in the adaptive equalizer. Burst noise mitigation limits the effect of noise impulses.

Implementation Considerations

In this section, block-diagram-level designs are described for the four corners of the cable modem system: modulators and demodulators in the upstream and downstream directions.

Upstream Burst Modems

Upstream burst modems are arguably the most challenging design problem in the cable television (CATV) system,[29] since the channel effects are severe and the TDMA burst structure places tight synchronization requirements on the modem. This section describes the basic signal-processing blocks used to implement upstream burst modems according to the MCNS specification.

Upstream burst modulator implementation: *Figure 3* is a block diagram of a typical subscriber upstream burst modulator, such as may be found in a cable modem or set-top box with radio frequency (RF) return. Data is input at the upper left of the figure from the MAC chip into a modulator chip. For the purposes of this implementation example, the modulator chip contains all digital processing functions as well as the output digital-to-analog converter (DAC).[30]

The modulator chip accepts input data for one upstream TDMA burst, which may contain several FEC code words. The data is formatted into one code word block at a time and encoded by the Reed-Solomon encoder, which appends a parity tail of length $2t$ bytes to the data in each code word. The code word length is programmable, and may be shortened on the last code word of a data burst if insufficient data is available to fill a whole code word. Next, the data is scrambled. The scrambler is restarted with a programmable seed at the beginning of each burst, and continues advancing its state across all the code words in a burst. The preamble, which has been selected under head end control as a substring of the stored superstring, is then inserted at the beginning of the burst. The data burst is now a bit stream of finite length awaiting transmission.

The bitstream is mapped into QPSK or 16-QAM symbols by the symbol mapper. As part of this process, differential encoding may be applied. Differential encoding converts absolute data into transitions between quadrants of the transmitted symbols. The preamble is differentially encoded as part of the data. Thus to obtain a specific desired preamble pattern, with required autocorrelation and spectral characteristics, the preamble must first be differentially decoded before storage. Differential encoding then restores the original preamble string as received by the transmitter output and by the demodulator.

At this point the data to be transmitted is precisely timed, both in its symbol rate and in the instant at which the upstream burst is initiated. The symbols, which have in-phase (I) and quadrature (Q) components, are applied to identical I and Q Nyquist filters with square-root raised-cosine response.

Either before or after the Nyquist filter, a transmit equalizer may optionally be included. The purpose of the transmit equalizer is to predistort the signal to compensate for effects of the channel, including filtering, multipath, and interference. The transmit equalizer coefficients are computed at the head end and sent downstream to the subscriber. One powerful predistortion technique is Tomlinson-Harashima precoding.[31] If TH precoding is used, the equalizer is located ahead of the Nyquist filter, as shown in *Figure 3*. The digital bus width at the input of the Nyquist filter is increased by TH precoding, adding complexity.

Nyquist Filter Implementation Considerations in Burst Modulators The Nyquist filters are generally implemented as finite impulse response (FIR) filters. In a burst modem, the FIR filters must have the capability to represent symbols with an algebraic zero amplitude (quiescent or OFF state) at their input and through the FIR shift register. The zero state may require an extra bit in the data representation. Before and after each burst, the symbols input to the filter have zero value. This causes the impulse response of the filter to ring up at the beginning of the burst and to ring down at the end of the burst, preventing on/off transients which would cause spurious frequency components in the transmitted waveform.

The span of the transmit FIR filter is typically twelve to sixteen symbols in order to meet the transmit-spurious requirements. The receive filter may be of comparable length, so the combined response is as long as twenty-four to thirty-two symbols. Since the minimum interburst gap is five symbols, there is considerable time overlap between successive TDMA bursts, both in the transmit filter and in the overall response. In the transmit filter, the last few symbols of the previous burst, followed by at least five zeros, and then by the first few symbols of the next burst, may coexist concurrently in the FIR shift register. The output of the Nyquist filter is then the superposition of the tail of one burst and the precursor of the next burst.

The MCNS specification requires a large guard time between consecutive bursts of different modulation types (QPSK and 16-QAM), permitting the FIR filter shift register to flush out between modulation types. However, by proper design of the burst modulator, it is possible to abut QPSK and 16-QAM bursts with arbitrarily small guard times. This is done using the following novel method. In 16-QAM mode, the full constellation is transmitted with I and Q coordinates (+/– 1, +/– 3). In QPSK mode, a four-point subset of the 16-QAM constellation is transmitted, consisting of the off-axis point (3,1) and its four 90-degree-multiple rotations. The average power of the 16-QAM and QPSK constellations is $1^2 + 3^2 = 10$. Thus QPSK and 16-QAM symbols can coexist seamlessly in the FIR shift register without a power discontinuity. The fact that there is a phase offset in the off-axis QPSK points relative to the full 16-QAM constellation makes little difference in the design of the burst receiver, since each burst is generally treated independently with regard to carrier phase.

Upstream Burst Demodulator Implementation

Upstream Demodulator Chain: *Figure 4* is a block diagram of a typical head end upstream burst demodulator chain. The input signal is carried in fiber-optic cable originating at an upstream fiber node, typically located up to tens of kilometers from the head end in a service neighborhood, either on a pole or in a ground pedestal. The optical signal consists of the aggregate upstream , AM modulated band (5 MHz to 42 MHz, plus other bands if stacking is used). This signal terminates in a fiber receiver, which demodulates the AM optical envelope and converts the signal to the electrical coaxial medium. An optional splitter may send replicas of the upstream signal to other equipment. If frequency stacking is used, each one of the stacked upstream bands may be down-converted to 5 MHz to 42 MHz, or to whatever frequency band is required for interface to the upstream receivers. After down-conversion, a splitter routes the signal to a bank of upstream receivers, typically one for each TDMA channel in use. It is also possible for several fiber receiver outputs to be summed. This would be done early in deployment when the take rate is low, and is not shown in the figure. The signal is bandpass-filtered to reject noise below 5 MHz, and possibly also to limit the frequency range tuned to by a single demodulator. The gain of the upstream link is trimmed by amplification and/or attenuation in order to meet the dynamic range interface of the demodulators. An analog tuner is used to select the signal of interest; this tuner may be replaced by a digital filter bank in alternate implementations. After anti-alias filtering, the signal is applied to a burst demodulator ASIC for digital signal processing and data recovery.

Burst Demodulator ASIC: *Figure 5* is a block diagram of a typical head end upstream burst demodulator ASIC. For the purposes of this example, the analog-to-digital (A/D) and digital-to-analog converters are included in the ASIC. The demod ASIC performs all digital processing of the signal.

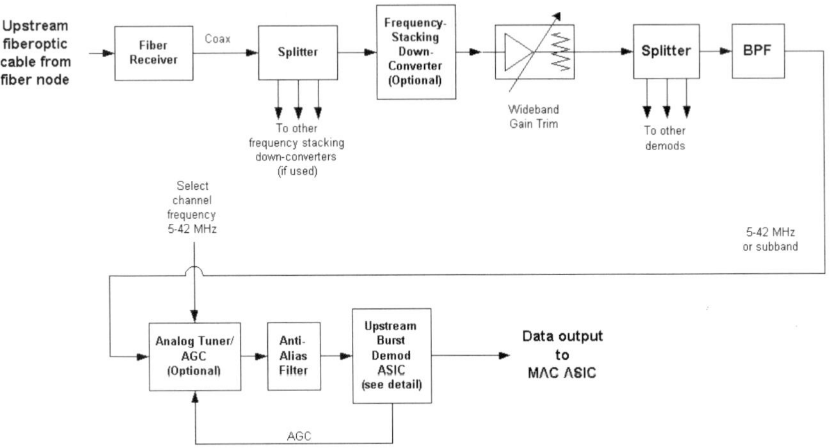

Figure 4: *Block Diagram of Head End Upstream Demodulator Chain (example)*

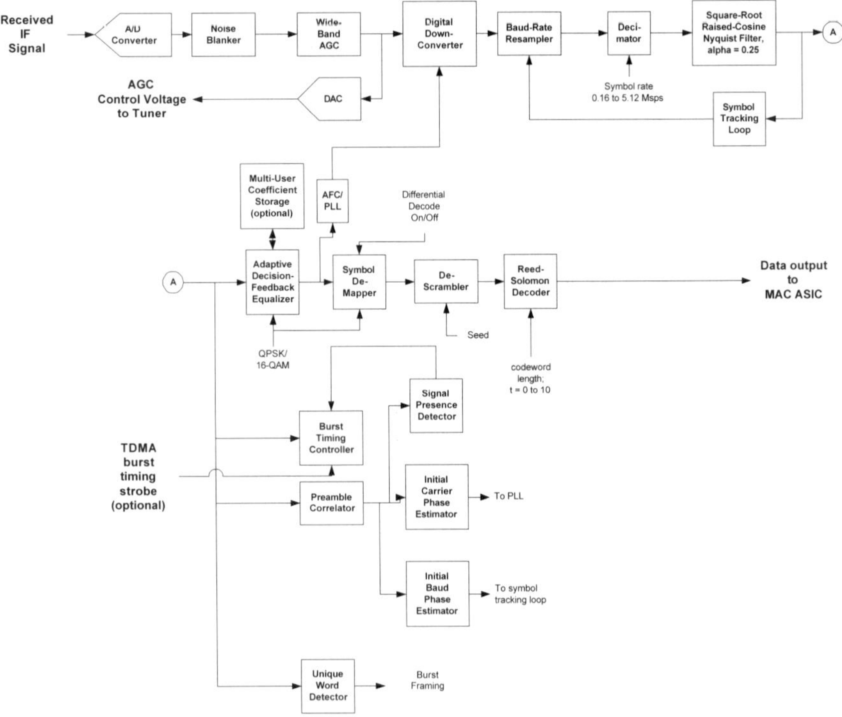

Figure 5: *Block Diagram of Head End Upstream Demodulator ASIC (example)*

The signal is A/D–converted, and noise spikes above a selected threshold are replaced by zeros. A wideband AGC controls the external amplification through a DAC. A down-converter translates the signal to complex baseband representation, and may provide fine-tuning control. A resampler, which is controlled by the symbol tracking loop, adjusts the sampling rate to track the signal's symbol transitions. In alternate implementations, the symbol tracking loop adjusts the frequency of an analog or digital oscillator, which controls the frequency and phase of the A/D converter clock. The correctly sampled signal is filtered and decimated according to the variable symbol-rate requirement, 0.16 Msps to 5.12 Msps (IEEE 802.14 range). A digital Nyquist filter (square-root raised cosine, $\alpha = 0.25$) serves as the matched filter and half the pulse-shaping filter, as well as rejecting adjacent channels and ingress. At this point the signal represents a single TDMA channel at the selected symbol rate.

Continuing the description of *Figure 5,* the filtered signal is processed in an adaptive decision-feedback equalizer. The equalizer adjusts the channel response to compensate for filtering effects and microreflections, and provides some ingress interference rejection capability. In some implementations, active interference cancellation may be used, though this is not shown in the figure. Adaptive equalization is usually required for 16-QAM upstream signals, and may also improve the robustness of QPSK channels. An automatic frequency control (AFC) and PLL aids the equalizer in tracking phase noise and frequency offsets. After the equalizer recovers the signal constellation, the symbols are recovered in a symbol demapper, which differentially decodes the signal if enabled.

The signal at this point consists of a bit stream demarcated in FEC code blocks. The bits are descrambled, then decoded by the Reed-Solomon decoder. The data is then output to the MAC chip for higher-level-layer processing.

Burst Synchronization Circuits: Synchronization is performed on each TDMA burst as follows. Continuing the description of *Figure 5*, the filtered signal is applied to a burst timing controller whose purpose is to determine the starting and ending samples of the burst. After a subscriber has been ranged, a TDMA burst timing strobe may be available so that on each subsequent burst from the same subscriber, timing does not have to be reacquired. For initial network entry before ranging, or if a timing strobe is not implemented, the burst may be detected by observing the preamble with a correlator. The output of the correlator is processed to provide an indication of signal presence, an estimate of the carrier phase (used to initialize the PLL), and an estimate of the baud clock phase (used to initialize the symbol tracking loop). In some implementations, additional "unique word" synchronization symbols are included after the preamble. The unique word resolves timing ambiguities in the preamble and/or provides additional information enabling the demodulator to reliably determine which symbol is the first information symbol in the burst.

Downstream Continuous Modems

Downstream Modulator Implementation: Figure 6 is a block diagram of a typical head end downstream modulator (for an MCNS application). The modulator chip shown in this example performs all digital signal processing functions up to and including an intermediate frequency digital to analog converter (IF DAC). Input data from the media access control (MAC) chip enters into a motion pictures experts group (MPEG) framer, which blocks the data into 188-byte blocks to emulate an MPEG-II video source. Other video sources may then be multiplexed into the data stream, so that the downstream may be time-shared between cable modem data and video data. If no data is available, idle cells are sent in the downstream. The resulting data stream is encoded, interleaved, and scrambled. The interleaver has selectable latency. The data stream is then coded with an inner trellis code and differentially encoded. The signal is mapped to QAM symbols, and the symbols are shaped in a Nyquist filter. In this example implementation, the filtered signal is interpolated and digitally modulated to an intermediate frequency, then converted through a DAC and output from the modulator chip.

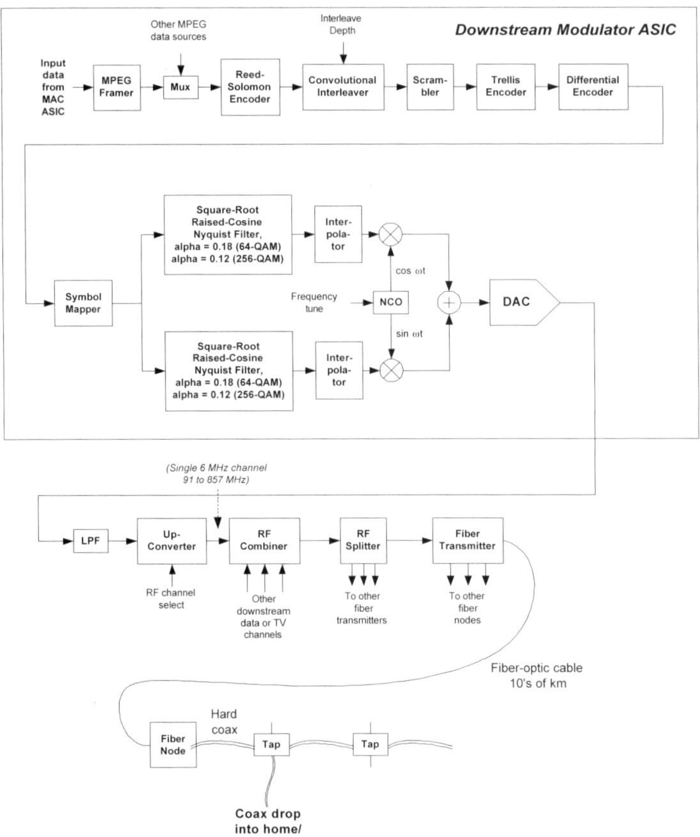

Figure 6: *Block Diagram of a Head-End Downstream Modulator Chain (example)*

The output of the modulator chip is filtered and upconverted to the desired RF channel. This channel is combined with other channels to produce an aggregate FDMA spectrum. The aggregate signal is split and sent to downstream fiber transmitters for transport to fiber nodes in the subscriber neighborhoods. At the fiber node, the signal is converted to electrical coax, and tapped off onto subscriber drops to enter homes and businesses.

Downstream Demodulator Implementation: *Figure 7* is a block diagram of a typical subscriber downstream demodulator, such as may be found in a cable modem or set-top box.[32] The RF signal from the diplexer (shown in *Figure 4*) is input to an analog tuner. The tuner selects a single 6-MHz channel and provides down-conversion to 36- or 44-MHz IF, filtering, and gain. The IF signal may be further down-converted to in-phase and quadrature baseband signals, or, as in *Figure 7*, it may be sampled directly at IF by an A/D converter. A noise blanker replaces noise spikes that are above a threshold with zero samples. An AGC circuit controls the gain of the analog tuner. In some tuner implementations, separate RF and IF gain controls are provided. A digital down-converter converts the signal to complex baseband format. A resampler synchronizes the samples to the symbol transitions. The signal is then filtered in a Nyquist filter and processed by an adaptive decision-feedback equalizer.[33] The equalizer corrects for channel impairments such as non-ideal filtering and microreflections; with 64- and 256-QAM, even slight impairments can cause significant degradation. Some implementations use a linear equalizer (without the decision-feedback section).[34] An AFC/PLL loop aids the adaptive equalizer to track out frequency offset and phase noise; the latter can be of considerable concern given the often inexpensive oscillators in the tuner and fragile 256-QAM modulation. A symbol-tracking loop controls the sampling phase of the resampler. In some implementations, symbol tracking is done via an analog or digital oscillator that controls the frequency and phase of the A/D sample clock.

The equalized constellation is mapped into bits by the symbol demapper, which operates in conjunction with a differential decoder. A Viterbi decoder processes the inner trellis code. The result at this point is a continuous bit stream. The signal is then descrambled, de-interleaved, and sent to the Reed-Solomon block decoder. The resulting error-corrected bit stream is synchronized by the MPEG deframer and output to the MAC chip. Those MPEG frames having a cable modem ID are processed further in the cable modem, while the MAC ignores or routes other frames, such as MPEG video, to other destinations.

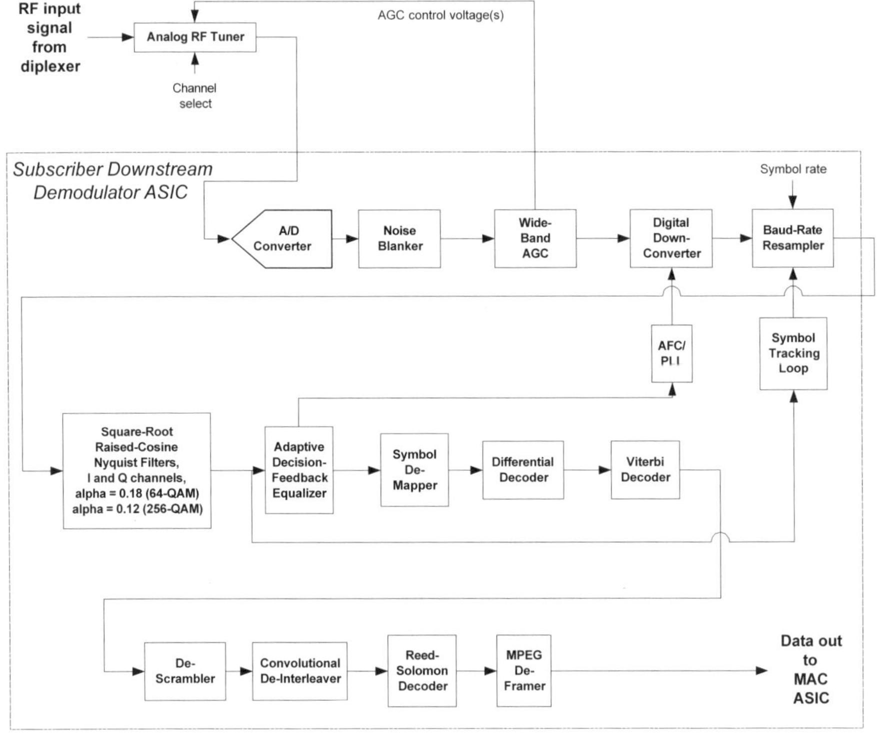

Figure 7: *Subscriber Downstream Demodulator Block Diagram*

Conclusion

This paper has given a brief overview of the physical layer technology implemented in today's cable modems. Acknowledgment is given to the reviewers and to the many contributors to the IEEE 802.14 and MCNS specification process.[35]

Notes

1 Gagen, P. F., and Pugh, W. E., "Hybrid Fiber-Coax Access Networks," *Bell Labs Technical Journal*, Summer 1996, 28–35.

2 DAVIC: Digital Audio Video Council. ETSI: European Telecommunications Standards Institute. IEEE: Institute of Electrical and Electronic Engineers. MCNS: Multimedia Cable Network System. SCTE: Society of Cable Telecommunications Engineers.

3 MCNS interim specification: Cable Television Laboratories, "Data-Over-Cable Service Interface Specifications: Radio Frequency Interface Specification," SP-RFI-104-990724, July 24, 1998.
4 IEEE 802.14 standard: IEEE Project 802.14/a Draft 3, Revision 2, "Cable-TV Access Method and Physical Layer Specification,"August 1, 1998.
5 Chen, W. Y., Valenti, C. F., and Edmon, E., "Performance Estimation of an HFC DMT System," IEEE 802.14-95/111, September 11, 1995.
6 Gross, R. et al., "An Overview of DWMT for HFC–based Telecommunications Services," IEEE 802.14-95/023, March 7, 1995.
7 Stoneback, D. A., and Beck, W. F., "Designing the Return System for Full Digital Services," SCTE Cable-Tec Expo '96, Nashville, TN, June 1996.
8 Kolze, T., "Error Performance Degradation Due to Differentially Encoded Modulation when Combined with Forward Error Control (FEC) Coding," GI IEEE 802.14-96/195, Enschede, Netherlands, July 8, 1996.
9 Ungerboeck, G., and Karaguz, J., "Formal Proposal: Frequency Agile Multi-Mode (FAMM) Single-Carrier Modems for Upstream Transmission in HFC Systems," IEEE 802.14-95/131, November 7, 1995.
10 Ungerboeck, G., "Channel Coding with Multilevel/Phase Signals," *IEE Transactions on Information Theory*, vol. IT-28, No.1, pp. 55–67, January 1982.
11 Original analysis by Leo Montreuil (Scientific Atlanta) for DAVIC case of α = 0.3 and QPSK.
12 General Instrument proposal: Thomas Kolze, "802.14 PHY Layer Proposal," IEEE 802.12-95/130R1, November 1995. This proposal included multiple symbol rates, burst lengths, FEC options, and modulation formats.
13 Basic analogy outlined by Isaac S. Wingfield, private communication.
14 Hou, Victor T., et al., "TDMA System Synchronization Using Timebase Marker Messages," IEEE 802.14-96/090, March 11, 1996.
15 Lockwood, L. W., "VSB and QAM," *Communications Technology*, December 1995, 16–23.
16 ITU-T Recommendation J.83, "Television and Sound Transmission: Digital Multi-Programme Systems for Television, Sound and Data Services for Cable Distribution," October 1995 (and revisions).
17 Heegard, C., "A Comparison of ITU J.83 Annex A & B Forward Error Correction (FEC)," IEEE 802.14-96/094, March 11, 1996.
18 Forney, Jr., G. D., Eyuboglu, V., and Ungerboeck, G., "On Downstream Forward Error Corrction," IEEE 802.14-96/231, September 17, 1996.
19 Cable Television Laboratories, "Digital Transmission Characterization of Cable Television Systems," November 1994.
20 Cable Television Laboratories, "Two-Way Cable Television System Characterization," April 12, 1995.

21 Currivan, B., "Upstream Channel Model, Rev. 1.0," IEEE 802.14-95/133, November 1995.
22 Kolze, T., "HFC Channel Model," IEEE 802.14-97/xxx, July 27, 1997.
23 IEEE 802.14 draft standard, section P.12.2, "HFC Channel Model," section written by Thomas Kolze.
24 Kolze, T., "Upstream HFC Channel Modeling and Physical Layer Design," *Broadband Access Systems: Voice, Video, and Data Communications*, SPIE Photonics East Conference, Boston, MA, 18–22 November 1996.
25 Campos, L. A., and Krinsky, D., "HFC Upstream Plant Impairments and Transmission Performance Implications," IEEE 802.14-95/159R1, October 19, 1995.
26 Williams, T., Chelehmal, M., and Prodan, R., "Field Evaluation of Reverse-Band Channel Impairments," Cable Television Laboratories, December 1995.
27 MCNS specifications [3], Table 9-1, "Channel Burst Parameters for Advanced Modems."
28 The IEEE 802.14 downstream channel model was developed by Alan Gatherer of Texas Instruments.
29 Currivan, B., and Xenakis, W., "Considerations in Implementing Upstream Channel Communications in CATV Systems," SCTE Cable-Tec Expo '96, Nashville, TN, June 1996.
30 See, for example, Stanford Telecommunications, Inc. (Sunnyvale, CA), STEL-1109 burst QPSK/16-QAM modulator data sheet.
31 Forney, G.D., and Eyuboglu, M.V., "Combined Equalization and Coding Using Precoding," *IEE Communications Magazine*, December 1991, 25–35.
32 Keller, S., "Understanding 256-QAM Demodulator Power Cable Modes," *Engineering Design News*, Products Edition, February 12, 1997.
33 Joshi, R. B., and Samueli, H., "A 100 MHz, 5 Mbaud QAM Decision-Feedback Equalizer for Digital Television Applications," 1994 IEEE International Solid-State Circuits Conference *Digest of Technical Papers*, Paper WP 4.2, 68–69.
34 Lane, F., et al., "A Single Chip Demodulator for 64/256 QAM," *IEEE Transactions on Consumer Electronics*, vol. 42, no. 4 (November 1996): 1003–1010.
35 Tom Kolze (ComStar Communications, formerly with General Instrument) and Jeyhan Karaoguz (Motorola, formerly with IBM) acted as reviewers of this paper and made major contributions to the cable modem specification process.

TDMA System Synchronization Using Timestamp Messages

Victor T. Hou
Robert Mullins
General Instrument

Bernardo Paratore
Sarnoff Digital Communications

Abstract

This paper discusses an upstream hybrid fiber/coax (HFC) synchronization technique using timestamps to synchronize upstream time division multiple access (TDMA) transmissions. This method involves sending timestamp messages in the downstream channels to subscriber cable modems (SCMs) in order to synchronize the transmission of the bursts in upstream TDMA slot times. This synchronization method is different from one in which the TDMA timebase is derived from the downstream framing. This method has the advantages of being independent from the downstream modulation scheme, downstream symbol rate, and the downstream framing structure. As a result, this synchronization method does not require an association of particular upstream channels to a downstream channel, and thus, an SCM can be assigned to any upstream channel no matter what downstream to which the SCM is tuned. In addition, this TDMA synchronization scheme does not require the upstream frame structure and downstream frame structure be linked. For example, this synchronization scheme does not require that the duration of an upstream frame be an integer multiple of the downstream frame time.

Timestamp messaging for synchronization of upstream burst transmissions has been incorporated into the Multimedia Cable Network System (MCNS) Radio Frequency Interface (RFI) specification and also into working drafts of IEEE 802.14.[1,2] Descriptions of TDMA structure in this paper are intended to be generic and do not necessarily reflect specific details of either of these two specifications.

TDMA

Requirements

The requirement is to distribute on the downstream channels a time reference. The media-specific access control (MAC) uses this reference in the subscriber cable modem (SCM) at the user's premises, to derive the operational parameters for the current TDMA framing as conveyed by the appropriate MAC messages.

The high-level system requirements are as follows:

- must be independent of the downstream transmission rate
- must be independent of the downstream modulation scheme
- must operate independently from the downstream channel that is used
- must not constrain the upstream transmission rate
- must facilitate an adequate ranging scheme
- must allow for adjustment of ranging offset

TDMA Scheme

TDMA is the baseline access scheme for upstream transmission. The following sections provide a detailed description of an upstream TDMA structure. The TDMA scheme supports a high degree of programmability, easy reconfigurability, and allows variable data rates.

The head-end controller (HC) MAC management entity controls the agile upstream frequencies. At a given frequency, the HC MAC management entity relays TDMA configuration parameters to the SCM. The upstream TDMA is fully configurable and programmable in terms of the parameters outlined in the following sections. *Figure 1* illustrates a TDMA framing structure. The slots that are illustrated can be generalized to include mini-slots, in which a transmission opportunity can consist of a mini-slot or a variable number of consecutive mini-slots.

Slot Size The slot size parameter of the TDMA scheme is defined in terms of the number of timebase reference counts or "ticks" the slot occupies. The reference tick is utilized to measure the passage of time for upstream transmission for a bit or a group of bits. The head end processor MAC management entity provides the slot size information to the SCM during the initialization and provisioning state.

The slot size does include guardtime to allow for nondeterministic delay variations in the TDMA timing system. Guardtime is composed of null bit times that accommodate system timing variations to prevent overlap of a burst in one time slot into the previous or the following time slot. In a ranged system, the guardtime does not

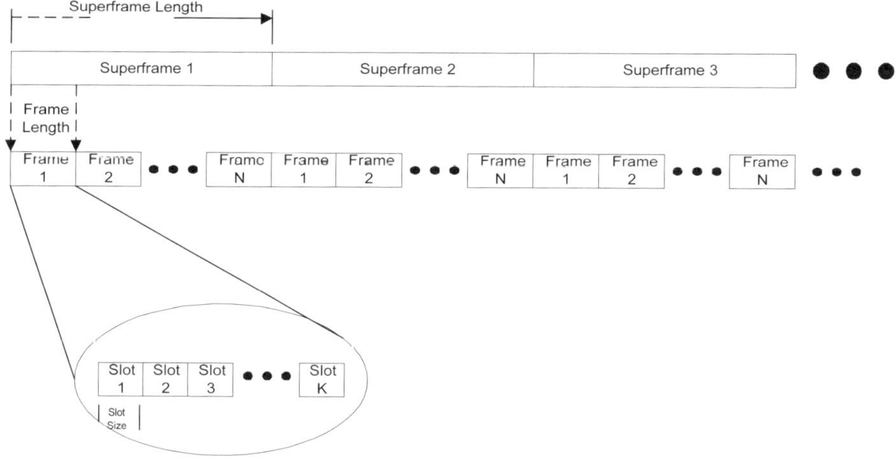

Figure 1: *TDMA Structure*

need to be allocated to compensate for the propagation delays of various SCMs that are varying distances from the head end, but it needs to be allocated for sources of timing variation in which delays are not fixed.

Frame Size The head end processor MAC management entity can specify the frame size associated with an upstream channel. Frame size is expressed in the number of slots per frame. The frame size value is configurable across different frequency channels, allowing a multitude of upstream data rates for individual users based on slot assignment.

Superframe Size The head end processor MAC management entity can specify the superframe size associated with the upstream TDMA channels. This is included in the timestamp messages. The length of a superframe in time ticks is M+1, where M is the terminal count value of the TDMA timebase counter, or in other words, the highest counter value until the counter rolls over back to zero.

The length of a superframe on all upstream channels is the same, and superframes begin at the same time and end at the same time on all upstream channels. Because the timestamp messages are sent down all downstream channels, and all the upstream channels share a common superframe structure, there is no limitation that only certain upstream channels are associated with a particular downstream channel. An SCM can be assigned bandwidth on any upstream channel. Furthermore, downstream framing is not required to indicate the upstream framing structure.

Timebase Synchronization

The first step required by a SCM for TDMA operation is timebase synchronization. This means that the local timebase contained within each SCM counts in step with the master timebase at the head end. This is achieved by distributing a common timing tick in the form of a timing marker (TB_MARKER). This marker is broadcast regularly at a rate which is a multiple of the timebase clock resolution (TB_CLOCK). TB_MARKER is an N-bit number representing the modulo M timebase count (TB_COUNT). The broadcast rate is determined by the characteristics of the locking mechanism used in the subscriber cable modem to compensate for clock drifts and inaccuracies.

The time marker, or timestamp, is contained in a downstream MAC message, which will be called the synchronization timebase message. The transmission of this message from HC to the SCMs is shown in *Figure 2*.

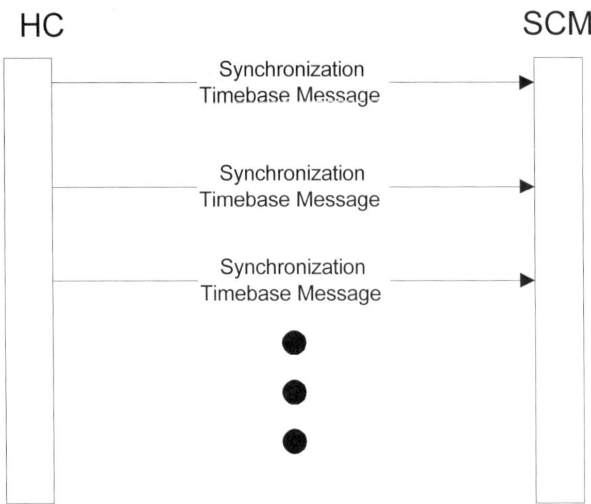

Figure 2: *Synchronization Signaling Sequence*

This process can be visualized by thinking of TB_CLOCK as the second hand of a clock. The count modulo in this case is sixty. If we decide to send TB_MARKER every five ticks starting at zero, TB_MARKER would take on the following values: 0, 5, 10, 15, ..., 55, 0, ..., etc. On the other hand, if we decide to send TB_MARKER every nine ticks starting at twelve, TB_MARKER would take on the following values: 12, 21, 30, 39, 48, 57, 6, 15, etc.

The nominal frequency in which to send a timestamp message depends on the overall system synchronization requirements and the design parameters of the control loop used in the SCM to replicate the head end timebase. To give an idea of the nominal frequency, the MCNS RFI specification states that the nominal frequency shall be greater than or equal to five timestamp messages per second, or in other words, the maximum value of the nominal interval between timestamp messages is 200 ms.[1]

Note the use of the word *nominal* in the previous paragraph. That implies that the timestamp messages do not have to be sent with strict periodicity. Any reasonable recovery circuit can tolerate a range of non-periodicity, and therefore, nominal periodicity is adequate. This attribute of timestamp messaging is useful when statistically multiplexing because a data packet (fixed or variable length) does not have to be held back from transmission just because a timestamp message is required to be sent at a specific time.

Local Timebase Design

There are two issues to deal with. First, the SCM must track the head end time base frequency over relatively short time intervals (tracking over long intervals would not work because of the short-term frequency drift error). To achieve this, a relatively stable clock must be generated that replicates the head end timebase unit clock. Second, the SCM must be able to align the internal reference, in time, to the head end assigned slot clock in order to accurately transmit within an allocated time slot. A downstream MAC message supplies the required information (which includes a timing offset determined through the ranging procedure) to the SCM so that the TDMA slot designation can be aligned in phase with the head end.

Synthesis of the Master Timebase Clock

When the SCM is connected to the cable plant and powered on, the SCM will tune to an active downstream channel and begin acquiring timestamp messages. The messages will contain the value of the head end timebase counter at the instant the message is inserted into the data stream. By using these time markers, the SCM can synchronize an internal clock to the head end timebase clock frequency. *Figure 3* shows timestamp messages being distributed over a downstream channel and the timing relative to the TDMA frame clock.

Note that in *Figure 3* the time marker transmissions [TB_MARKER(n)] are asynchronous to the upstream TDMA frame sync clock. Also, note that the actual frame sync clock does not exist; it may merely be a number in a processor memory. The time markers are loosely coupled to the upstream TDMA rate in the sense that the

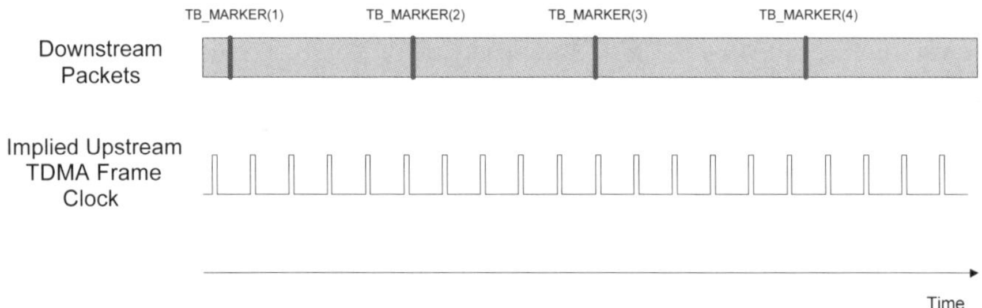

Figure 3: *Relationship Between Time Markers and TDMA Frame Markers*

duration between the messages and the short term drift over that duration dictate the required guard band on each TDMA slot.

At initial power-up, the SCM tunes to a default downstream channel or finds a channel bearing time marker MAC messages. Upon receiving the first timestamp message TB_MARKER(1), the local timing subsystem loads the time marker value TB_MARKER(1) into a programmable counter. The programmable counter is incremented by a clock that is derived from a variable rate synthesizer. The synthesizer is initialized to the nominal frequency of the head end timebase increment clock; however, it is locked to the local oscillator. This means that the timebase counter will be incremented at the following rate:

$$F_{\text{SCM Timebase Counter Clock}} = F_{\text{TB}} + F_{\text{error}}$$

where F_{TB} is the actual increment rate of the head end timebase clock, and F_{error} is the frequency difference between the SCM and HC oscillators. This error consists of constant offset and a short-term drift.

$$F_{\text{error}} = F_{\text{offset}} + F_{\text{drift}}$$

At reception of the second time marker message TB_MARKER(2), the count value in the local timebase counter (LTBC) denoted by TBL_MARKER, is compared to the received message TB_MARKER(2). During the time interval from TB_MARKER(1) to TB_MARKER(2), the LTBC will have attained a value of

$$\text{TBL\_MARKER}(2) = \text{TB\_MARKER}(1) + \text{CNT}_{\text{local}}(2)$$

where $CNT_{local}(2)$ is the number of $F_{\text{SCM Timebase Counter Clock}}$ clock cycles that occurred during the time interval from TB_MARKER(1) to TB_MARKER(2). Ideally, if there was no frequency error over the time interval, TB_MARKER(2) would be equal to TBL_MARKER(2). However, there may be an error F_{error}. Therefore,

$$\text{TB\_MARKER}(2) = \text{TBL\_MARKER}(2) + \text{CNT}_{\text{error}}(2) \text{ or,}$$
$$\text{TB\_MARKER}(2) = [\text{TB\_MARKER}(1) + \text{CNT}_{\text{local}}(2)] + \text{CNT}_{\text{error}}(2)$$

Thus, the error count may be calculated from the known terms:

$$\text{CNT}_{\text{error}}(2) = \text{TB\_MARKER}(2) - [\text{TB\_MARKER}(1) + \text{CNT}_{\text{local}}(2)]$$

The term accurately represents the constant error and short-term drift over the interval. Since this error is stable over a short interval of time (less than a few seconds), the next time marker value can be estimated as

$$\text{TB\_MARKER}(3)_{\text{est}} = \text{TB\_MARKER}(2) + \text{CNT}_{\text{local}}(3) + \text{CNT}_{\text{error}}(2)$$

or more generally

$$\text{TB\_MARKER}(N)_{\text{est}} = \text{TB\_MARKER}(N\text{-}1) + \text{CNT}_{\text{local}}(N) + \text{CNT}_{\text{error}}(N\text{-}1)$$

assuming that the inter-timestamp message time is constant within a small tolerance.

This means that the clock of the SCM timebase counter must be corrected by a count of $\text{CNT}_{\text{error}}(N)$ over the time marker interval. Ideally, this is done by adjusting the frequency of the clock driving the time base counter.

Figure 4 presents a second order loop for varying a digitally controlled oscillator (DCO), which tracks the time stamp synchronization message in frequency and phase. The gain values G_p and G_i are adjusted to vary the acquisition, tracking, and jitter reduction properties of the loop. The DCO is sized to meet the required digitally controlled oscillator (DCO) quantization error. Also, note that what is illustrated is just one particular type of integrator. For example, a weighted exponential integrator can also be used.

When a timebase synchronization message is received by an SCM, the difference between the timestamp value in the message and the local timebase counter is computed. The proportional-integral controller is updated, and the result is latched into the phase increment register.

The digital feedback loop in *Figure 4* was analyzed and simulated. As an example, based upon a 10-Hz rate of timebase synchronization messages, a 1.024-MHz TDMA system clock and a high-speed 12.8-MHz clock were used to clock the DCO. The loop settling time was 3.5 seconds with no preloaded accumulator increment value. A jitter of +/– 50 ms was applied to nominal 100-ms intervals between timebase synchronization messages, and no degradation was noticed. Timestamp jitter attenuation of the loop was 6.3 dB. The loop did not lose stability until the rate of the timebase synchronization messages went below 1.25 Hz. The jitter due to DCO quantization is a function of the size of the accumulator, the rate of the timebase synchronization messages, the system clock frequency, and the high-speed clock that times the DCO. For example, with a 23-bit accumulator, the jitter is 149 ns. Since this is a second order loop with two integrators, it operates with zero steady state error.

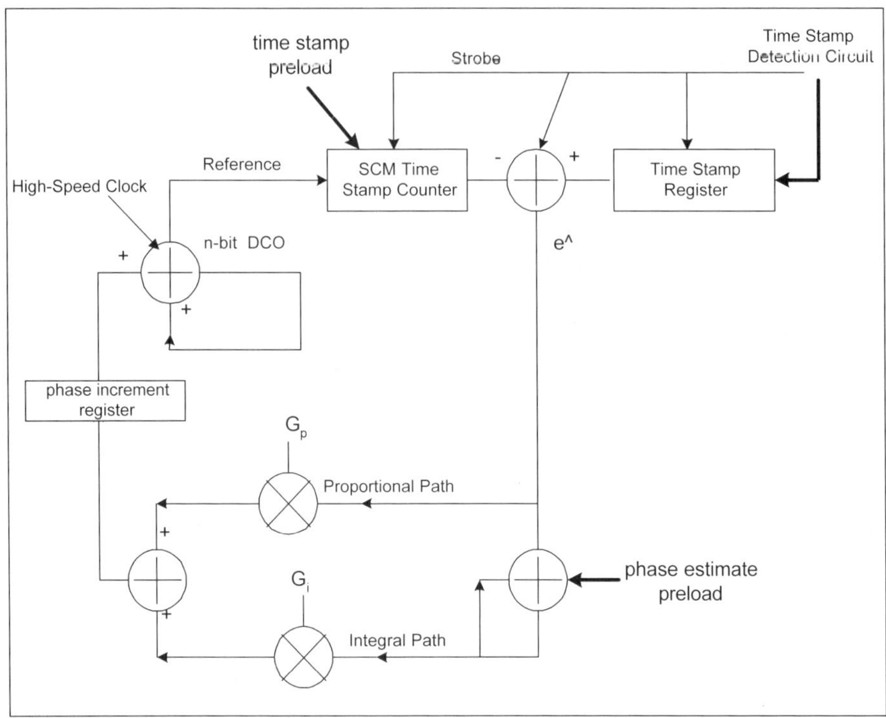

Figure 4: *SCM Synchronization Circuit*

Upstream Synchronization

From the SCM point of view, upstream synchronization is coordinated by the MAC subsystem in the SCM. It uses the local time marker (TBL_MARKER) in conjunction with the TDMA operating parameters provided by the head end processor via MAC messages, to determine the time slot boundaries on which transmission is allowed.

At the head end or central office, the key element in the synchronization system is the burst timing entity within the upstream demod unit.

The burst timing entity generates a time reference t_{slot} for each slot in the form of the aperture signal, as shown in *Figure 5* and *Figure 6*.

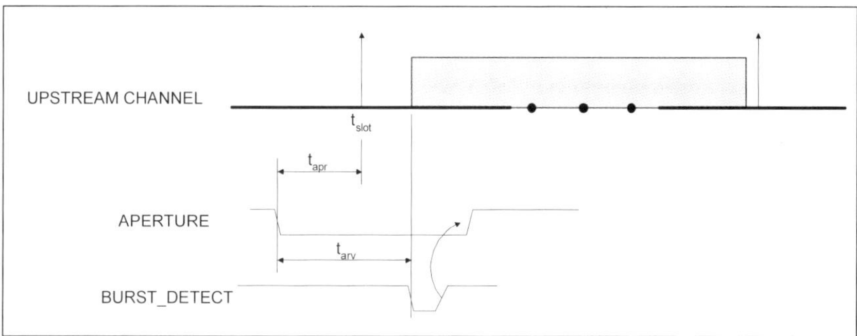

Figure 5: *Late Burst Timing*

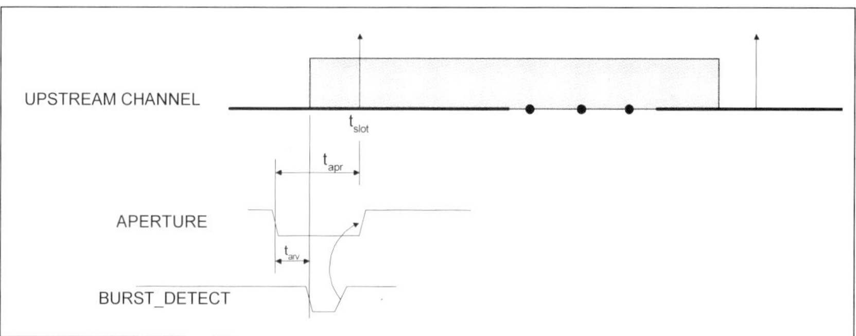

Figure 6: *Early Burst Timing*

The slot time t_{slot} is a function of TB_MARKER and the associated channel parameters loaded by the head end processor (HC) entity. The position of the aperture signal with respect to t_{slot} is programmable by the HC.

If one defines the burst arrival timing error as $t_{err} = t_{apr} - t_{arv}$., if $t_{err} < 0$, the burst is late, and if $t_{err} > 0$, the burst is early. Normal operation requires $|t_{err}| < t_g/2$, where t_g is the guardtime between bursts. For systems containing elements with negligible processing delay variation, t_{err} is the correction factor used by the ranging algorithm.

System Error Presentation

Figure 7 summarizes the source of timing error in the typical cable system. Bandwidth limiting delays have been assumed to be zero or of negligible values. Transmission and fixed delay will also not be considered as these delays are measurable and can be nulled out by the ranging procedure or by calibrated measurement. The nondeterministic system timing jitter, which should define the guard band spacing on the upstream channel, is outlined in this diagram.

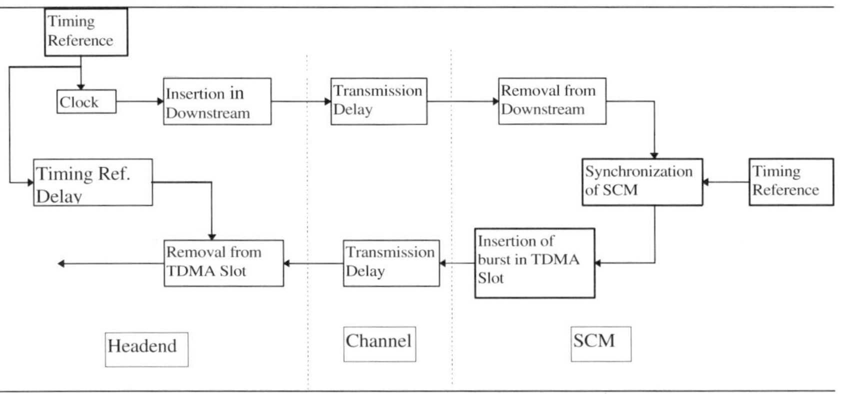

Figure 7: *System Timing Uncertainties*

A description of the various errors and the timing resolution of each are presented in *Table 1*. Differences in the QAM channels are assumed to be negligible due to utilization of a common clock in the HC. The remaining items in the table are discussed as follows:

1) Insertion of timebase into QAM data stream: the resolution of this is one clock tick of the master timebase clock. This should be implemented in hardware at the head end.

2) Reception of timebase message: again, one clock tick of the master timebase clock and implemented in hardware at the SCM.

3) SCM DCO frequency resolution: this is primarily due to the frequency resolution capability of the DCO determined by the number of bits in the phase accumulator, which is an implementation issue of the SCM. For example, the DCO can be sized for 4 ppm performance.

4) Insertion of the message into the return channel: again, implemented in hardware and one clock tick.

5) Message detection in head end processor: assumes approximately 1/2 symbol accuracy depending on which timing epoch is selected in the demod corresponding to one of multiple transmission modes.

6) Measurement Resolution: one clock tick, implemented in hardware.

Description	Resolution
Insertion of time base into downstream channels	1 clock tick
Reception of time base at SCM	1 clock tick
SCM DCO Frequency Resolution	e.g., 4 ppm
Insertion of message into data stream (during access)	1 clock tick
Message detection in demod	±.5 symbol
Measurement resolution	1 clock tick

Table 1: *Sources of Timing Variation*

In general, fixed processing delays can be voided by the ranging algorithm. Channel delay, which is determined by the propagation of a signal through the cable plant, can be voided by the ranging algorithm. Transmission delay, transmitting a known number of symbols onto the physical medium, is also a fixed delay that is nulled out by the ranging algorithm. Packet jitter occurs when the data clock and the transmission clock are asynchronous with each other. This occurs at the head end where the master timebase clock and the downstream data clock on the modulator board are free-running in respect to each other. In the SCM, the local timebase and the transmit clock are also free-running in respect to each other. If the data clock frequency is greater than the timebase clock frequency, then the jitter is +/– 1/2*DATA_CLOCK. In the reverse case, the jitter is +/– 1/2*TB_CLOCK. Packet jitter cannot be compensated for and must be accounted for as part of the guard band. Thus, in general, the ranging procedure seeks to null out all fixed delays, and guardtime allocated in the upstream time slots is for the purpose of handling delay variations throughout the TDMA system.

The ranging algorithm takes place in two different scenarios: 1) a range upon power-up and 2) an ongoing range during packet transmission under normal upstream communications. It should be noted that guardtimes during power-up ranging must be adequate to cover the delay variation encountered with an unranged unit, or the transmission from an unranged unit may overlap into a time slot allocated to another user.

Conclusion

This paper discusses a TDMA timebase synchronization scheme that involves distributing the master TDMA timebase clock located at the head end. This synchronization scheme operates independently from the downstream modulation scheme, the downstream data rates, and the downstream framing structure. It also synchronizes all users tuned to any of multiple downstream channels, so that an SCM can be commanded by the HC to tune to another downstream channel emanating from the same HC without having to acquire TDMA timebase synchronization. In addition, the superframe structure of the upstream TDMA channels is all synchronized. On each upstream channel, superframes begin and end at the same time. Thus, a reassignment by the HC of the upstream channel does not require the SCM to reacquire the superframe structure. The superframe structure relates to the TDMA timebase because the terminal count value of the TDMA timebase is one time tick less than the length of a superframe in timebase ticks. Finally, this TDMA synchronization scheme provides additional flexibility by not requiring that upstream framing and downstream framing be locked together.

References

1 MCNS, *Data over Cable Service Radio Frequency Interface Specification*, SP-RFI-I02-971008.

2 IEEE Project 802.14, *Cable TV Access Method and Physical Layer Specification*, 802.14 Draft2, R2, 31 July 1997.

Synchronized Discrete Multitone Modulation for Upstream Transmission on Cable Networks

Krista S. Jacobsen
Senior Systems Engineer, VDSL Department
Amati Communications Corporation

John A. C. Bingham
Chief Scientist
Amati Communications Corporation

Abstract

This paper describes synchronized discrete multitone (SDMT), a technique that enables the use of discrete multitone (DMT) in the reverse channel of cable or hybrid fiber/coax (HFC) networks. The advantages of SDMT are enumerated, and a procedure to achieve the required remote modem synchronization is presented. An example configuration is described to show that SDMT is a cost-effective physical layer solution that enables efficient, robust, high-speed upstream communications on cable and HFC networks.

Introduction

First-generation cable modem standards such as IEEE 802.14 require use of quaternery phase shift keying (QPSK) for upstream transmission. Although QPSK is robust, inexpensive, and well-understood, it is also inefficient, supporting spectral efficiencies of less than 2 bps/Hz, which limits both the number of subscribers that can be supported in the shared reverse channel bandwidth and the maximum burst rate at which those subscribers can send data. For this reason, the IEEE 802.14 Working Group has established a study group whose charge is to investigate and propose for standardization a more efficient, higher-performance modulation scheme for upstream transmission. This paper describes one such technique called synchronized discrete multitone (SDMT).

SDMT is a powerful technique that enables a number of DMT–based remote terminals to transmit to a single central-site receiver. SDMT maintains in multipoint-to-point environments all the advantages that DMT provides in point-to-point environments. These advantages include the ability to support high data rates on channels with a variety of transmission impairments such as passband ripple, spectral nulls, impulse noise, and radio-frequency interference, all of which are expected in the reverse channel of cable and hybrid fiber/coax (HFC) networks.

Section 2 reviews DMT and its advantages. SDMT is described in section 3, and a procedure to synchronize remote modems to enable SDMT is presented. In section 4, an example SDMT configuration for cable or HFC networks is described. Section 5 concludes that SDMT enables high-performance, cost-effective multipoint-to-point communications on cable and HFC networks.

Discrete Multitone (DMT) Modulation

This section briefly reviews discrete multitone modulation, which is the basis of SDMT. For additional details, the reader is referred to other resources.[1–7]

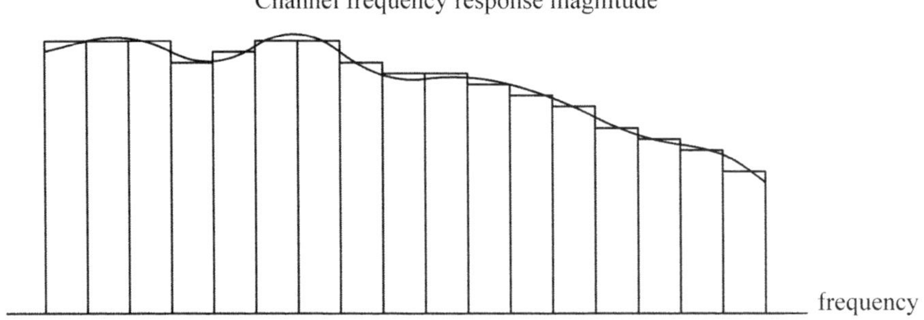

Figure 1: *A Channel Partitioned into Subchannels*

Discrete multitone (DMT) modulation uses an inverse discrete Fourier transform (IDFT) to partition a transmission channel into a set of orthogonal, equal-bandwidth subchannels so that the frequency response of each subchannel is roughly constant across its bandwidth, as illustrated in *Figure 1*.

Because the frequency response is nearly flat across each subchannel, the subchannels are almost without memory, and the small amount of intersymbol and inter-subchannel interference caused by each subchannel's non-unity impulse response length can be eliminated by use of a cyclic prefix. As its name implies, the cyclic prefix precedes each symbol. It is a copy of the last ν time-domain (that is, post–IDFT) samples of each DMT symbol. Hence, the cyclic prefix carries redundant information and is transmission overhead. If there are no more than $v + 1$ non-zero samples in the channel impulse response, then intersymbol interference (ISI) caused by each symbol is confined to the cyclic prefix of the following symbol. By discarding the cyclic prefix samples in the receiver, ISI is eliminated completely. Therefore, a traditional equalizer is not required in the DMT receiver. If the channel impulse response length exceeds the cyclic prefix length, then the received signal incurs some distortion.

In general, the channel can be partitioned into any convenient number of subchannels, as long as the condition of nearly-flat frequency responses over the subchannel bandwidths is met. However, one cost-effective implementation of DMT uses the fast Fourier transform (FFT) to compute efficiently the DFT and IDFT required in the receiver and transmitter, respectively. To exploit the computational efficiency offered by the FFT, the number of subchannels should be an integer power of two. Cost constraints and performance requirements then dictate the precise number of subchannels into which a channel is partitioned.

In the DMT transmitter, a bitstream is encoded as a set of quadrature amplitude modulated (QAM) subsymbols, where each QAM subsymbol represents a number of bits determined by the signal-to-noise ratio (SNR) of its associated subchannel. During each DMT symbol period, a set of subsymbols is input as a block to a complex-to-real IDFT. The cyclic prefix is prepended to the output, and the result is converted from digital to analog format. If passband transmission is required, the resulting signal, a sum of real time-domain sinusoids, is then upconverted to the desired carrier frequency and transmitted over the channel. In the receiver, after downconversion and analog-to-digital conversion, the cyclic prefix is stripped, and the samples are input to a DFT. Each output value is then scaled by a single complex number to compensate for the magnitude and phase of each subchannel's frequency response, and a memoryless detector decodes the resulting symbols. The set of complex multipliers is known as the frequency domain equalizer (FEQ). Changes in the channel magnitude or phase are accommodated by updating the FEQ taps as the system operates. A generalized passband DMT transmitter and receiver pair is shown in *Figure 2*.

Advantages of DMT

DMT provides a number of advantages for transmission on difficult channels such as the reverse channel of cable or HFC networks. First, DMT can maximize the bit rate supported on any channel. In contrast to single-carrier techniques, which support a fixed number of bits within a particular frequency band, the number of bits supported by DMT varies with frequency. Subchannels are assigned as many bits as they can support reliably. As a result, the aggregate bit rate on the channel is generally higher than it would be using a single-carrier technique. For example, QPSK

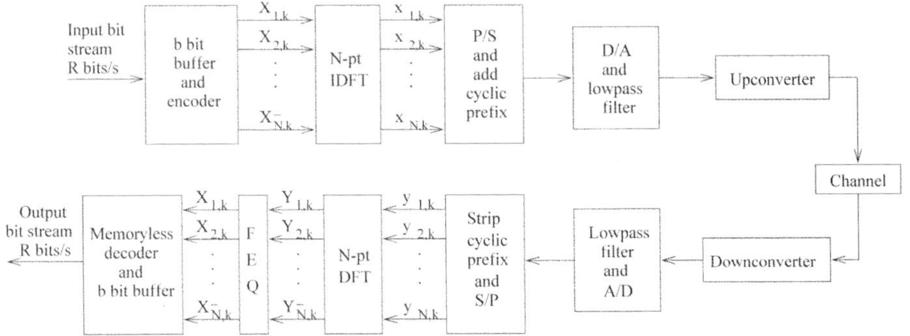

Figure 2: *Passband DMT Black Diagram*

transmits 2 bits per symbol. Taking into account the required excess bandwidth, the spectral efficiency of QPSK is less than 2 bps/Hz, irrespective of the channel capacity. In contrast, the spectral efficiency of DMT is closely tied to the channel capacity: more bits per symbol are supported on channels that have higher capacities. It is expected that the SNR in some spectral regions of the reverse channel will be high enough to support 5 or even 6 bits. DMT takes advantage of these high SNR bands by assigning the appropriate number of bits to subchannels in those regions, thereby maximizing the channel bit rate.

DMT also automatically avoids those regions of the spectrum in which reliable data transmission cannot be supported. During initialization, the bit distribution is computed based on the observed channel and noise conditions. Those subchannels with SNRs insufficient to support any bits are simply turned off. By tailoring the bit distribution to the channel characteristics, DMT optimizes the transmit bandwidth. During steady-state operation, the bit distribution is adapted as necessary to maintain a desired performance. The subchannel SNRs are monitored in a data-driven manner by the receiver. Upon detecting a degradation in a subchannel SNR, the receiver computes a modified bit distribution that better achieves the desired error performance. Depending on the SNR of a degraded subchannel, some or all of its bits may be moved via a bit swap algorithm to one or more other subchannels that can support additional bits. The bit distribution change is reported to the transmitter, which then notifies the receiver when the new bit distribution will be activated. The adaptivity afforded by bit swapping enables DMT to alleviate problems caused by changes in the channel frequency response or noise while a system operates. DMT can also accommodate substantial changes in the channel capacity by decreasing (or increasing, if the channel capacity increases) the number of bits transmitted during each DMT symbol period. Thus DMT is inherently rate-adaptive.

Another advantage of DMT is its immunity to impulse noise, which is generally better than that of single-carrier systems. Because the DMT receiver processes received samples in blocks, after the receiver transformation the energy of any impulse noise that corrupted some of the received samples is spread evenly among the subchannels. Hence, impulse noise causes a temporary decrease in each subchannel SNR. If the subchannel noise margins are large enough, impulse noise may not affect the system error performance. If, however, the expected impulse noise is severe enough to exceed the noise margin, then forward error correction with interleaving can be applied to mitigate errors.[8]

Synchronized DMT (SDMT)

The reverse channel of an HFC system must be shared by many independent and physically separated transmitters. Their transmissions may be multiplexed into a common channel using frequency-division multiplexing (FDM) or time-division multiplexing (TDM), or, for greatest efficiency in bandwidth usage and flexibility of data rate for many types of users, a combination of the two. This section describes how DMT can use its subchannels to implement very efficient FDM; the combination of frequency-division multiple access (FDMA) with a more traditional time-division multiple access (TDMA) system is discussed briefly in section 4 but is mostly beyond the scope of this paper.

The DMT system described in section 2 is point-to-point: all sub-carriers are modulated in the same transmitter (using a single IDFT), and, except for a small variation of delay across the used frequency band, their modulation envelopes arrive at the receiver at the same time. Therefore they can be processed using a single DFT.

For the greatest efficiency of an FDMA system it is desirable to allocate a contiguous frequency band to each remote. The allocation can be either permanent, as in a conventional FDMA system, or temporary in a combined FDMA/TDMA system. Using either approach, maximum efficiency is achieved if guard bands and band-separating filters are not required to isolate transmissions from different remote units. This maximum efficiency can be achieved with complete flexibility in a DMT system by allotting contiguous sets of any number of subchannels to the remotes. When the modulated sub-carriers emanate from different remote transmitters, those transmitters must be both frequency-locked and timed so that their modulation envelopes also arrive at the head end at the same time. One method of meeting the time and frequency requirements is shown in *Figure 3*. The procedure is as follows:

1. When a cable system is first installed the operator must know (or conservatively estimate) the round-trip delay from the head end to and from the furthest remote.

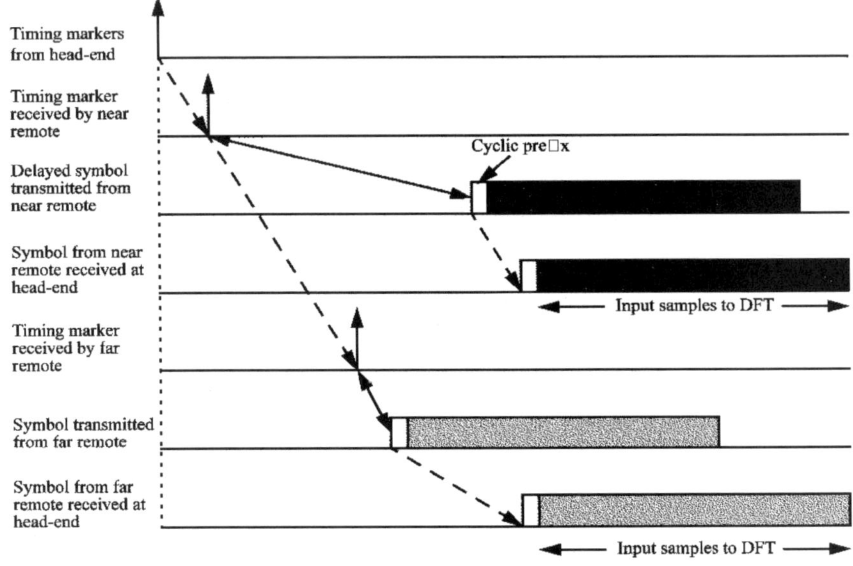

NOTES:

1. The dashed arrows represent propagation delays.
2. The double-headed arrows represent the transmit delay programmed into each remote.
3. The far remote is almost at the maximum distance programmed into the system.
4. For the sake of clarity, only one symbol is shown for each remote; in practice, consecutive symbols will be contiguous.

Figure 3: *Synchronization of a Near and a Far Remote Using a Downstream Timing Strobe*

Thereafter the downstream signal must carry:

(a) a frequency reference (a common one is the 8.0-kHz network timing reference, which is used for all voice digitization);
(b) a timing strobe whose repetition period is longer than the possible variation in the round-trip delay (that is, maximum minus minimum).

2. When a remote modem is first installed, it has no a priori knowledge of how far it is from the head end. Therefore it must complete a ranging procedure, which enables it to learn its round-trip delay from the head end, and hence the complementary amount by which it must delay its transmissions. This procedure, which is not unique to networks using SDMT, is beyond the scope of this paper. *Figure 3* assumes the ranging procedure has been completed.

3. After completing the ranging procedure, each remote locks to the strobe and begins transmitting each DMT symbol (beginning with its cyclic prefix) the

appropriate number of DMT sampling periods after receipt of the strobe. The symbol and subchannel assignments are coordinated by the access protocol, a discussion of which is beyond the scope of this paper.

When the remotes are synchronized using this procedure, the full wideband signal can be demodulated at the head end using one DFT. The demultiplexing can be done subchannel by subchannel at the output of the decoder shown in *Figure 2* by applying the appropriate mixture of FEQ taps.

Example System for Upstream HFC Transmission

This section presents an example SDMT configuration to support reverse channel communications. Assuming a sub-split cable or HFC configuration, the entire reverse channel spans the frequency band from 5 MHz to 42 MHz. However, designing remote terminals that use the entire 37-MHz band is impractical for a number of reasons. First, the computational complexity of a transceiver operating in 35+ MHz bandwidth is quite high. Second, it is unlikely that the data rate requirements of the remote terminals warrant use of the entire reverse channel at any time. Based on discussions within IEEE 802.14, the maximum expected instantaneous bit rate of the services considered is 10 Mbps. If a spectral efficiency of just 2 bps/Hz can be achieved in some part of the reverse channel, then a maximum of 5-MHz bandwidth is required to support 10 Mbps (neglecting any excess bandwidth). Hence, none of the remote terminals is ever likely to need access to the entire reverse channel. A more practical solution than designing wideband transceivers is to segment the reverse channel bandwidth into smaller bands, each of which supports an independent SDMT signal.

Description	Resolution
Insertion of time base into downstream channels	1 clock tick
Reception of time base at SCM	1 clock tick
SCM DCO Frequency Resolution	e.g., 4 ppm
Insertion of message into data stream (during access)	1 clock tick
Message detection in demod	$\pm$.5 symbol
Measurement resolution	1 clock tick

Table 1: *Example System Parameters*

Assuming that minimizing the cost of the remote terminals is a design goal, dividing the entire reverse channel into 1024 34.5-kHz-wide subchannels appears to be a good compromise between performance and complexity. To ensure the complexity of the remote terminals is not prohibitive, the set of 1024 subchannels can be segmented into eight disjoint groups of 128 consecutive subchannels.[9] Given this partitioning, any remote terminal can use a maximum of 128 subchannels that span 4.416-MHz bandwidth at any time. As a result, the maximum sampling rate of the remote terminals must be only 8.832 MHz, and the required DFT size is 256. *Table 1* summarizes the example system parameters.

Each remote terminal configured according to the example parameters has a computational complexity that can be handled easily by a single DSP chip. Furthermore, even less expensive remote terminals can be designed. For some cost-sensitive applications, it may be desirable to design remote terminals that use fewer than 128 subchannels. For example, a video conferencing signal may require 384 kbps. Assuming a minimum of 2 bits per subchannel, then at a symbol rate of 32 kHz only 6 subchannels are required to support this bit rate. To protect against subchannels that are too noisy or attenuated to support data, a conservative design for video conferencing signals might use, say, 32 subchannels. The head end controller could then assign these smaller bandwidth units to a *subgroup* within a particular group, as shown in *Figure 4*. The main advantages of using fewer than 128 subchannels are the resulting reduction in transmitter computational complexity and a reduced digital-to-analog converter complexity. When only 32 subchannels are used by a remote terminal, the required DFT size is 64, and the sampling rate, assuming the example configuration environment, is only 2.208 MHz. To maintain a subchannel spacing of 34.5 kHz, the cyclic prefix length is then 5 samples.

In addition to reducing remote terminal costs, the flexible eight-group configuration offers another advantage. As numerous cable operator studies have revealed, some parts of the reverse-channel bandwidth are poorer quality than other parts. In particular, it is generally agreed that the bandwidth from 5–20 MHz is significantly more susceptible to ingress than is the bandwidth from 20–42 MHz.[10] By designing the remote terminals with frequency-agile carriers, a remote terminal can transmit using the group of subchannels that is best able to support its data rate requirements. Hence, groups that can support high aggregate data rates can be used either by some number of high bit rate users or by a larger number of lower bit rate remote terminals. Those groups that are affected by severe attenuation, passband ripple, or ingress can then be more lightly loaded. Furthermore, even though multiple independent DMT signals can be transmitted simultaneously in the eight-group configuration, a single head end receiver, spanning the entire reverse channel bandwidth, can be designed. The receiver performs a 2048-point FFT to demodulate incoming

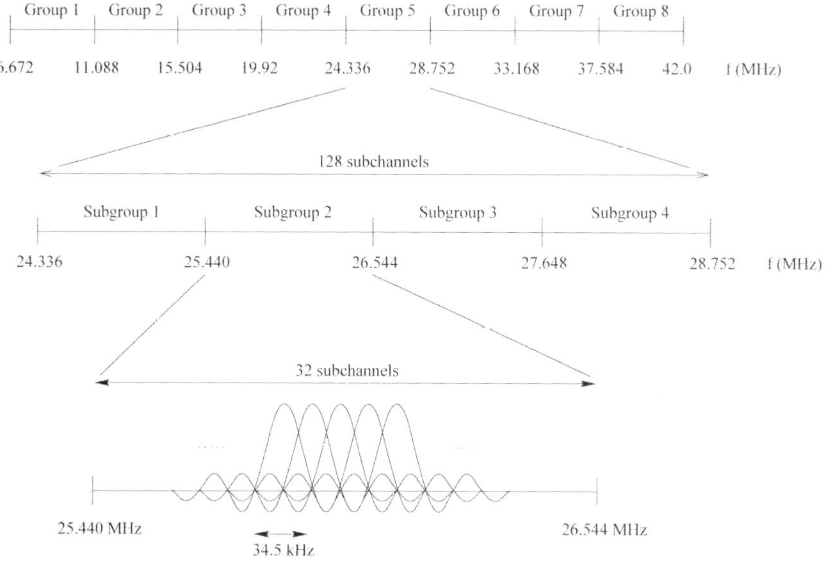

Figure 4: *Partitioning of Reverse Channel into Groups and Subgroups*

signals, and a controller then ensures signals are decoded correctly. If the complexity of the wideband receiver is too high for a particular installation, separate head end receivers can be designed for each of the eight groups.

Summary

In this paper, synchronized discrete multitone (SDMT) has been presented as a technique for coordinating multipoint-to-point reverse channel digital communications on cable or HFC networks. Although DMT has been applied traditionally only in point-to-point environments, this paper illustrates that, by appropriately synchronizing remote terminal transmissions, DMT can also be used effectively in multipoint-to-point environments. An example configuration was presented to show that a low-cost, low-complexity SDMT design can be implemented for cable or HFC networks. Thus, SDMT enables cost-effective multipoint-to-point reverse channel communications while preserving the well-known advantages of DMT, including its efficient use of the available bandwidth and its robustness to channel impairments and noise.

References

1 Bingham, J. A. C., "Multicarrier Modulation for Data Transmission: An Idea

Whose Time Has Come," *IEEE Communications Magazine*, vol. 28, no. 5, May 1990, pp. 5–14.

2 Cioffi, J. M., "A Multicarrier Prime,." *ANSI T1E1.4 Committee Contribution,* November 1991, pp. 91–157.

3 Saltzberg, B. R., "Performance of an Efficient Parallel Data Transmission System," *IEEE Transactions on Communication Technology*, COM vol. 15, December 1967, pp. 805–811.

4 Aslanis, J. T., "Coding for Communication Channels with Memory," PhD thesis, Stanford University, November 1989.

5 Kasturia, S., "Vector Coding for Digital Communication on Spectrally Shaped Channels," PhD thesis, Stanford University, March 1989.

6 Tu, J. C., "Theory, Dcsign and Application of Multichannel Modulation for Digital Communications," PhD thesis, Stanford University, June 1991.

7 Jacobsen, K. S., "Discrete Multi-Tone-Based Communications in the Reverse Channel of Hybrid Fiber-Coax Networks," PhD, thesis, Stanford University, August 1996.

8 Aslanis, J. T., Tong, P. T,. and Zogakis, T. N., "An ADSL Proposal for Selectable Forward Error Correction with Convolutional Interleaving," *ANSI T1E1.4 Committee Contribution*, August 1992, pp. 92–180.

9 Bingham, J. A. C., and Jacobsen, K., "CATV Reverse Channel Transmission Using Synchronized DMT, Part 1. Overview: Network, Data Rates and Services, Protocol," *IEEE 802.14 Working Group Contribution number 95 001*, January 1995.

10 Rogers Cable Systems, "Two-way Plant Characterization," 1994.

An Approach to Upstream HFC Channel Modeling and Physical-Layer Design

Thomas J. Kolze, Ph.D
President
ComStar Communications, Inc.

Abstract

To facilitate economic development of robust upstream communications systems over hybrid fiber/coax (HFC) plants, the characteristics of the HFC plant must be captured and presented in a form meaningful to communications systems/modem designers. These channel characteristics, along with performance goals for information rate and error rates, must drive the selection of the appropriate choice of modulation and error control coding.

This paper presents a set of parameters that characterize the upstream HFC cable plant in which upstream signaling is expected to perform. Furthermore, the paper presents quantitative values for these parameters that represent a reasonable impairment level for designers to accommodate; doing so will guarantee reliable communications with wide availability.

Single-carrier frequency and time division multiple access (F/TDMA) provides a low-risk, high-capacity approach that offers the best choice for upstream modulation over the characterized HFC system. This modulation technique is a mature technology; it easily accommodates proven, effective mitigation techniques for combating the HFC upstream channel impairments, and is bandwidth efficient in the HFC channel.

Frequency agility of the carrier, with both quaternery phase shift keying (QPSK) and 16-quadrature amplitude modulation (QAM), multiple symbol rates, and flexible forward error control (FEC) coding and frame and preamble structure are advocated.

Introduction

This paper is organized into two parts: 1) modeling of the HFC channel to facilitate communications system design trade-offs and performance assessment, and

2) presentation of an upstream modulation approach that is at once low-risk, robust, and capable of high data rate/low error rate performance in the HFC upstream. The paper presents the HFC channel model in Section 2 and an upstream modulation approach in Section 3.

HFC Upstream Channel Model

This section contains a model for the upstream HFC cable plant. The impairments of the HFC plant are included in a form useful for both 1) evaluating cable modem and system performance, and 2) determining whether specific cable plants are compliant. The quantitative values are provided with two criteria in mind: 1) satisfactory performance in the presence of these impairment levels is achievable in a practical and economical cable modem product, and 2) the impairment levels represent a reasonably severe, but not atypical, bounding case. In regard to the latter, it is believed that designing to the impairment levels of the model will ensure that the resulting system will successfully communicate in the existing and proposed HFC cable plants.

The channel model closely follows the channel model generated for NextLevel Systems' (formerly General Instrument) cable modem design,[1] and is based on an amalgam of data gathered by NextLevel,[2-4] published in Cable Lab's papers and reports,[5-8] IEEE 802.14 Working Group's discussions and filings,[9-12] or other sources.[13] Much of this information is not available to the public (yet), and there is still a need for more data due to the diversity of HFC plants in operation. However, it is not premature to present a channel model at this time that captures the essence of the HFC upstream plant in a fashion to commence modem design, and places a stake in the ground for the impairment levels that a successful modem should combat.

Upstream Channel Model Block Diagram

Figure 1 shows the block diagram of the upstream HFC channel impairments introduced between the output of a modulator in a subscriber premises and the output of the fiber-optic receiver at the cable plant's head end. Signal filtering, nonlinear devices, and additive and multiplicative interferences are encountered in the upstream HFC plant. The various blocks are described later in this paper.

Filtering Distortion

This block is a linear filter and accounts for passband amplitude and group delay nonflatness; it includes tilt, reflections, and amplifier filtering. The model includes discrete reflections, and appends additional amounts of tilt, group delay variation (GDV), and amplitude variation to these reflections.

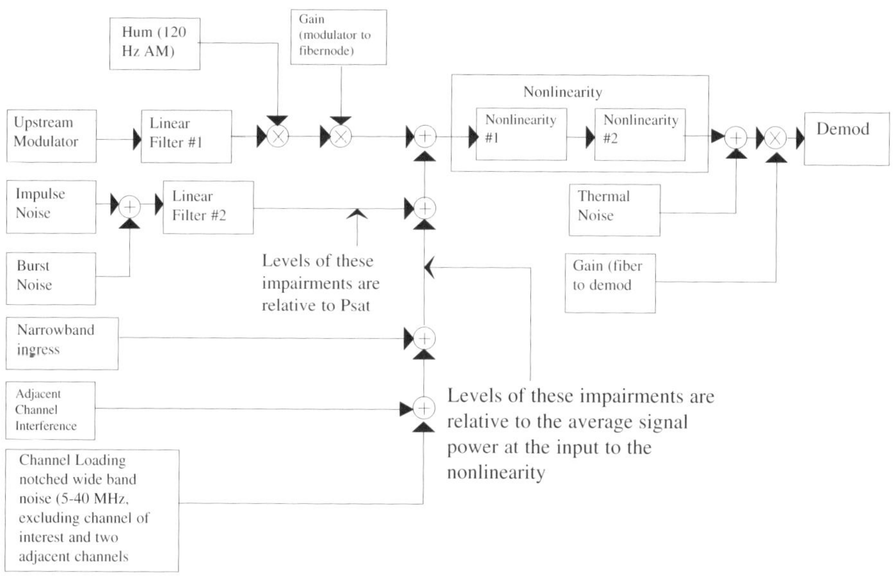

Figure 1: *Upstream HFC Impairment Model Block Diagram*

Total Echo Power

The reflection scenario accounts for multiple reflections of various delay and power levels. The table below shows eight different delay bins, with the corresponding amount of reflection power accumulated in each bin (see *Table 1)*. The model allows for one or more discrete reflection components within each time bin, with randomly placed delay and random distribution of reflection power within the bin. However, the total reflection power in each bin should be as given in the table (for the limiting case—less if a less severe scenario is to be modeled, of course).

0 to 200 nsec delay	**-10 dB**
200 to 400 nsec	**-12 dB**
400 to 800 nsec	**-14 dB**
800 to 1200 nsec	**-18 dB**
1200 to 2500 nsec	**-24 dB**
2500 to 5000 nsec	**-30 dB**
5000 to 15000 nsec	**-35 dB**
>15000 nsec	**none**

Table 1: *Eight Different Delay Bins*

Note that placing the echo power at the maximum of the range yields the largest GDV and smallest period of oscillations in the frequency domain. Breaking the echo power into multiple echoes within the range lowers the GDV (using RSS as a combining technique). But heuristically, this will have a worse effect on performance since the worst-case ISI voltage is larger for the same amount of echo power (due to higher peak-to-RMS ratio with multiple signals combining than with a single echo).

Tilt

The tilt is 0.3 dB/MHz.

Additional Filtering Amplitude Ripple and GDV

The amplitude is 0.3 dBp-p/MHz of signal bandwidth. The wider the channel, the more amplitude variation to be added to the reflection profile and the tilt (see *Table 2*).

GDV:	
5 MHz to 35 MHz	slope: 50 nsec/MHz quadratic: 10 nsec p-p GDV per MHz of signal bandwidth (e.g., 40 nsec p-p in 2 MHz)
35 MHz to 40 MHz	slope: 100 nsec/MHz quadratic: 20 nsec p-p quadratic GDV per MHz of signal bandwidth

Table 2: *Filtering Amplitude Ripple and GDV*

Dynamic Variation

To account for dynamic variation of the plant, of the composite –6.5 dBc of reflection power of the Total Echo Power section of this paper, components comprising –14 dBc can instantaneously change phase, time delay, and power level, while requiring that the composite reflections remain within the constraints of this section. Such a change is a relatively rare occurrence. Sustained periods with a rate of one event per five minutes may be unduly pessimistic.

Nonlinear Distortion

The nonlinear distortion is dominated by the upstream laser, but the amplifiers may also contribute. The nonlinear model consists of a two-stage concatenation of nonlinearities. The first stage accounts for small signal distortion:

$$y = x + ax^2 + bx^3$$

The second stage accounts for the asymmetric laser saturation or clipping:

$$z = \min(y, A_{sat})$$

The value A_{sat} (in volts for example) corresponds to the power level of a sinusoid that begins exhibiting significant harmonic generation in the upstream laser, and is referred to as Psat dBmV. (See the Gain section for value of P_{sat}.) The values for 'a' and 'b' still need to be determined, but based on current data it is believed that they are negligibly small. There should be occasional monitoring to assess if intermodulation (IM) distortion is getting too severe, resulting in backing off all transmitters appropriately, as a technique to combat gain changes in the HFC from the upstream modulators up to the laser transmitter.

For improved model accuracy when fine-tuning performance estimates, a more rounded model of the saturation nonlinearity is in order, based on the measurement of the particular laser to be employed. The precise Pout versus Pin shaping can appreciably impact the resulting intermodulation performance.

Hum

This is multiplicative noise, introducing 3.5% amplitude modulation on the upstream signal. The shape of the multiplying waveform is trapezoidal with a frequency of 120 Hz (i.e., period = 1/120 seconds), with 1-msec transitions.

Adjacent Channel Interference (ACI)

This is a 25% square-root-raised-cosine Nyquist-shaped spectrum, with various symbol rates (160 ksym/sec, twice this rate, four times, eight times, and sixteen times this rate for 2.56 Msym/sec), and spacing of 200 kHz, and twice, four times, eight times, and sixteen times, respectively, according to symbol rate. The channel of interest is between two other digital channels with 6 dB higher density in each. Note that a reasonable worst case is for the adjacent channels to be the widest mode. Frequency uncertainty and symbol rate uncertainty are negligible for this ACI.

Channel Loading

This signal emulates all the upstream carriers impinging upon the upstream laser, except the channel-of-interest and its adjacent channels. The noise–loading waveform is wideband (5 MHz to 40 MHz) additive Gaussian noise, notched to eliminate the channel of interest and the two adjacent channels. The noise density of the loading noise shall be 2 dB higher [3 dB–10log(5/4)] than the density of the signal of interest. Another scenario uses only 9 MHz of notched, wideband noise, which models the case where there is only 25% of the upstream spectrum in use.

Primarily, the channel-of-interest shall be in the middle of the wideband loading noise. This provides for more IM distortion from the wideband noise falling into the channel-of-interest than if the channel is at a band edge. However, second order products, and even higher order products, from carriers in the 5-MHz to 40-MHz band are capable of falling in other channels in the band, so channel-of-interest placement at the high end of the band should be investigated. With 25% spectrum utilization, investigation should include noise loading the spectrum at half the channel-of-interest frequency.

If a simulation must be executed using complex baseband implementation of the channel-of-interest (due to time-bandwidth product considerations), then the combination of the wideband loading noise and the saturating nonlinearity may be modeled via addition of pulses generated with a Poisson arrival time. The pulse amplitude probability distribution shall be the translated tail distribution of a Gaussian random process (saturation threshold translated to zero). The pulse statistics will depend on the value of the ratio (multicarrier signal power)/(upstream laser clip threshold) being modeled.

Thermal Noise

The carrier to noise ratio (CNR) is to be varied parametrically. A CNR greater than or equal to 25 dB may be used for a value representative of a "typical" fairly good portion of the upstream spectrum. However, this does not include head end demod thermal noise.

Narrowband Ingress

The narrowband ingress is created by 20 kHz narrowband additive Gaussian noise interferers. The interferer power is to be varied parametrically, adding one interferer for each 1.5 MHz of signal bandwidth; e.g., 200 kHz, 400 kHz, and 800 kHz—one interferer; 1.6 MHz—two narrowband interferers located randomly in channel; 3.2 MHz—three interferers located randomly in channel. While the interferer power is to be varied parametrically, a target value for an impaired channel would be –7 dBc for the 200-kHz and 400-kHz channel bandwidths, and –10 dBc for the 800 kHz bandwidth, 1.6-MHz bandwidth (total for the two ingress bands in the latter case), and 3.2-MHz bandwidth (total for the three ingress bands in the latter case).

Note that this model is representative of the very disruptive carrier-related components that couple into the upstream from the downstream composite waveform, due to even-order nonlinearities and coupling in the common path components of the downstream and upstream, in cable plants experiencing degraded operation.[12] Unfortunately, this is a condition that occurs often enough that it should be accounted for in the design. For radio broadcast ingress, the interference is often due to modulation types with more controlled (less fluctuating) amplitude levels. For a given relative power level, they are less difficult to operate with (since the extreme amplitudes of the AWGN interference will tend to drive performance). Also, much of the radio broadcast ingress is less than 20 kHz in bandwidth. Still, the model provided here is a challenging yet reasonable characterization for the narrowband ingress upstream impairment.

Dynamic appearance and disappearance of narrowband ingress, such as with amateur radio operators, may occur. A model for dynamics of narrowband ingress is an on/off cycle every two minutes.

Burst Noise

Disagreement exists on the frequency of occurrence of bursts longer than 1 microsecond. In general, such bursts occur only a handful of times in a 24-hour period, and yet there may be times and conditions and plants wherein a much higher frequency is sustained for a relatively long duration. Some of these conditions are described in the references.[8] The following items describe the burst noise model, consistent with the philosophy of capturing the dominant features of impairments in as simple a fashion as possible useful to modem designers:

- Events are Poisson-distributed with upstream rates of one per ten seconds (but this is very subjective, with limited available data).

- Each burst of the event is AM–modulated, zero-mean, Gaussian noise, 1-MHz RF bandwidth (0.5-MHz lowpass baseband noise), with standard deviation equal to one-quarter of carrier amplitude.

- The carrier of each burst of an event has the same amplitude and frequency.

  A) The carrier frequency of each event is selected from a uniform distribution between 5 MHz and 40 MHz (captured data in the references seem to imply a range of 5 MHz to 15 MHz for this oscillation.[3,7] But other burst mechanisms may occur, and other conditions leading to higher oscillation frequencies should not be discounted).

  B) The amplitude of each event is selected from a sawtooth distribution, which is zero at 0 Vrms and is peaked at 1 Vrms (equal to 60 dBmV across 75 ohms).

- The number of bursts per event occur with the following probabilities (approximately a binomial distribution):

one burst: 0.25
two bursts: 0.42
three bursts: 0.25
four bursts: 0.072
five bursts: 0.008

- With multiple bursts per event, the duration of each burst is selected from a uniform distribution between 1 microsecond and 30 microseconds. With a single burst, the burst length is exponentially distributed with mean 30 μs.

- The space between a burst and the following burst (if there is another in the event) is selected from a sawtooth distribution peaked at 10 microseconds and down to zero at 130 μs.

Impulse Noise

The impulse noise shall be a filtered (5 MHz to 40 MHz) impulse at laser's P_{sat} + 15 dB, at a rate of approximately 360 Hz. Every 1/360 seconds, a window with a duration of 1/4 * 1/360 seconds opens. During this window, there are Poisson-distributed impulse events with a mean rate of 4 * 360 per second (thus averaging one per open window).

This impulse noise model is based on power line discharges in the vicinity of the cable plant.[12] This impulse is also filtered through a linear filtering segment, but differing from that of the signal of interest. The lingering reflections of extremely large impulses could account for the claim (of some) of frequent bursts longer than 1 microsecond (but still relatively short).

Gain

Gain range for the two distinct segments of the upstream path are given based on the work of Stoneback and Beck (see *Table 3*).[13]

Segment	Max. Gain	Min. Gain
Modulator to Fiber Node	+2 dB	-3 dB
Fiber Node to Demod	+3 dB	-3 dB

Table 3: *Gain Range for the Upstream Path*

The range for the gain from the fiber node to the demod constitutes two path segments, broken at the F/O Rcvr, and the range is reduced from 14 dB to 6 dB by performing trimming at installation.

Of further importance upstream are the power levels for the individual carriers at the transmitter and demod, and the total carrier power at the laser transmitter and demod. These values actually may vary from plant to plant, depending on how the operators choose to set the signal levels, for example, at the input to the upstream fiber node. An example of signal levels within the HFC plant is given below, assuming a particular 10-dB range for the Psat level of the upstream fiber-optic transmitter as seen at the input to the fiber node. The Psat of most plants is likely to fall within this range (see *Table 4*).

Power at Laser Input	**Min. Power**	**Max. Power**	
Laser Clip Power, P_{sat}	+25 dBmV	+35 dBmV	
Backoff from P_{sat}	-10 dB	-6 dB	
Total Power into Laser	+15 dBmV	+29 dBmV	*range of carrier power into laser
Frequency Utilization Factor	0 dB	+6 dB	
Fractional Carrier Power	-15 dB	-15 dB	*example for 1MHz bandwidth signal
Power Control Accuracy	-2db	+2 dB	
Single Carrier Power into Laser	-2 dBmV	+22dBmV	
Demod Total Input Power			
Total Power into Laser	+15 dBmV	+29 dBmV	
Fiber Node to Demod	-3 dB	+3 dB	
Total Power into Demod	+12 dBmV	+32 dBmV	
Demod Input Power Per Carrier			
Carrier Power into Laser	-2dBmV	+22 dBmV	
Fiber Node to Demod	-3 dB	+3 dB	
Carrier Power into Demod	-5 dBmV	+25 dBmV	
Modulator Output Power			
Carrier Power into Laser	-2dBmV	+22 dBmV	
Modulator to Fiber Node	-(+2 dB)	-(-40 dB)	
Modulator Output Power	-4 dBmV	+62 dBmV	

Table 4: *Signal Levels within the HFC Plant*

Note that these values are based on the segment gain ranges implied in the references,[14] assuming a system with 27 dB maximum tap.

The modulator output carrier power dynamic range is very large, even assuming a 27-dB system. The minimum modulator carrier power is –4 dBmV, and the maximum is +62 dBmV. The low value is due to the assumptions for this calculation that 1) there is a 2-dB gain from the modulator to the fiber node, 2) all of the 35-MHz frequency band is used, 3) a 10-dB backoff is employed for the laser, and 4) the power control algorithm is 2-dB low for that modulator. The high value assumes a 1-MHz bandwidth mode, +2 dB of power control error, a Psat up to +35 dBmV, only 6 dB backoff, and that only 25% of the upstream frequency band is used. A modulator may not have to transmit at these extreme low or high levels, depending on the parameters of the system to which it is deployed, but this calculation provides an example of derived transmit power requirements that might occur without careful deployment of cable modems into HFC plants. The references present examples of dynamic range requirements for an upstream cable modem transmitter and for an upstream receiver at the cable head end.[14–16]

Flexible F/TDMA Provides Robust and Efficient Upstream Approach

Conventional F/TDMA provides a low-risk, high-capacity approach that offers the best choice for upstream modulation over the characterized HFC system. This modulation technique is a mature technology, which easily accommodates proven, effective mitigation techniques for combating the HFC upstream channel impairments; and is bandwidth-efficient in the HFC channel.

The use of F/TDMA is readily combined with frequency agility of the carrier, both QPSK and 16-QAM modulation, multiple symbol rates, and flexible FEC coding, frame and preamble structure.[17] Since the basic techniques have long been in practice, these flexible features provide a low-risk modulation format. The flexibility also allows operation in the most difficult HFC upstream environments, using longer preambles, more powerful FEC, lower symbol rates, and even falling back to QPSK as a last resort. For less challenging environments, highly efficient, high-capacity data transfer is available. Even within a given plant, robust modes may be used at one portion of the upstream spectrum while higher capacity modes are used in other portions of the spectrum.

For a thorough comparison of F/TDMA with other schemes, one may refer to the references;[18] other reference reach the same conclusion: frequency agile F/TDMA is the wisest choice for upstream modulation available today.[19]

The remainder of this section describes the proposed frequency agile modulation, multiple symbol rates, FEC coding, and data stream framing structure. The scheme is based on delivering ATM cells at rates per user ranging from tens of kbps to greater than T1, with error rate performance across the HFC upstream link acceptable to ATM networks. Mini-slots on each frequency channel are also considered, which can be used for improving MAC performance or for user-defined purposes such as synchronous transfer mode (STM).

Processing

Figure 2 shows the layered block diagram of the upstream transmission processing. The data entering the transmitter is formatted as individual packets intended for communication upstream through a hybrid fiber/coax (HFC) plant in a burst F/TDMA system. In order to support the transmission of this bursted data through the HFC plant, appropriate data conditioning, modulation, acquisition, demodulation, and data restoration must be performed.

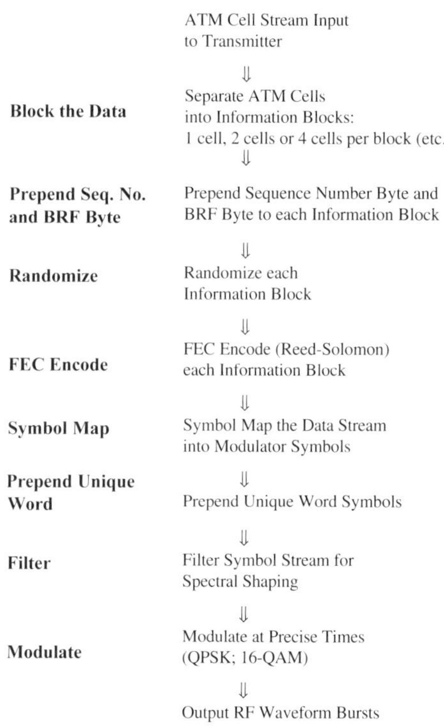

Figure 2: *The HFC F/TDMA Upstream Transmission Processing Block Diagram*

Depending on the modulation mode selected, the ATM cell stream is grouped into information blocks of one, two, four, or more 53-byte ATM cells (up to some maximum limit, such as eighteen). To support an automatic repeat request (ARQ) at a higher layer, a sequence number byte is added to each information block, and to support dynamic bandwidth allocation, a bandwidth request field (BRF) byte is added.

The resulting information block is randomized by modulo-2 addition with a data pattern in an attempt to provide balanced utilization of the symbols and symbol transitions in the data stream. At this point FEC encoding is performed wherein the randomized ATM cell(s), with prepended BRF byte and sequence number, are encoded with a Reed-Solomon code over GF (256) (i.e., eight bits per Reed Solomon symbol). Each mode uses its own particular Reed-Solomon code. The TDMA bursts contain one, two, or more codewords, depending on the selected mode. The resulting data stream for each TDMA burst is augmented with a prepended unique word sequence, which is crucial to the acquisition and synchronization processes at the receiver. This data stream is prepared (symbol-mapped) for either QPSK or 16-QAM modulation, depending on the mode (and there is also the option for differentially encoded modulation). This sequence of symbols is filtered (pulse-shaped) for 25% square-root-raised-cosine spectral shaping and QPSK or 16-QAM modulated and transmitted at the prescribed time and carrier frequency allocated in the TDMA/FDMA system. The carrier frequency assignments are agile, with center frequencies selectable from 5 MHz to 40 MHz in 40 kHz steps; this allows the system to assign channels in a fashion to "step around" particularly bad interference entering the HFC system.

Figure 3 shows the frame formatting for three example modes: A) a single ATM cell per TDMA burst, B) two ATM cells in one codeword per burst, and C) a mode with four ATM cells per burst, divided into two equal length codewords.

The receiver is tuned to the appropriate frequency, and continually scans the received signal at this frequency for the occurrence of bursts of data. A demodulator in the receiver processes all detected bursts of data appearing in its tuned channel; note that the burst data in the time slots will originate at different transmitters in general. After detecting the presence of a burst of data via special filtering tuned to the unique word pattern, the demodulator will further process the received waveform to determine the end of the unique word sequence and the beginning of the data field, and will perform precise carrier acquisition and begin carrier tracking to facilitate demodulation and detection of the data. The demodulator will perform estimation of the time of arrival of the burst relative to a reference timing signal it is provided, and will estimate the power of the received burst and the signal-to-noise ratio (with good resolution in the range of SNR usable for signaling). In the absence

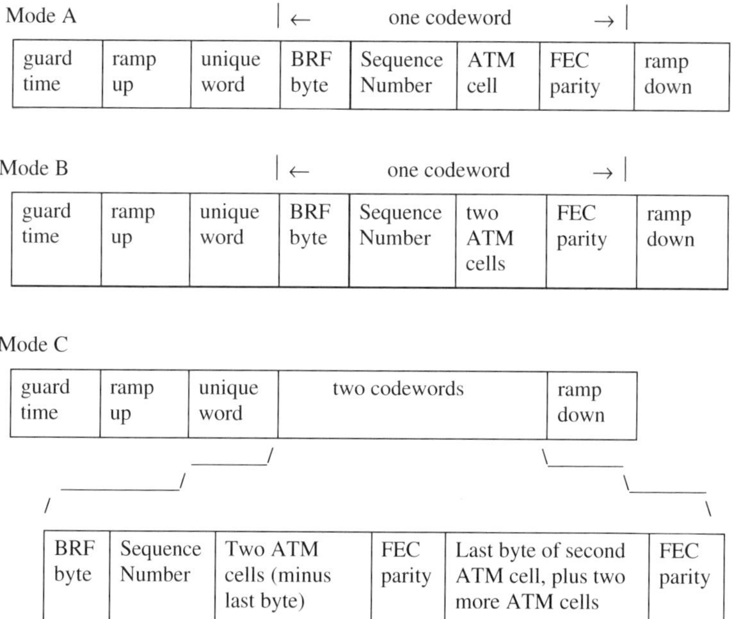

Figure 3: *Example Frame Structures with Flexible Burst Length Modes*

of a detected burst, the demodulator estimates the power in the tuned channel (so that a noise power estimate for that channel is available). Further processing at the receiver consists of decoding the differential encoding induced upon the data portion of the cell (if the differential encoding option was selected). The data is then passed to the Reed-Solomon decoder and finally derandomized.

QPSK and 16–QAM

Both QPSK and 16–QAM modulation are advocated. Different portions of the frequency band can be operated in different modes. The majority of the frequencies in the 5-MHz to 40-MHz band in the HFC plants are expected to have sufficiently low noise-plus-interference and sufficiently high passband fidelity to allow successful 16–QAM operation. QPSK modulation is provided for those bands where the passband distortion and/or available C/N is not capable of supporting 16–QAM. QPSK has less throughput capacity than 16–QAM but is more robust and allows excellent service to be widely available at low risk.

Differential and Non-differential Modulation

Providing capability in the transmitter for both differential and non-differential modulation is advocated. Differentially encoded modulation provides easier implementation of the receiver and can offer performance advantages in a long burst modulation with severe channel impairments, but it can also offer significant performance loss relative to FEC non-differential modulation.[20] The references present the analysis approach and an example computation for assessing the performance loss of differentially encoded modulation compared to non-differential modulation with the same FEC.[20] With the non-differential modulation, carrier–tracking loop cycle slips in the receiver are considered in the analysis.

Multiple Symbol Rates

Multiple symbol rates are offered in the flexible modulation scheme. Rates of 160 ksym/sec, 320 ksym/sec, 640 ksym/sec, 1.28 Msym/sec, and 2.56 Msym/sec are advocated. These rates have been adopted in the MCNS Radio Frequency Interface Specification and in the IEEE 802.14 draft standard Draft 2 Revision 2.[17,21]The latter also includes provision for 5.12 Msym/sec, though the detailed requirements for this highest rate are yet to be determined. These rates are preferred due to both the extended lower end of the range for risk mitigation and the high end of the range, which offers high burst rate with an achievable level of implementation difficulty in an economical modem.

The slower rates occupy less bandwidth, and thus offer the potential of "fitting" between severe ingress interferers situated such that the wideband modes will not operate. The slower rates also offer less latency without sacrificing throughput when many low-rate users vie for service.

Further, for a given reflection scenario, the slower symbol rates will suffer less intersymbol interference (ISI) degradation than the wideband modulation. Finally, with burst noise duration between 1 and 10 microseconds, the slower symbol rates will suffer fewer symbols impacted by each event and thus will be more robust in such a scenario, while the shorter symbol duration modulation will have fewer symbols (proportionately) hit by truly impulsive noise. The symbol rate flexibility offers many opportunities to find a successful operational mode in a challenging HFC plant.

Frequency Agility

Frequency agility of some fashion goes without saying in an F/TDMA scheme, since assignment to any of the multiple channels is a possibility. In this instance, taking the frequency agility to a fraction of the narrow channel bandwidth (e.g., 40 kHz),[12] for any of the channel widths, allows avoidance of narrowband ingress without sacrificing much of the upstream spectrum. While operation with some amount of narrowband ingress inband is highly desirable, to salvage as much of the spectrum as possible it is still far better to simply avoid large narrowband ingress. The avoidance strategy combined with a range of symbol rates allows the HFC system operator to avoid the severe ingress without losing much of the plant's spectrum.

Symbol Shaping and Channel Bandwidth

25% square-root-raised-cosine Nyquist filtering or pulse shaping is advocated.[17] This represents a good compromise between spectral efficiency (only 25% more than the theoretical minimum bandwidth), burst spacing (minimal impact), implementation complexity, and performance loss due to channel impairments and modem implementation.

Flexible Burst Length and FEC

Flexible burst length allows the bundling of multiple ATM cells into longer codewords and TDMA bursts. This offers the advantages of increased coding gain and reduced TDMA overhead per ATM cell (thus providing higher throughput). For optimum (minimum) latency performance and for optimum cell delay variation (CDV) performance, at both the "slow" and "high" symbol rates, modes with a single ATM cell per TDMA burst are available.

Even for a given number of ATM cells in a burst, varying the number of parity bytes in the codeword(s) increases the robustness of the communications, minimizing the risk of not being able to operate in the most difficult scenarios that may be encountered.

The flexible burst length would not be possible (or meaningful) without the accompanying flexible FEC, wherein the coding error correction can range from minimal (two parity bytes included allowing for one byte error correction) to powerful (twenty parity bytes appended allowing up to ten bytes error correction). The FEC consists of Reed-Solomon coding over GF (256), with codeword lengths of up to 255 bytes. Error correction of one to ten bytes is available.

Flexible Preamble Length

A flexible preamble length of one to over 128 symbols allows for short preambles in the case of high-fidelity HFC plants in pristine portions of the spectrum. In more challenging scenarios longer preambles are available. Long preambles may be needed and useful for channels with narrowband ingress or severe ISI, where the long preambles are useful in training the equalization used to mitigate the plant impairments. Longer preambles may also be useful with 16-QAM. Furthermore, techniques are available to enable short preambles in some impairment scenarios.[22]

Mini-Slots

In addition to the bursts with FEC and ATM cells, there is benefit in allowing shorter duration bursts of five to eight bytes of information and CRC instead of a Reed-Solomon FEC. A given frequency channel must normally have all users of that band operating with the same burst format (same modulation, burst length, preamble, and FEC). However, the single exception of allowing mini-slots may prove worthwhile. Such mini-slots improve the efficiency of the MAC by placing contention requests in short rather than normal-sized bursts, and can possibly be utilized for other user-defined purposes such as STM.

Interleaving

Interleaving in conjunction with TDMA seems incongruent, but with impulses possibly "pinging" around in the HFC plant for microseconds—a relatively long time at the highest symbol rates under consideration—the benefit of interleaving is substantial.[17,23] In this approach, several ATM cells are bundled together into a burst (still a short duration since the symbol rate is assumed high), and multiple codewords are formed.[23] These codewords are then interleaved (within the one burst). At the receiver deinterleaving precedes decoding, and the short bursts of errors due to "pinging" impulse events are spread among the several codewords rather than concentrating in a single codeword.

Trellis Coded Modulation (TCM)

TCM could provide significant performance advantage to HFC upstream communications in some cases, in combination with Reed-Solomon outer coding. The TCM would simply overlay the existing QPSK or 16-QAM modulation with 8PSK or 32-QAM; after the convolutional decoding of the modulation, the formats of the bursts are identical to those with conventional QPSK and 16-QAM.[24]

Tomlinson-Harashima (TH) Precoding [25–27]

The use of TH precoding in the modulator should be investigated as an approach to mitigate the propagation of errors in decision feedback circuits of the receiver in conjunction with equalizers employed for mitigation of ISI and narrowband interference in the HFC plant. TH precoding is possible for such scenarios since 1) narrowband interference is expected to be relegated to the same small set of frequencies unique to each HFC plant, and can thus be "known" and shared with the modulators, and 2) the ISI for each modulator is not expected to vary dramatically in a dynamic fashion, so the receiver may "learn" the tap weights most suitable for equalization. In each case, the feedback taps of the equalization structure can be communicated to the modulator so that TH precoding can be invoked. By executing the feedback taps at the modulator, wherein there is no possibility of feeding the wrong data into the feedback structure, error propagation associated with the powerful decision feedback taps is largely mitigated.

Upstream Frame Synchronization

Initialization and synchronization procedures are necessary to bring the upstream modulators "on-line." A time-stamp delivered in the downstream signal is recommended as the vehicle for conveying the head end frequency standard/frame clock to the upstream modulators.[28,29] After an initial ranging burst by a modulator, burst guard times of less than 12 microseconds can be allowed in the frame design; these can be even shorter in the higher symbol rate modes.

Leveraging off the frame clock operating in the modulator (for keeping the frame timing), the upstream carrier and clock frequencies can be locked to the head end frequency standard. The frame clock is appropriately smoothed and used as the reference in synthesis loops to synchronously generate the modulator carrier and clock frequencies. The resulting upstream carrier and clock uncertainty at the receiver are then negligible.

Conclusion

There exists a flexible F/TDMA modulation scheme that simultaneously provides the possibility of high data rate upstream and minimizes the possibility that a user simply will not be able to achieve sufficiently low error rate in the upstream HFC plant. The techniques described in this paper provide for this dual capability, as they are readily implementable today and have been proven in many previous applications. The techniques presented later in the paper are also capable of providing performance enhancement in the HFC upstream, but the increased system complexity involved makes application of these techniques less clear-cut.

References

1 Kolze, T., "PCLinX Performance Model of Downstream and Upstream Signal Distortion, Interference, and Noise Introduced by the HFC Cable Plant," General Instrument Corporation, Communications Division—Phoenix, Document Number: TJK.ISC.030a-96, 8 March 1996.
2 Kolber, M., "Data Path Channel Characteristics System Level Design Specification, Revisions 1.5," General Instrument Corporation, Communications Division—Hatboro, PA, 3 March 1994.
3 Himayat, N., "Return Path Characterization," preliminary report, General Instrument Corporation, Communications Division—Hatboro, PA, 12 November 1994.

4 Eldering, C., Himayat, N., and Gardner, F., "CATV Return Path Characterization for Reliable Communications," *IEEE Communications Magazine*, August 1995: 62–68.
5 Digital Transmission Characterization of Cable Television Systems; Cable Television Laboratories, Inc., November 1994.
6 Two-Way Cable Television System Characterization; Cable Television Laboratories, Inc., 12 April 1995.
7 Request for Proposals for a High Speed Cable Data Service, Cable Television Laboratories, Inc., 12 April 1995.
8 Williams, T., Chelehmal, M., and Prodan, R., "Field Evaluation of Reverse-Band Channel Impairments," SCTE 1996 Conference on Emerging Technologies, San Francisco, January 1996.
9 Kolze, T., "Proposed HFC Channel Model," General Instrument IEEE 802.14 contribution, IEEE 802.14-96/196, Enschede, Netherlands, 8 July 1996.
10 Currivan, B., editor, "CATV Upstream Channel Model," Rev. 1.0, IEEE 802.14-95/133, 3 November 1995.
11 Currivan, B., "CATV Upstream Channel Model Update," IEEE 802.14-96/091, 10 March 1996.
12 Discussions, without specific related written contributions, at IEEE 802.14 Working Group Meetings, and reflector traffic on the IEEE 802.14 Working Group reflectors.
13 Stoneback, D. and Beck, W., "Designing the Return System for Full Digital Services," SCTE Conference on Emerging Technologies, San Francisco, January 1996.
14 Kolze, T., "Proposed Upstream Transmit Power Specification," General Instrument IEEE 802.14 contribution, IEEE 802.14-96/191, Enschede, Netherlands, 8 July 1996.
15 Kolze, T., "Upstream PHY Transmitter Dynamic Range," General Instrument IEEE 802.14 contribution, IEEE 802.14-97/039, Irvine, CA, 9 March 1997.
16 IEEE Project 802.14/a Draft 2 Revision 2, Cable-TV access method and physical layer specification, approved at IEEE 802.14 Working Group meeting, Maui, Hawaii, July 1997.
17 Kolze, T., "Flexible Upstream Format," General Instrument IEEE 802.14 contribution, IEEE 802.14-96/137, Los Angeles, 2 May 1996.
18 Ghaibeh, G., "Is CDMA a viable choice for the upstream path?" Communications Engineering & Design, August 1995: 40–50.
19 Kolze, T., "Upstream PHY Modulation Technique Recommendations," General Instrument IEEE 802.14 contribution, IEEE 802.14-96/086, San Diego, 12 March 1996.

20 Kolze, T., "Error Performance Degradation Due to Differentially Encoded Modulation when Combined with Forward Error Control (FEC) Coding," General Instrument IEEE 802.14 contribution, IEEE 802.14-96/195, Enschede, Netherlands, 8 July 1996.
21 MCNS, *Data Over Cable Service Radio Frequency Interface Specification*, SP-RFI-I02-971008.
22 General Instrument Response to: Phase 3 RF Interfaces—Line Items WM-PH3RFLIW07-960715; General Instrument Corporation, San Diego, Document No.: NET-096-073, 4 September 1996.
23 Kolze, T., "802.14 PHY Proposal," General Instrument IEEE 802.14 contribution, IEEE 802.14-95/130, San Diego, 30 October 1995.
24 Ungerboeck, G. and Karoaguz, J., "Frequency agile multi-mode (FAMM) single-carrier modems for upstream transmission in HFC systems," IBM IEEE 802.14 contribution, IEEE 802.14-95/104, Atlanta, 12 September 1995.
25 Tomlinson, M., "New automatic equalizer employing modulo arithmetic," Electronic Letters, Vol. 7, March 1971: 138–139.
26 Forney, G. D. and Eyuboglu, M. V., "Combined Equalization and Coding Using Precoding," *IEEE Communications Magazine*, December 1991: 25–34.
27 Ghoshe, M. and Hulyalkar, S., "Method of combating ingress and multipath in a CATV return channel," Philips IEEE 802.14 contribution, IEEE 802.14-96/111, Los Angeles, 1 May 1996.
28 Hou, Victor T., "TDMA System Synchronization Using Timebase Marker Messages," General Instrument IEEE 802.14 contribution, IEEE 802.14-96/090, San Diego, 11 March 1996.
29 Ungerboeck, G. and Karoaguz, J., "Standard Protocol for Cable-TV Based Broadband Communications Network," IBM IEEE 802.14 contribution, IEEE 802.14-96/209, Enschede, Netherlands, 8 July 1996.

Frequency-Agile TDMA System for Upstream Cable-Modem Applications

Jonathan S. Min
Senior Staff Scientist
Broadcom Corporation

Henry Samueli
Chief Technical Officer
Broadcom Corporation

Abstract

This paper presents a frequency-agile, time division multiple access (TDMA) burst transmission system based on 16–quadrature amplitude modulation (QAM) or quaternary phase shift keying (QPSK) modulation schemes suitable for next-generation cable modem upstream applications. Later, the paper describes advanced techniques to improve existing TDMA systems and handle various channel impairments in a hostile upstream cable plant. The techniques employed in the proposed system include fine carrier frequency agility, truly variable symbol rates, programmable constellation size, dynamic channel allocation based on ongoing channel quality monitoring, programmable forward error correction (FEC) with an optional interleaver, and adaptive pre-equalization. Implementation of such an upstream receiver is one of the key factors for robust and efficient two-way cable data communications. Finally, the paper presents the results of simulations of critical issues regarding a combined, coherent 16–QAM or QPSK burst receiver.

Introduction

The hybrid fiber/cable (HFC) network employs a point-to-multipoint topology between many subscriber cable modems (CM) and the head end or cable modem termination system (CMTS). Typically in North America, the upstream band covers between 5 MHz and 42 MHz, while it extends to 65 MHz in Europe. The downstream band can reach up to 1 GHz in new cable plants. For the Multimedia Cable Network System (MCNS)/Date-over-Cable Service Interface Specification (DOCSIS) and Institute of Electrical and Electronic Engineers (IEEE) 802.14 cable modem system, the available frequency band is divided into frequency channels. The multiple access scheme that enables users to share channels is time-division multiplexing (TDM) for the downstream channel and TDMA for the upstream. For better bandwidth efficiency, high-level constellation modulation formats such as 64–QAM and 256–QAM with adaptive equalization are employed for the downstream transmission. Digital cable TV transmission systems have employed these technologies for a few years. They have proven to be robust and bandwidth-efficient (5–7 bits/sec/Hz), providing up to 30–40 Mbps over the 6 MHz channel.

The modulation scheme used in the upstream has typically been QPSK. Severe channel impairments in the cable return path have been preventing higher modulation techniques. Dominant upstream impairments include ingress, impulse or burst noise, microreflections, and plant response distortion. A combination of microreflections and plant response phase distortion results in intersymbol interference (ISI) at the receiver. Ingress, also known as radio frequency interference (RFI), is the source for many narrowband interferences, especially in the lower band of the upstream spectrum (between 5–20 MHz). Impulse noise or burst noise with longer duration of each event, caused by electronic coupling at home, is of serious concern for the return plant since all the noise funnels into the head end receiver.

This paper focuses on the architecture of the upstream physical layer (PHY) cable modem transmission system. Advanced techniques to deal with channel impairments using full 16–QAM burst modulation, thus also increasing the overall upstream channel rates, are discussed in section two. Here, a set of advanced features is proposed to provide an evolutionary path from the existing TDMA system. Some of the simulation results of a 16–QAM burst receiver, which is a key factor for implementing robust two-way cable communications, are then presented in section three.

Proposed Upstream PHY System: An Evolutionary Path to the Next-Generation Cable Modem System

For robust upstream transmission, QPSK modulation with relatively low symbol rates (typically less than 1.5 Mbaud) has been used in today's TDMA systems. One of the main issues to resolve in future cable modems is robustness at higher data rates and channel loading. Advanced features that address these issues in the upstream are proposed here to provide a graceful evolution to the next-generation TDMA cable modem system.

16–QAM Modulation and Programmable Constellation Format

To increase data rates, 16–QAM is chosen as a baseline modulation format with a QPSK fall-back option. It improves the channel bandwidth efficiency by a factor of 2, from 1.6 bits/Hz to 3.2 bits/Hz with 25% of excess bandwidth. For example, it provides up to 20.48 Mbps at 5.12 Mbaud. However, burst detection of 16–QAM is much more difficult than QPSK alone since its decision is amplitude-sensitive. Moreover, adaptive equalization is required for multipath mitigation. One of the requirements of the advanced burst receiver should also be to provide channel status information. Depending on the channel status, the constellation format can be dynamically changed to optimize the overall channel efficiency. For instance, when the channel is good (e.g., with an SNR greater than 20 dB), the head end can use 16–QAM as a default modulation format. As shown in section three, more than 20 dB of SNR can be achieved after the initial equalizer training even in the worst case of ISI predicted in the IEEE 802.14 channel model;[1] thus, 16–QAM modulation format can be used in many channels in the upstream. Of course, periodic training should follow to adapt to slowly varying channel conditions. When the channel SNR drops below 20 dB, the head end should command all the subscribers in that particular frequency to switch to QPSK. This programmability in constellation format, with dynamic channel allocation, will improve the channel bandwidth efficiency significantly.

Channel Quality Monitoring

The upstream channel for the HFC network is a scarce resource. Not only can it be shared by other services such as cable telephony, but also channel impairments like ingress noise pose serious challenges for reliable burst transmission. Thus, it is important for the head end to provide some sort of spectrum management functions so that it can provision an RF channel that is not impaired or in use by other services, or move the subscribers' upstream transmitters to a "cleaner" RF frequency.

Ongoing channel monitoring based on packet-based statistics plus signal-to-noise ratio (SNR) and channel noise power should be incorporated into the upstream receiver system (see *Figure 1*). In a way, this feature enables upstream spectrum analysis to be performed locally for optimal use of the available spectrum without resorting to an expensive external spectrum management unit. Information such as lost packet ratio, bad packet ratio, and correctable packet ratio can be derived from the statistics provided by the head end receiver, as listed in *Table 1*. Long-term SNR averaged over many bursts or channel (noise) power level measurement should also be available from the burst receiver for this purpose. These statistics based on the packet reception, FEC status, and channel SNR are passed to the head end spectrum management or allocation routine to determine the need for possible carrier frequency change. The noise power level in each idle channel can indicate its quality.

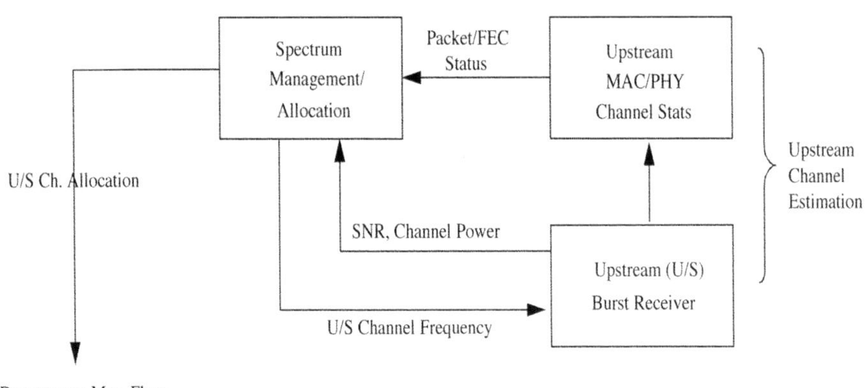

Figure 1: *CMTS Spectrum Management Functional Block Diagram*

Statistics
Number of packets (total)
Number of packets undetected (no unique word)
Number of packets with corrected errors
Number of packets with uncorrectable errors
Number of FEC blocks (total)
Number of FEC blocks with corrected errors
Number of FEC blocks with uncorrectable errors

Table 1: *Upstream Channel Quality Statistics*

High-level channel adaptation algorithms can use upstream channel quality information to change physical layer parameters in real time. This feature is especially important when the channel conditions are not too severe to change but require more FEC coding gain or longer guard time for robust transmission. It is well understood that the more FEC parity bytes used, the more the coding gain derived for a given size of the information bytes (see *Figure 2*). To switch the burst parameters dynamically, each upstream transmitter and the head end receiver should support multiple burst profiles, which are more than simply a number of different slot types. A burst profile refers to a group of PHY parameters relating to a certain burst slot type. An example of MCNS upstream burst types is shown in *Table 2*.

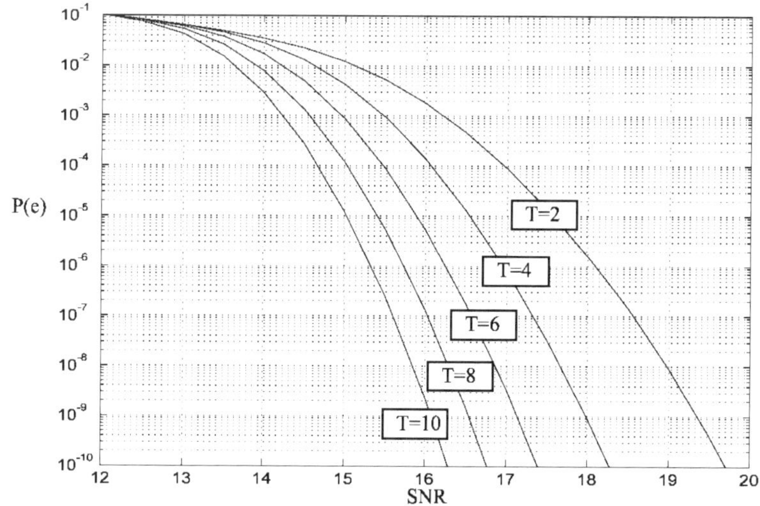

Figure 2: *RS Coding Gain over Various T's (16–QAM, K=64 bytes fixed)*

Burst Name	Burst Type Code	Description
Request	1	contention slot for requesting data bandwidth
Request/Data	2	contention slot for immediate data
Initial Ranging	3	contention slot for initial maintenance
Periodic Ranging	4	reservation slot for periodic maintenance
Short Data	5	reservation slot for smaller data packets
Long Data	6	reservation slot for longer data packets
Reserved	9-14	-------

Table 2: *MCNS Upstream Burst Types*

Fine Carrier Frequency Agility and Truly Variable Symbol Rates

The upstream transmitter (in the subscriber) and the burst receiver (at the head end) for the MCNS/IEEE modem system already have the carrier frequency agility to change its carrier frequency channel if commanded by the head end during the initial RF channel provisioning stage or during reprovisioning. In order to use the scarce upstream bandwidth more effectively, the next-generation upstream transmitter/receiver should utilize the fine frequency agility feature to tune carrier frequencies within a few Hertz of resolution. This digital fine tuning capability is especially useful for the low side of the upstream band, where the narrowband ingress is dominant (see *Figure 3*). As shown in Appendix F of the MCNS Specification,[2] the narrowband interference gaps can vary substantially. This implies that uniform channelization is typically prevented. The ability to fine tune the carrier frequency to avoid the interference region in a non-uniform fashion will improve the channel utilization efficiency much more than will avoiding interference regions in uniformly spaced channels.

The current MCNS/IEEE specifications allow powers-of-two variations in upstream symbol rates. For advanced modems, the system operator should be able to switch between any symbol rates in a near-continuous manner without physically changing the master crystal reference. This feature is important where multiple data rates, often required by modem applications, are not in powers-of-two. In the upstream, where the ingress in the channel typically prevents uniform channelization, the ability to vary the data rates in an arbitrary manner allows

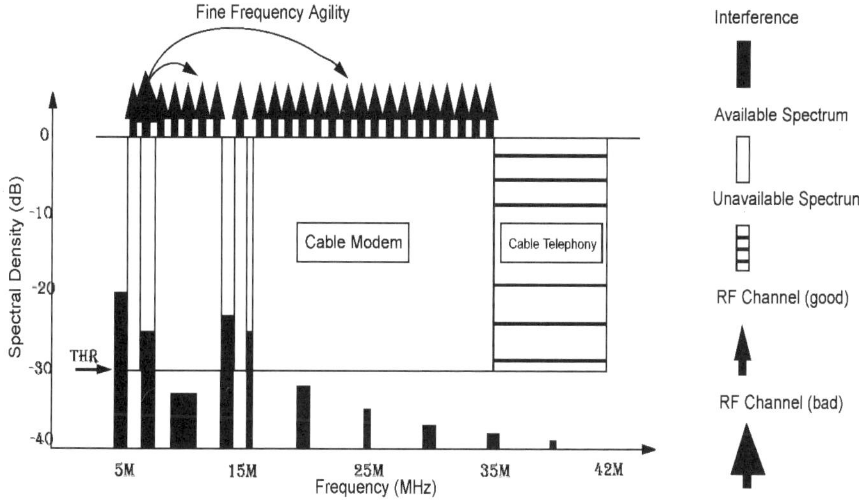

***Figure 3:** Upstream Fine Frequency Agility Example*

data carriers to be placed in between narrowband interferences, an important feature for utilizing the upstream spectrum. *Table 3* shows an example of variable rates between 100 kBaud to 5.12 Mbaud that can be supported in the upstream channel.

TDMA with Dynamic Channel Allocation vs. CDMA for Ingress Handling

Unlike a code-division multiple access (CDMA) system where narrowband interference is averaged out by the processing gain (PG), a frequency-agile TDMA system can be seen as an avoidance system where an interference channel is entirely avoided. Fine carrier frequency agility and truly variable symbol rates allow dynamic channel allocation of a single carrier TDMA system based upon ongoing channel quality estimation. This dynamic feature allows an optimal utilization of the upstream channel with ingress without enforcing every user in the channel to maintain the worst-case processing gain that limits the maximum data rate. For example, 0 dB of narrowband interference with respect to a signal can only be suppressed when each user has a PG of 20 dB or higher (for 16–QAM systems), either in bandwidth or in power. A frequency-agile TDMA system allows symbol rates and thus channelization to be more flexible, allowing ingress to be avoided while maintaining higher bit rates.

Symbol Rate (ksym/sec)	Channel Width (kHz, α=25%)	QPSK Data Rate (kbits/sec)	16-QAM Data Rate (kbits/sec)
128	160	256	512
160	200	320	640
256	320	512	1,024
320	400	640	1,280
512	640	1,024	2,048
640	800	1,280	2,560
1,024	1,280	2,048	4,096
1,280	1,600	2,560	5,120
2,048	2,560	4,096	8,192
2,560	3,200	5,120	10,240
4,096	5,120	8,192	16,384
5,129	6,400	10,240	20,480

Table 3: *Variable Rate Example*

An adaptive ingress cancellation technique can also be applied to cancel ingress noise, rather than to avoid it. Due to the code orthogenality issue, however, it will be much more difficult to implement this technique in CDMA.

Adaptive Pre-Equalization to Mitigate Micro-Reflections and Plant Response

Both microreflections and cable plant response phase distortion in the upstream channel result in ISI at the receiver. With higher symbol rates and amplitude-sensitive 16–QAM, adaptive equalization should be mandatory to mitigate inter-symbol interference due to these distortions (see *Figure 4*). Two ISI channel models have been used for equalizer trade-off studies for upstream transmission: one from IEEE 802.14 and the other from MCNS.[1, 2] For the IEEE model, the total echo power that stretches up to 1.5 μs is -6 dBc with respect to the signal energy and its coherence bandwidth, loosely defined as the bandwidth separated by two nulls in its frequency spectrum, is roughly 2.5 MHz (see *Figure 5*), while the signal echo power is -10 dBc and the coherence bandwidth

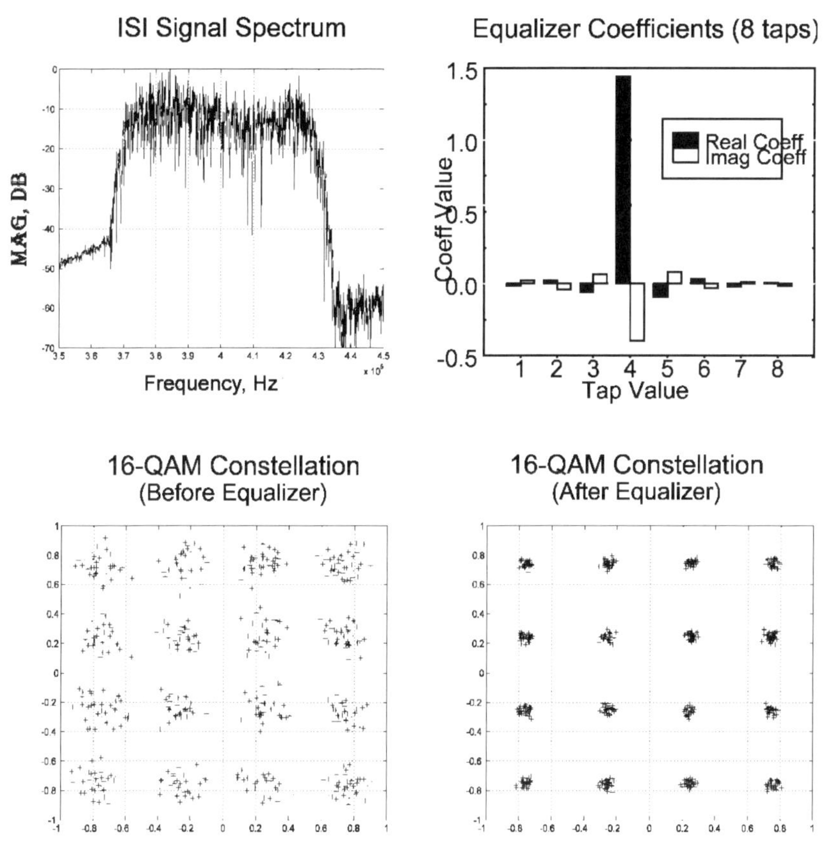

***Figure 4:** Effect of Adaptive Equaization in ISI Channel*

is about 2 MHz for the MCNS model. The IEEE channel model is the worst of the two, and thus is used as the basis for our equalizer studies. The original response at 100 MHz is resampled at F_{clk} = 17 MHz and renormalized to preserve the original channel characteristics (see *Figure 6*). This sampling rate has been chosen primarily to reduce the simulation time.

The measured SNR (E_s /N_o) values at the demodulator for various IF carrier frequencies (with a step size of 0.5 MHz) are shown in *Table 4*. Different symbol rates (0.5 Mbaud, 1 Mbaud, and 2 Mbaud) were simulated to find the relations between the baud rates and the achieved SNR for the given ISI channel. The median SNR value is 23.5 dB for F_{baud} = 0.5 MHz, 18 dB for F_{baud} = 1 MHz, and 14 dB for F_{baud} = 2 MHz. The worst performance is obtained near an IF of

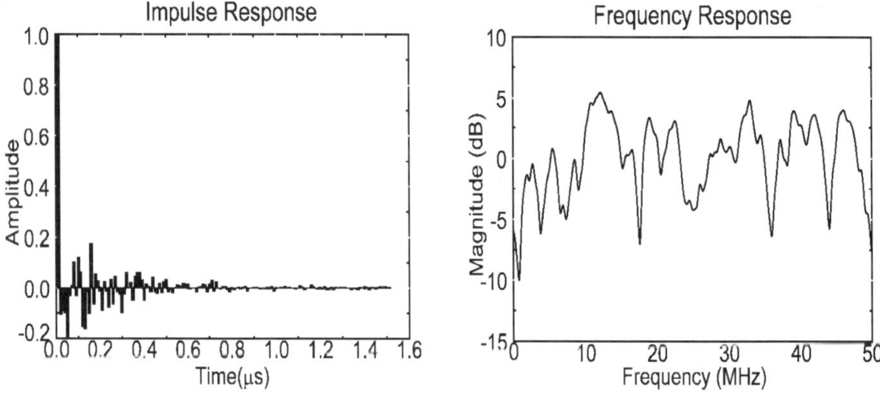

Figure 5: *Upstream ISI Channel Model*[1]

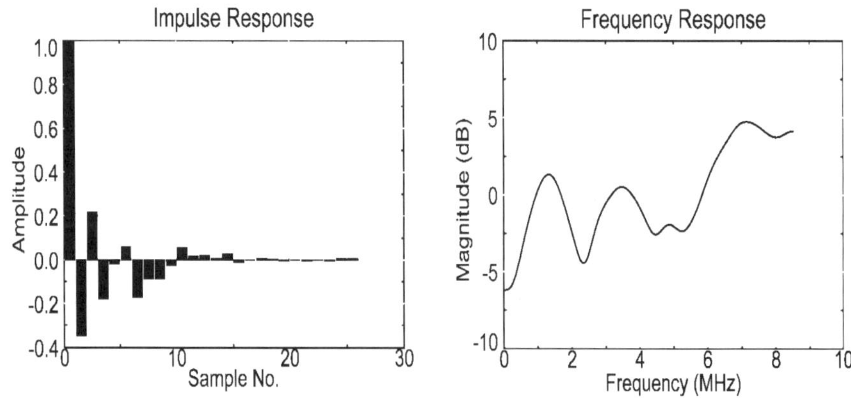

Figure 6: *Equivalent IF Channel Model (@F_s = 17 MHz)*

2 MHz, as can be deduced from the channel frequency response (see *Figure 6*). This IF carrier frequency was chosen to measure the performance improvements due to equalization, and the simulated results are summarized in *Table 5*. To achieve a BER = 10^{-8} without forward error correction (FEC), the required SNR is 15 dB for QPSK and 22 dB for 16–QAM. Therefore, equalization would be required at high symbol rates (greater than 1 Mbaud), especially for 16–QAM, to mitigate impairments due to microreflections. With a total of four symmetrical taps, an improvement of more than 10 dB is achieved. For the simulated case at 2.56 Mbaud, for example, an equalizer with at least eight taps would be required to guarantee robust upstream transmission.

Carrier Freq.	M_{baud} 0.5 MHz	M_{baud} 1.0 MHz	M_{baud} 2.0 MHz	Carrier Freq.	M_{baud} 0.5 MHz	M_{baud} 1.0 MHz	M_{baud} 2.0MHz
1.0 MHz	21.2 dB	15.0 dB	---	4.5 MHz	29.2 dB	24.8 dB	17.0 dB
1.5 MHz	22.2 dB	17.3 dB	12.0 dB	5.0 MHz	30.0 dB	27.6 dB	20.0 dB
2.0 MHz	18.4 dB	12.5 dB	11.0 dB	5.5 MHz	23.5 dB	18.0 dB	14.0 dB
2.5 MHz	19.5 dB	15.3 dB	14.0 dB	6.0 MHz	21.0 dB	15.5 dB	13.0 dB
3.0 MHz	24.0 dB	17.0 dB	15.0 dB	6.5 MHz	25.2 dB	19.0 dB	14.0 dB
3.5 MHz	34.0 dB	26.0 dB	16.0 dB	7.0 MHz	31.0 dB	25.2 dB	20.0 dB
4.0 MHz	23.5 dB	17.8 dB	16.0 dB	7.5 MHz	31.5 dB	25.6 dB	---

Table 4: *Measured SNR over Various Carrier Frequencies*

Adaptation speed of equalization with different step sizes (μ) was also simulated. Even with the largest step size allowable for stability, the minimum equalizer coefficient acquisition time is approximately fifty symbols, implying that the equalizer may need a dedicated training packet at the beginning of a session or during an initialization process. Relying on the preamble to train the equalizer may be too costly in terms of the bandwidth efficiency. Equalizer training was performed using random QPSK symbols. The example shown in *Figure 7* had a

Feedforward/ Feedback Tap	SNR (1.0 MBaud)	SNR (2.0 Mbaud)	SNR (2.56 Mbaud)
0/0	12.5 dB	11.0 dB	9.0 dB
0/4	22.5 dB	17.5 dB	15.0 dB
2/2	25.5 dB	22.0 dB	19.0 dB
2/4	28.0 dB	25.0 dB	23.0 dB
4/4	36.0 dB	34.0 dB	30.0 dB
4/6	39.0 dB	38.5 dB	32.0 dB
8/0	36.0 dB	32.0 dB	29.0 dB

Table 5: *ISI Effects and Equalizer Performance at Worst IF*

symbol rate of 2.56 Mbaud and with a null of the ISI spectrum appearing in the passband of the signal. Once trained, the demodulator can continue to track the equalizer coefficients on a per modem basis. This eliminates the need for retraining unless the user is assigned to a new channel.

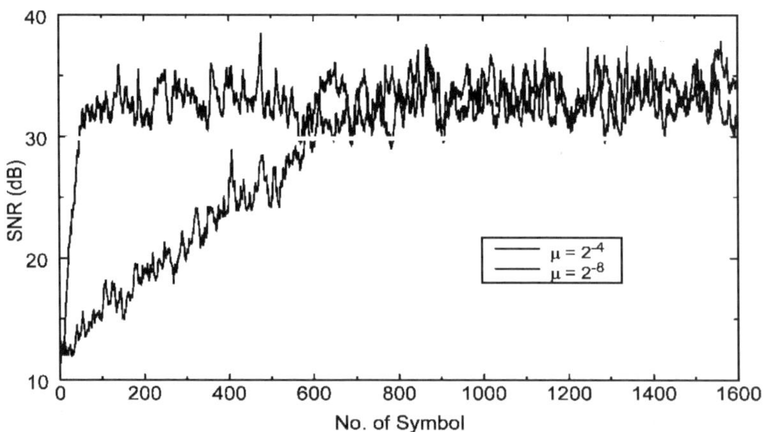

Figure 7: *Equalizer Convergence Performance*

Although adaptively equalizing at the head end receiver alone is feasible, it is complicated because of the switching required between users at different time slots. Thus, training each user at the upstream receiver and transferring the equalizer coefficients to the corresponding burst transmitter (pre-equalization) is the preferred approach. It has been verified that both Tomlinson-Harashima precoding and linear equalization provide sufficient correction gain for the IEEE upstream ISI channel model. After initial equalizer training, achieving more than 20 dB of SNR for normal 16–QAM operation would not present a problem.

In a CDMA system, a RAKE receiver can theoretically be implemented to handle the multipath problem caused by microreflections. However, implementation of such a receiver with already complex multiple correlators corresponding to all possible users in the channel would not be practical. Without the RAKE architecture, the CDMA system requires some sort of pre-shaping to flatten the wide spread band; however, it requires a careful consideration when the ingress is present in the channel. Learning the channel at the receiver and pre-shaping at the transmitter in the presence of ingress will actually worsen the transmitter shaping since the ingress noise is additive. In the proposed TDMA system, user-based pre-equalization is meant to handle ISI, not to deal with ingress.

Programmable Reed-Solomon Channel Coding with an Optional Interleaver to Improve Burst Noise Immunity

Burst or impulse noise is one of the most serious impairments present in the upstream channel. To improve the burst noise immunity and the SNR performance, Reed-Solomon coding with variable correction factors (T = 0-10) has been implemented. Increasing T not only improves the coding gain but also enhances the burst noise correction capability (T_c):

$$T_c = (T \times I) / (R_{sym}) \qquad \text{(EQ 1)}$$

where T = RS error correction factor, I = Interleaving depth, and R_{sym} = RS symbol rate.

Interleaver will increase burst noise immunity but with increased latency; therefore, interleaving should be optional. The majority of the burst noise length is about 1 μs.[3] Depending on an application's needs (for example, voice vs. data) and symbol rates, one should choose proper T's and switch on/off interleaving in the upstream path.

More bandwidth-efficient trellis-coded modulation (TCM) as a replacement to a block code or the inner code of a more powerful concatenated code has been suggested. Since the upstream uses burst transmission in which each packet needs to be reacquired, a simpler RS block code approach is recommended. Moreover, when burst or impulse noise is dominant as in the upstream, the trellis-based coding can actually propagate the burst error and thus worsen the problem without a proper interleaver.

Combined 16–QAM/QPSK Burst Receiver Architecture

Due to amplitude sensitivity, burst detection of 16–QAM signals is more difficult compared to QPSK–only burst detection. Coherent detection is mandatory, and actually performs better than differential detection even for the QPSK case. Some important assumptions have been made to support 16–QAM burst reception. First, it is assumed that the carrier frequency offset, coarse slot timing error, and coarse amplitude error are compensated for during the initial ranging process. Therefore, during a particular burst data transmission, only timing phase, carrier phase, and small amplitude errors need to be recovered. Again, for 16–QAM, amplitude errors result in significant performance degradation, unlike in QPSK where small signal level variations are not critical.

Second, the minimum SNR required for reliable transmission in QPSK and 16–QAM is assumed to be different. For example, at least 20 dB of SNR (E_s/N_o) is required for 16–QAM, while 15 dB is sufficient for QPSK. These channel conditions, with the help of forward error correction, will guarantee data transmission with a BER below 10^{-8}. With initial equalizer training, obtaining these operational SNRs is not a problem. Finally, the tracking loops are assumed to follow the initial preamble processing block. With this scheme, the preamble acquisition errors in amplitude, time, and carrier phase are just initial degradation sources that will be reduced after tracking.

The average SNRs over various payload sizes, including 16–byte minislots, 64-byte slots, and slots longer than 64 bytes have been measured by simulation. After preamble acquisition, the modulation parameters can be acquired or estimated to an initial condition, as exemplified in *Table 6*. The specified initial estimation errors are kept above three times the RMS deviation predicted by the Cramer-Rao bound at a given SNR.[4] The aggregate degradation due to the initial errors without tracking has been simulated to be 4.2 dB for QPSK at 15 dB of SNR and 4.8 dB for 16–QAM at 20 dB of SNR. However, when tracking loops are enabled during data slots, the average degradation is reduced significantly (see *Table 7*), thereby making robust transmission possible even for 16–QAM signalling. As expected, the longer slots perform better than the shorter ones. As far as the burst receiver is concerned, therefore, there is no limit in the length of the burst packet due to this continuous tracking. In this type of scheme, the overall BER is dominated by the performance at the beginning of the payload where the phase and amplitude errors are the largest. These simulations were carried out at the minimum SNR for acceptable performance and with the large initial errors (more than 3 sigma deviation). As the channel SNR increases, the initial parameter estimation errors are significantly reduced.

	QSPK (15 dB SNR)	16-QAM (20 dB SNR)
Timing Error	$T_{sym}/10$	$T_{sym}/16$
Carrier Phase Error	10 deg.	5 deg.
Amplitude Error	---	1 dB
Overall Degredation	4.2 dB	4.8 dB

Table 6: *Initial Preamble Acquisition Errors and Degredation Example*

Payload Size		$50\ T_{sym}$	$250\ T_{sym}$	$500\ T_{sym}$	$750\ T_{sym}$
QPSK (15 dB SNR)	no tracking	10.8 dB	10.8 dB	10.8 dB	10.8 dB
	with tracking	13.5 dB	14.0 dB	14.3 dB	14.5 dB
16-QAM (20 dB SNR)	no tracking	15.2 dB	15.2 dB	15.2 dB	15.2 dB
	with tracking	17.5 dB	18.5 dB	19.0 dB	19.2 dB

Table 7: *Measured SNR over Various Slot Sizes*

A system block diagram of the burst receiver is shown in *Figure 8*. It consists of an analog front-end including an ADC, a quadrature mixer, a preamble processing unit, tracking loops, a ranging parameter measurement unit, and a medium access control (MAC) interface. The role of the preamble processing unit is to detect the presence of a burst of data and achieve fast acquisition and estimation of modulation parameters such as symbol timing, carrier phase, and amplitude information. This block also includes a unique word (UW) detector that determines the beginning of the payload. The UW detector also resolves the four-fold ambiguity of the carrier phase, thereby removing the need for differential encoding and decoding. Differential decoding results in doubling of symbol errors that significantly reduces the effective performance of forward error correction. The tracking loops are implemented using a phase-locked loop (PLL) architecture with digitally programmable loop bandwidths.

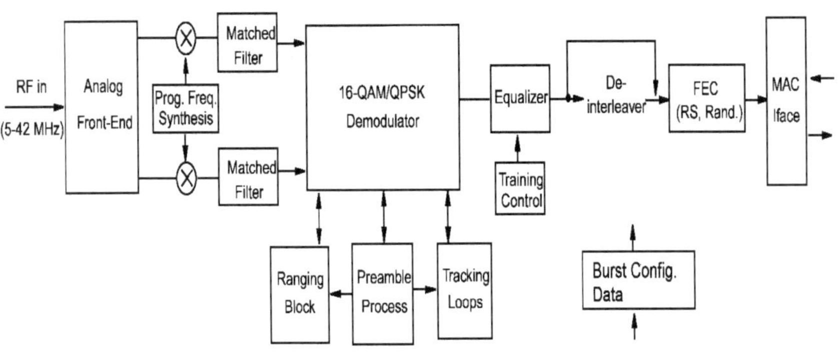

Figure 8: *Block Diagram of Upstream Burst Receiver*

In a TDMA system, the fast acquisition process is performed on the preamble symbols. The size and pattern of the preamble should be chosen carefully so that the burst demodulator can acquire relatively quickly without sacrificing performance or overhead efficiency. An upstream receiver that can acquire burst data transmissions in as few as sixteen symbols has been implemented (see *Figure 9*). As shown in *Figure 10*, each burst contains PHY overhead that includes a preamble sequence and an optional equalizer training sequence for ranging packets. Each packet should also include a guard time to compensate for ranging errors and ramp up and down time. Both short and long unique words are supported, the former for normal burst packets and the latter potentially for ranging packets. Even for 16-QAM modulation, QPSK–like signalling is used for the preamble.

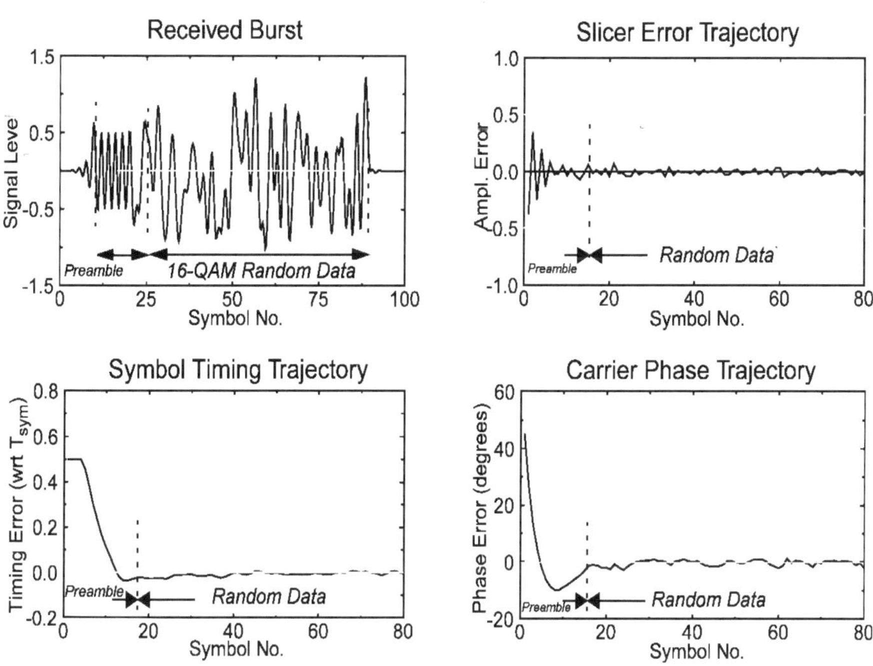

Figure 9: *Burst Acquisition Performance*

Figure 10: *TDMA Burst Format*

To handle the amplitude variation problem between bursts for 16–QAM demodulation, open-loop digital gain correction and adaptive equalization must be performed in each burst. Amplitude estimation for the slicer is based on either noncoherent power estimation, which is performed by taking the square root of the power of the preamble to estimate the amplitude level, or coherent amplitude estimation, which is possible after initial carrier/symbol timing acquisition (see *Table 8*). The latter approach is simpler, since wordlength increases in square for the former case. It does, however, require a longer preamble.

Method	CNR (E_S/N_0)	N (avg. window)	3σ Amp. Error	16-QAM SNR Degradation
Power Estimation	20 dB	12T	± 0.65 dB	0.9 dB
	15 dB	12T	± 1.1 dB	2.2 dB
Coherent Amplitude Estimation	20 dB	4T	± 0.9 dB	1.75 dB
	20 dB	8T	± 0.65 dB	0.9 dB
	20 dB	12T	± 0.51 dB	0.7 dB
	15 dB	8T	± 1.1 dB	2.2 dB

Table 8: *Amplitude Estimation Measures*

New requirements of fine carrier frequency agility and truly variable symbol rates for the upstream receiver mandate digital implementation using direct digital frequency synthesis and digital resampling techniques, respectively.[5] Adaptive equalization and programmable RS FEC blocks also contribute to a fully integrated digital architecture for the future head end receiver implementation. Channel status information as well as the ranging function should be incorporated at the burst receiver system as well.

"Ranging" refers to an iterative process in which the initial uncertainties in slot timing, transmit power, and carrier frequency of a TDMA subscriber are compensated. Ranging occurs during the initialization or channel re-provisioning process through head end receiver feedback. For instance, the different cable distance in an HFC network introduces a relatively large timing uncertainty for the upstream traffic (up to 800 μs of one-way delay in a worst-case MCNS cable plant of 100 miles). The complete ranging algorithm must be accomplished in combination with the MAC protocol since it needs iterative feedback between the subscriber modem and the head end receiver. The PHY portion of ranging can be implemented with the same preamble processing unit as is used for fast recovery of the normal data packet. A special ranging sequence with a longer preamble in a ranging control slot is reserved for ranging purposes. The duration of the ranging control slot should be long enough to handle a maximum round trip delay of 1.6 ms in the MCNS HFC plant (see *Figure 11*). The difference between the actual sampling phase of the ranging sequence and the desired timing phase gives the ranging error. Power ranging is accomplished in a similar manner as time ranging. The received signal strength value, estimated by the power estimation block, is sent back to the subscriber modem for the transmitter to adjust its transmission power level in order to create a constant power at the receiver. A relatively large frequency offset can also be recovered during the ranging process and sent back to the transmitter for correction. This allows efficient channel utilization, preventing spectrum leakage into adjacent frequency channels.

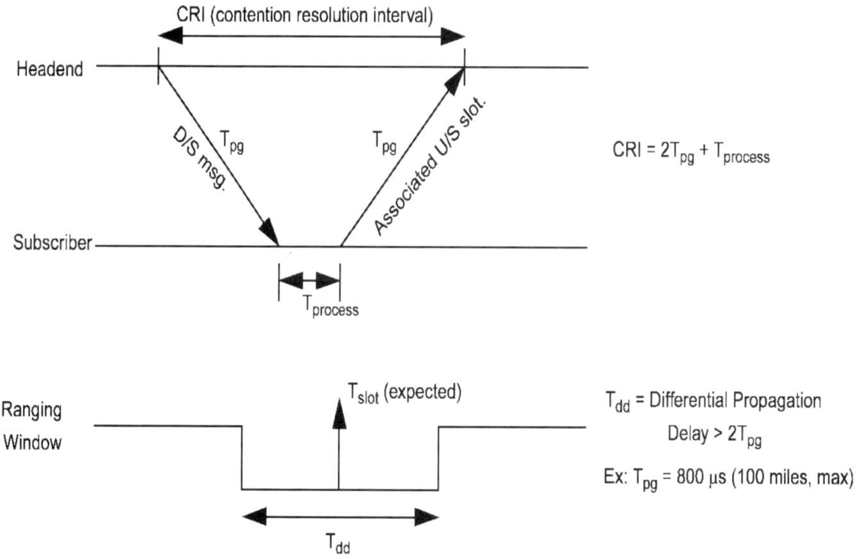

***Figure 11:** Time Ranging Process*

In summary, the key features of the advanced burst receiver are as follows:

- Fully frequency-agile, TDMA burst receiver
- Coherent 16-QAM/QPSK burst detection
- Fast burst acquisition (preamble size as small as sixteen symbols)
- Adaptive equalization
- Fully programmable modulation parameters: constellation size (16–QAM/QPSK), truly variablesymbol rates, and fine carrier frequency
- Multiple burst configuration banks with programmable preamble size/patterns and RS FEC parameters
- Channel quality monitoring and ranging error measurements fully integrated

Conclusion

An evolutionary path to improve the existing TDMA upstream system in terms of overall throughput and robustness has been proposed. For 16–QAM, reliable burst transmission can be achieved at a channel SNR greater than 20 dB when amplitude, timing, and carrier phase loops accompany the preamble processing unit. It has been shown from the simulation results that after initial equalizer training, many channels can provide greater than 20 dB of SNR and thus increase the data rates by a factor of two, compared to QPSK–only systems. Dynamic channelization based on continuous channel quality monitoring at the head end, coupled with fine carrier frequency agility and truly variable symbol rates anywhere from 100 kBaud to 5.12 MBaud, improves the overall channel utilization significantly even in the presence of ingress and microreflections. Burst/impulse noise can be handled by adaptively changing the correction capability (up to T=10) of the RS code at a given symbol rate and possibly turning on the optional interleaver. A fully integrated digital burst receiver architecture with all these techniques integrated also has been presented. Such an implementation is already available in the marketplace making more reliable TDMA data transmission based on a single carrier system possible for next-generation cable modem applications.

References

1 B. Currivan, "CATV Upstream Channel Model, Rev. 1.0," IEEE 802.14-95/133.

2 MCNS/DOCSIS Radio Frequency Interface Specification, SP-RFI-I02-971008, October, 1997.

3 R. S. Prodan, M. Chelehmal and T. Williams, "Analysis of Two-Way Cable System Transient Impairments," NCTA Proceedings, 1996.
4 Leo Montreuil, "DAVIC Preamble and Unique Word for the Upstream Transmission Synchronization," IEEE 802.14-96/142.
5 H.T. Nicholas and H. Samucli, "A 150-MHz Direct Digital Frequency Synthesizer in 1.25-1m CMOS with -90 dBc Spurious Performance," IEEE Journal of Solid-State Circuits, vol. 26, 1959-1969, December 1991.

Part III: The Medium Access Layer and Performance Issues

UniLINK as a Media Access Protocol for Community Cable TV

Chris Grobicki
Director Product Marketing, Broadband Technology
Nortel Networks

John M. Ulm
Principle Engineer, Broadband Technology
Nortel Networks

Abstract

UniLINK is a media access control (MAC) protocol optimized for community-area networking over new and existing hybrid fiber/coax cable (HFC) infrastructures. It is designed to operate on the tree and branch topology prevalent in the cable industry at distances up to 100 miles from the head end or central office. UniLINK is based on modified TDM concepts, supporting a mix of fixed, contention, and reservation bandwidth concurrently. It dynamically adjusts the mix as offered load changes. The protocol is used in the LANcity family of products. This implementation, with a frequency-agile quaternery phase shift keying (QPSK) modem, is deployed at more than 200 customer sites and provides a packet-oriented interface with a high-speed, symmetrical, 10-Mbps connection. This paper provides an overview of the UniLINK protocol, its operating environment, test results, and possible future enhancements.

Community Networking Using Cable TV

The cable operator is in a unique position to provide high-speed data communication services within the community. The infrastructure of copper coax and fiber that passes through more than 90% of the homes in the United States can be used for more than just the distribution of television programs. It is also becoming a valuable part of the national information highway. Computers and information sources can be accessed at speeds unheard of using the traditional phone modem, opening up bandwidth intensive multimedia applications across the community and around the country.

The types of applications that can be provided with the cable infrastructure are as varied as the homes and businesses passed by the cable. It can provide the services that a community of users require to run their businesses, teach their students, occupy their leisure time, and improve their productivity and quality of life. Typical applications at customer sites include Internet access services, local area network interconnectivity, working at home, multimedia information retrieval, and multimedia conferencing. These services are provided over large metropolitan areas at speeds much higher than the traditional telephone modem.

Challenges of Data Communication over HFC Networks

High-speed data communication using community-wide HFC networks presents the equipment provider with unique technical challenges. A MAC protocol operating in this environment must be designed for optimum efficiency, robustness, and fairness, addressing each of the environmental characteristics in the core of its functionality.

The following highlights some of the areas addressed by the UniLINK protocol.

Operation over Existing Cable TV Plants as Well as Evolving HFC Systems
More than 90% of the homes and businesses in the United States are passed by cable with an ever-increasing number becoming bidirectional capable. It is important that communication equipment operates within the constraints of the existing cable environment and does not require a rewiring or rearchitecting to become operational. This implies support for everything from pure coax tree and branch cable systems to the evolving hybrid fiber/coax systems at distances from one mile to 100 miles.

Supports a Large User Community
The cable infrastructure is a multidrop architecture with a single upstream and downstream channel shared among many users. Access to the upstream channel must be fair and stable, regardless of the network offered load. Because the type and number of applications in use at any time is indeterminate, the access protocol must dynamically adjust to optimize operating conditions.

Supports Extended Distances

The cable infrastructure covers distances that are much greater than those current LANs can support. Some cable networks in remote regions or in large metropolitan areas may have to span up to 100 miles from the head end or central office.

Supports LAN Applications in an Extended Network

The 10 Mbps performance level available with the Ethernet protocol has set the standard for network applications. Individuals have come to expect high-data content multimedia applications in real time. As data communication services move into homes and small businesses, the expectation follows.

Supports Multiple Types of Traffic Efficiently

The many applications that are run in a community of users require different types of data transfer mechanisms. Bulk data transfer, multimedia services, video conferencing, whiteboarding, and Internet access all have different performance and periodicity requirements, which should be serviced within the same communication channel to optimize the efficiency of the available reverse channel bandwidth. Channel spectrum in the upstream direction is at a premium, requiring mixed service support in a single channel.

Provides Guaranteed Bandwidth to Information Servers

Information servers that source high levels of data can be provided guaranteed bandwidth with assured performance levels. This allows data bandwidth to be apportioned between the information sources and users with a split based on local community requirements.

Provides LAN–Like Performance Levels

The computer paradigm of the 90s is a distributed client/server environment with information and data sources used throughout the network. The networking environment in the community should follow this model and provide the same types of service expected at the local level, with performance levels optimized for the upstream and downstream channels.

Simplified HFC Cable Architecture

The purpose of a data communication network is to provide the link that allows the movement of information and data between clients and servers. The HFC cable topology is different from that encountered with distributed LANs and point-to-point WANs. These differences must be accounted for in the design of a media access control (MAC) and physical layer to optimize the utilization of the bandwidth.

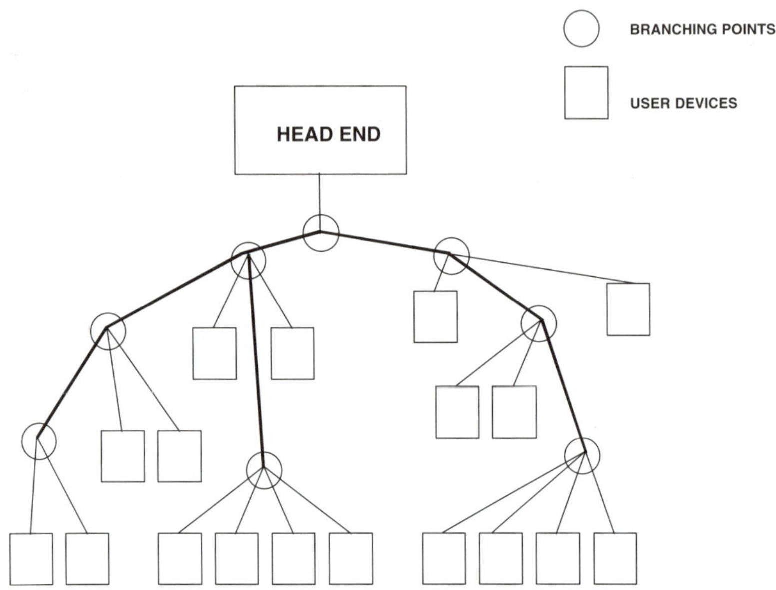

Figure 1: *Cable TV Tree and Branch Topology*

The topology of the cable-TV system is a tree and branch, or a collection of tree and branches, the root being at the head end or central office as shown in *Figure 1*.

The path from the home or end node is often called the upstream, reverse, or return channel, while the path from the head end to the end node is often called the downstream or forward channel. In most commercial cable plants the channels are on the same physical cable but at two different frequencies. The upstream frequencies tend to be in the 5 MHz to 30 MHz or 5 MHz to 42 MHz range. The downstream frequencies vary from 54 MHz to 350 MHz to 54 MHz to 750 MHz depending on the cable site. *Figure 2* is a simplified diagram of frequency allocation upstream and downstream with the frequency conversion at the head end.

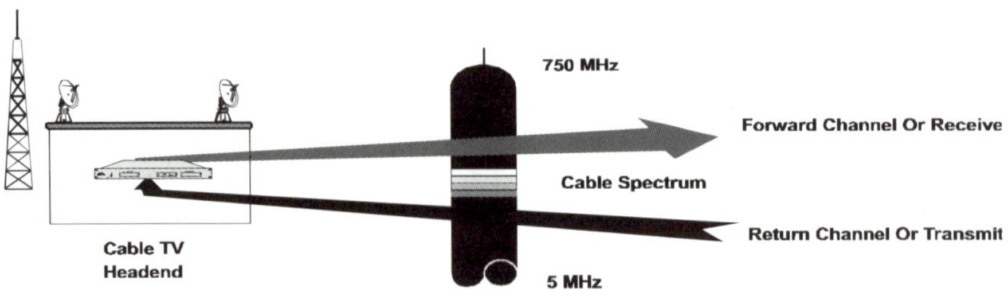

Figure 2: *Frequency Spectrum Allocation*

A topology-independent representation of a cable plant that is data communication-ready is shown in *Figure 3*, and will be used to present the key concepts of the UniLINK protocol.

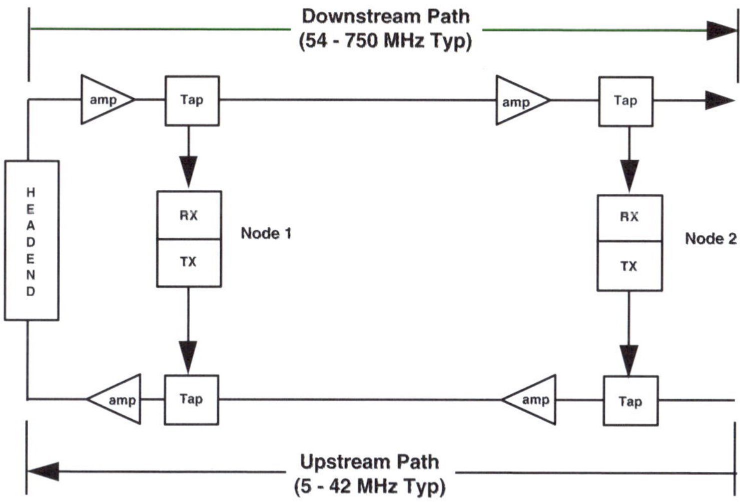

Figure 3: *Topology-Independent Representation*

Key points to note are the distributed, multidrop environment, with an upstream path that allows data to be carried to the head end, and the downstream path from the head end to the end nodes. The head end provides the connection point from the upstream path to the downstream path and in the case of a single coax system would contain a frequency translator.

UniLINK as a MAC Protocol for Cable Networks

The UniLINK protocol has been designed from the beginning to operate within the community-wide HFC infrastructure and address all of the issues cited previously. It is already implemented with a frequency agile QPSK modem. These are deployed at more than 200 customer sites and provide a packet-oriented interface with a high-speed, symmetrical, 10-Mbps connection.

The UniLINK protocol is based on modified TDM concepts and supports a dynamic mix of fixed (isochronous), demand-based reservation (dedicated), and random access (contention) assignments. These assignments are given for each available TDM slot, which actually represents a small amount of assigned bandwidth.

The random access assignments are available to all nodes. Under light network traffic loads, random access provides immediate access to the network and thus low latencies. However, under heavier loads, random access is subject to collisions, with a resulting lost in throughput and delay. Dedicated assignments are reservation-based and are granted as needed. Each node requests the amount of bandwidth it needs, and is assigned slots appropriately. These slots are contention-free and provide high throughput with predictable delays, even under heavy network traffic loads.

The isochronous assignments are relatively static allocations of slots to a particular node. This allows UniLINK to support applications that require fixed, guaranteed bandwidth and delays.

Three basic mechanisms are used to provide timing on a UniLINK network: periodic timing reference signals, TDM slot delineation, and compensation for each node's transmit to receive propagation delay.

Pacing and Block Synchronization

The UniLINK protocol uses a concept called pacing to provide a means for global network synchronization. The slot time synchronization as well as slot assignments are controlled through a single node called the pacer. Depending on implementation, the pacer could reside at the head end or be located elsewhere in the network. The pacer transmits a block sync packet at a fixed periodic rate. This is shown in *Figure 4*. The time from one block sync packet to the next is referred to as a block sync interval (BSI). The block sync interval is a network parameter and will typically have a value ranging from two milliseconds to thirty-two milliseconds.

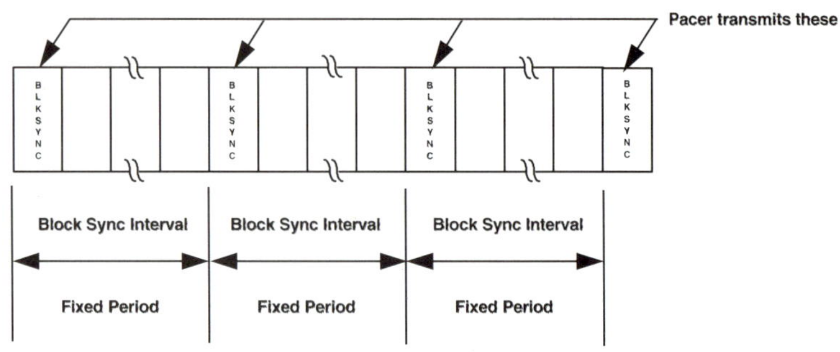

Figure 4: *Block Sync Interval Timing*

TDM Slots and Slot Assignments

UniLINK uses time division multiplexing (TDM) to partition data bandwidth into assignable entities called slots. Each node on the network synchronizes itself to the periodic block sync packets. *Figure 5* shows a block sync interval and its associated slots.

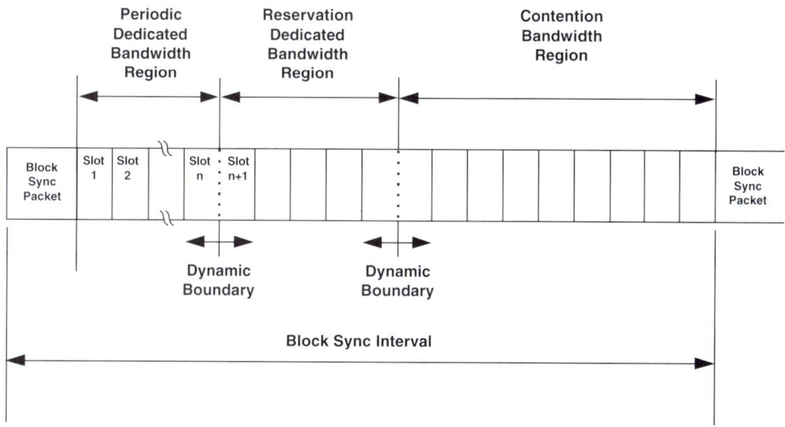

Figure 5: *Dynamic Allocation of Bandwidth*

All slot assignments are identified relative to the block sync packet. The pacer allocates these slots for each BSI. Each slot can be allocated to a single node and called a dedicated slot, or it may be allocated to multiple nodes and called a contention slot. Multiple slots may be assigned together to allow a node to transmit a variable sized packet. The protocol dynamically adjusts between contention and reservation slots on every block sync interval.

Compensating for Network Delays

Community area networks may stretch for tens of miles from their point of origin, whether this is a head end or a central office. In some remote locations or in a large metropolitan area, this distance may surpass seventy miles. These distances become significant when operating at LAN–type speeds, such as 10 Mbps.

Because each node on the network is at a different distance from the head end, a method must be established to assure that a transmission originating from any node on the network reaches the head end within the allocated slot. UniLINK uses a concept called loop delay compensation to accommodate these distances.

To line packets up that precisely in their assigned slots, a node needs only two critical pieces of information:

- a global timing reference (provided by block sync)
- knowledge of its own round-trip delay (referred to as loop delay)

The process of obtaining one's loop delay is often termed "ranging," and is done in a special region of the block sync interval known as the maintenance area. This region must be long enough to allow for the maximum loop delay on the network. Once a node obtains its loop delay, it can transmit its dedicated packets so they arrive at the precise time as determined by the allocation algorithm.

A transmission that compensates for the round-trip delay can be very precisely placed in a slot and, as all nodes perform ranging and compensate for their propagation delay, a very precise timing relationship is established for the network. From the block sync timing information, a node knows precisely when it will receive a particular slot. If it wishes to transmit in a slot, it simply starts the transmission early by its own loop delay to make it arrive at the exact time desired.

Data Bandwidth Allocation

Periodic dedicated bandwidth, reservation dedicated bandwidth, and contention bandwidth regions provide different service levels and efficiencies based on the requirements of the system. *Figure 5* illustrates the three types of data bandwidth available with the UniLINK protocol. Bandwidth is allocated by the pacer node.

The periodic dedicated bandwidth region provides dedicated and jitter-free TDM slot assignments that have a fixed interval-to-interval timing relationship to the block sync packet. An assignment uses a call set up/break down process that allows nodes to maintain an active and fixed assignment over multiple block sync intervals.

Bandwidth made available when a session becomes inactive is reallocated to the next requesting node and can be provided in one or multiple contiguous slots.

The reservation-dedicated bandwidth region provides dedicated slot assignments but with no interval-to-interval timing relationship maintained. This region is used to relieve congestion in the contention region because of high offered load and collisions, but bandwidth assigned to a node is kept only for the interval assigned and relinquished on subsequent block sync intervals. The number of slots within this region varies from interval to interval and, if not reserved, are available to the contention bandwidth region.

The contention bandwidth region contains slots that are available to multiple nodes on the network with a probability of collision based on offered load, which behaves similarly to a slotted carrier sense multiple access with collision detection (CSMA/CD) protocol. A block sync interval with low offered load and no periodic dedicated bandwidth assigned would have all contention slots.

The number of slots in each region can vary based on the offered load. Therefore the boundaries between the regions are dynamic. Each region can have a preconfigured maximum, and in the case of the contention area, a minimum number of slots available.

UniLINK is unique in that it allows either a single slot or multiple contiguous slots to be allocated to an individual node, allowing the efficient transport of LAN–type data packets throughout the community without necessitating fragmentation and reassembly. As might be obvious, future embodiments of the protocol could allow the efficient transport of both cell- and packet-based data as will be described later in this paper.

Fairness and Robustness

There are several levels at which fairness of access to the reverse cable plant is addressed:

- fair access regardless of distance from head end
- fair access regardless of offered
- lockout prevention by heavy users

The mechanisms used to allocate dedicated bandwidth and to resolve collisions are designed in such a way as to prevent unfair advantage or disadvantage to a node regardless of its offered load or placement in the network. Round-robin mechanisms are built in to assure that nodes denied a service will have highest probability of acceptance when the service is available.

The protocol does not become unstable at offered loads greater than 100% but instead continues to provide deterministic performance, as will be shown in the performance examples later in the paper.

Improving Efficiencies

The UniLINK protocol was designed from the beginning to efficiently overcome the distance limitation. The basic premise is that packets may be precisely lined up, with a minimal inter-packet gap in between, when they arrive at the head end. This provides a pipelining effect that breaks the performance bottleneck by allowing multiple packets onto the network at any instance in time.

The CSMA/CD–based protocols allow only one packet on the network at a time. They rely on the fact that the packet transmission time is greater than the maximum round-trip delay on the network to get efficient operation. When the transmission time equals the maximum loop delay, then the protocol reverts to aloha or slotted aloha performance, depending on its implementation. As the distance continues to grow, the CSMA protocols fall apart altogether. As a point of reference, a minimum-sized Ethernet packet requires 51.2 μs of transmit time. This corresponds to a distance of less than seven miles from a head end or central office, which is small by today's cable system standards.

In addition to allowing multiple packets from multiple nodes on the network, the UniLINK protocol also allows a single node to have multiple dedicated packets outstanding on the cable plant. This can greatly increase the throughput a single user can obtain on an idle but physically large network. Since dedicated packets are collision-free, there is no concern of re-ordering of packets.

Contention Support

The UniLINK protocol also has special features to improve the efficiency of contention-based packets. The allocation algorithm effectively restricts the start of all contention transmissions to a set interval, which is greater than or equal to the max loop delay of this network. This forces all contention packets transmitted with the same start opportunity to have the beginning of the packets exactly overlap. This helps make the collision detection much more robust.

Another feature is the length field at the beginning of the UniLINK frame format. With the contention start opportunities spaced at a maximum loop delay time, every node is guaranteed to hear a contention transmission before the next start opportunity arrives. A node waiting to transmit can now determine, using the length field, precisely when the current transmission ends. This effectively emulates a carrier sense capability. Because it knows its own loop delay and when the current transmit ends, a waiting node can now take advantage of this to start its transmission early (i.e. before the entire current frame is received). This means the next transmission may be placed immediately after the current transmission without a maximum loop delay of idle time in between frames. In order to ensure the ordering of packets, nodes are restricted to one outstanding contention transmission at any one time.

These two features effectively cause the contention region to behave as a slotted CSMA/CD mechanism. In an idle network, almost all of the available bandwidth is contention and this controls the network's behavior. A node wishing to transmit simply waits for the next available contention start opportunity. This delay is bounded by a maximum loop delay.

Dynamic Sharing of Dedicated and Contention

Once a network is very busy, the allocation algorithm assigns almost all of the available bandwidth to the dedicated region. In this mode, collisions are virtually eliminated and all nodes are fairly sharing the available bandwidth as determined by the allocation algorithm. As the network load changes, the UniLINK protocol gracefully and quickly adjusts to the changing conditions on each block sync interval.

Concatenation

Another significant UniLINK feature to improve network efficiency is concatenation. Within UniLINK, nodes that have multiple packets to transmit can concatenate them into a single transmission called a UniLINK frame. These concatenated packets are treated as a single entity and may be transmitted in either dedicated or contention modes. If a collision occurs on a concatenated frame, all packets are discarded and the entire concatenation is retransmitted. This ensures packet ordering is maintained.

Each packet within a concatenated frame still contains all of its addressing information. Thus, each packet may have its own unique destination. Receiving nodes decode the concatenated frame and accept only those packets that are destined for it, discarding the remainder. The concatenation remains completely transparent to all upper layers.

The first major benefit of concatenation is the reduction of transmission overhead. Only one IPG and one preamble are needed for many packets that are put together in the concatenation. This is most beneficial when the packet sizes are small.

The second benefit is increased efficiency from reduced collisions. For example, if ten small packets are concatenated, it appears to the network as a single transmission that is ten times as large. This helps in two ways: the effective tenfold increase in length greatly improves the slotted CSMA/CD performance, and the apparent offered load is reduced by a factor of ten, since only one transmission needs to be completed (instead of ten separate transmissions, each subject to collisions).

Measured Test Results in Real Systems

During 1992, LANcity implemented the UniLINK protocol in a gate array. The allocation algorithm was kept in software to provide flexibility and to allow for future enhancements. The added complexity for the UniLINK protocol was less than 10K gates more than an Ethernet controller chip with comparable functions. This technology has been deployed and thousands of units are running in several hundred sites worldwide. These are real cable plants running real applications. The following results were obtained from operational production units on real cable plants.

Network Parameters

The product was targeted for transparent interconnection with Ethernet networks, and implementation parameters for the UniLINK protocol were chosen for supporting primarily data applications. The product supports a full 10-Mbps data rate over a frequency-agile QPSK modem. A slot size of 25.6 microseconds (or 32 byte times) was selected along with a block sync interval of 31.6 milliseconds (or 1236 slots). At least 20% of every block sync interval is allocated for the contention region to allow nodes without access to the dedicated bandwidth some limited access. Additional network overhead, for things like reservation requests, may vary from <1% up to 10% overhead. The maximum concatenation size of 4576 bytes was chosen to allow up to three maximum-sized Ethernet packets or 65 minimum-sized Ethernet packets in a single concatenation. The maintenance area occurs once every block sync interval and is set to the maximum loop delay for that site. The overhead from this could range from less than 1% for networks up to fifty miles; to only 4% overhead for a network that spans up to 200 miles.

Data Throughput with Large Packets

With these production parameters, we ran several tests to determine the actual capabilities of the protocol. We ran maximum-sized Ethernet packets to determine maximum user data throughput and minimum-sized Ethernet packets (64 bytes) to determine the maximum packet carrying capability of the network.

For networks that cover up to fifty to sixty miles (which corresponds to maximum loop delays up to sixteen slots), the product was able to sustain transfers of 756 maximum-sized Ethernet packets per second (pps). When the maximum loop delay was increased to fifty slot times for a network spanning 150–200 miles, the throughput dropped to 655 packets per second.

Performance with Small Packets

To test the packet carrying capability, we set up eight traffic generators all transmitting at maximum possible offered load. Under these conditions, the UniLINK network supported a sustained throughput of more than 15,900 packets per second. Note that this is noticeably higher than the theoretical Ethernet maximum of 14,880 packets per second in a collision-free environment. Each individual traffic generator averaged about 2000 packets per second with a variation of +/– 300 pps when sampled every second. This increased throughput is directly attributable to UniLINK's concatenation feature. As the network size was increased to the 150-200 mile parameter used above, the network was still able to transfer almost 14,400 packets per second.

While this was happening, we performed some SNMP network management operations and ran some ping operations to get a qualitative feel for a network under stress. The results of the test were quite informative. The SNMP manager, which was configured to continually poll the devices, operated with the same interactive nature as in an idle network. The pings ran with typically 30–60 milliseconds of increased response time.

For comparative purposes, we ran the identical test with a short (<100m) Ethernet backbone replacing the UniLINK network. The Ethernet backbone was able to sustain a total throughput of approximately 14,000 packets per second. The individual throughput was more revealing. Even though all nodes had similar long-term averages, the one-second sampling showed that a node's throughput could vary from a few dozen packets to almost 5,000 packets. This had a disastrous effect on the SNMP manager. Since nodes were constantly being dropped due to time-outs, it was effectively not usable. The ping results were just as poor, with times varying wildly and some messages being lost due to time-outs. This shows that not only is UniLINK's dedicated bandwidth more efficient under heavy loads by eliminating collisions, but it improves fairness and allows low-bandwidth interactive applications to operate in almost a transparent fashion to the network load.

UniLINK Enhancements

The UniLINK protocol offers many features that are very desirable for multimedia applications. It can support the different quality of service (e.g. asynchronous transfer modes [ATM], constant bit rate [CBR], variable bit rate [VBR], and available bit rate [ABR]); it efficiently allocates scarce bandwidth on the reverse channels; and it is very stable under high loads, which is important in any community-area network. Future implementations of UniLINK will have to consider additional issues such as

ATM support, video, and telephony applications as well as increasing efficiency to get the most out of the reverse channel.

ATM Support

Since the protocol already supports ATM's quality of service, any future implementation must also provide for a seamless transport of ATM cells. In addition, allowances must be made for switching the cells at the head end equipment. These enhancements can be made without any changes to the basic concepts in UniLINK. The block sync timing and various control information (slot allocation, collision detection, ranging, etc.) must still be passed down in the forward direction. In fact, UniLINK could provide the connection across the infamous "Last mile" over existing cable lines to provide a virtual ATM connection to the home.

Voice and Telephony

Voice and telephony applications pose a different sort of problem because they are sensitive to delay and jitter. The block sync interval would have to be reduced to two or three milliseconds to be able to support the delay requirements for POTS. The jitter issue is handled quite easily with UniLINK. To increase the number of subscribers supported, it would also be desirable to support small (16- or 24-byte) transfers with a minimal of overhead.

Video telephony is another application with some rigid delay and jitter considerations. In addition, it requires much higher bandwidth than voice (e.g. 384 Kbps versus 64 Kpbs). Unlike video-on-demand, video telephony needs the bandwidth in both directions. UniLINK can support video telephony with the enhancements described above.

Asymmetric Operation

Another class of applications, video-on-demand and information retrieval services, for example, have a much larger bandwidth requirement in the forward direction than in the reverse. For these it is very desirable to support high bandwidth (30-40 Mbps) modulation schemes in the forward direction when the reverse channel may have limited bandwidth. As with ATM support above, the UniLINK protocol can support this with a small amount of control overhead in the forward direction, and no changes in the basic concepts.

The main issue is the size of the data pipes (what is the total bandwidth supported) and the services offered (e.g. support for ATM's QoS). These are the things that determine which applications can run. The need for additional bandwidth will always continue to grow.

Conclusion

This paper has presented UniLINK as a MAC protocol suitable for today's cable networks. UniLINK has proven itself in the field as a protocol that efficiently handles networks with long distances and high data rates. The protocol dynamically supports a mix of fixed, contention and reservation bandwidth concurrently. It also supports the quality of service necessary in today's world. As such, it is ideally suited to support the multimedia applications that run over the community area networks, both today and in the years to come.

Performance of MAC Protocols for Broadband HFC and Wireless Access Networks

Kotikalapudi Sriram
Distingulshed Member of Technical Staff
Bell Laboratories Lucent Technologies, Inc.

Abstract

This paper presents a brief overview of some key features that enable superior performance and efficiency in medium access control (MAC) protocols for hybrid fiber/coax (HFC) and wireless access networks. These features include the following: 1) minislots for bandwidth allocation granularity; 2) flexible transport capability for fixed-length asynchronous transfer mode (ATM) protocol data units (PDUs) and variable-length (VL) PDUs (e.g., IP, 802.3 frames, IPv6); 3) means for low delay transport of synchronous transfer mode (STM) voice information; 4) dynamically variable minislots for contention request resolution; and 5) piggybacking for additional transmission requests. This paper also includes a comparison of alternative MAC protocol features. The objective of the paper is to identify a set of MAC protocol features that results in significant performance and capacity gain, and (in conjunction with connection level bandwidth management algorithms) is capable of meeting different delay and bandwidth requirements for different traffic classes such as telephony, data, video, and multimedia. The resulting MAC protocol would enable efficient integration of various applications in a community access network employing HFC or wireless media. The IEEE 802.14 standards specification (draft) and the MCNS (a cable industry consortium) specification on cable modems have embraced several of the key features discussed here as part of their MAC protocols.

Introduction

Experiences over the last several years and projections into the future suggest an incipient revolution in the variety of communication services offered to the user and in the infrastructure used to support these services. Internet access, client

This paper appears in part with the same title in *Advances in Performance Analysis*, vol. 1(1), 1998, pp.1–37. A portion of this paper also appears in *Telecommunications Systems*: Special Issue on Multimedia, vol. 9, nos. 3&4, 1998, pp. 315–333.

applications with remote servers, video-on-demand, and interactive multi-media are among the new services that will supplement traditional telephony. Concomitantly, frame relay, ATM, and the synchronous optical network (SONET) are enlarging the capabilities of networks at all geographical levels. Wireless and cable-based access technologies are providing attractive alternatives to wireline loop and analog modem–based access. Collectively, these promise to change the means and price of future communication services.

The new services are characterized by a plethora of throughput rates and delay requirements, along with varying degrees of burstiness. Looking to the future, combinations of these transmission attributes will become even richer and will create new technical challenges for equipment designers and network operators. An important goal of the system architecture and protocol design for HFC and wireless networks is service integration and real-time statistical multiplexing of various traffic types. This will allow efficient use of scarce upstream (i.e., from home/mobile to network) bandwidth, amortization of equipment (switches, routers, operations systems, radio frequency (RF) modems, etc.), and plant upgrade costs across multiple services.

Currently, work is in progress to design MAC protocols that enable efficient utilization of bandwidth in upstream and downstream channels over broadband access media, (e.g., HFC and wireless).[1–16] The upstream and downstream transmissions take place in separate RF channels occupying different parts of the RF spectrum. The available total bandwidth is usually much more in the downstream direction than in the upstream direction. An HFC network typically has several dozen RF channels, each at a rate of about 10 Mbs to 40 Mbs in the downstream, and a dozen or so RF channels each at a rate of 2.56 Mbs in the upstream.[2,3] The downstream transmission takes place in a broadcast mode with each MAC protocol data unit (PDU) carrying the address of a station to which it is destined. The upstream bandwidth is shared by many stations using a multiple access protocol, operating under the control of a central controller that may be located in the head end of an HFC network or the base station of a wireless network. In the rest of this paper, we use HFC terminology for ease of communication (with an understanding that the corresponding wireless terminology can be substituted when the MAC protocol is discussed in the context of wireless networks). In the context of HFC networks, the central controller is referred to as a head end controller (HC). The stations are referred to as cable modems (CM).

Figure 1 shows how a MAC protocol fits into an overall systems architecture for HFC networks supporting a multi-service environment including voice, data, video, and multimedia applications. The architecture illustrated in *Figure 1* calls for a flexible MAC protocol capable of transporting information in STM, fixed-size ATM PDUs, or VL PDUs (e.g., IP, 802.3 packets). Such a MAC protocol would also be bandwidth efficient, taking advantage of statistical multiplexing of bursty traffic and providing low latency for delay-sensitive traffic. ATM and VL PDU sizes are integral multiples of a basic upstream unit such as a minislot, which will be discussed later. When VL PDUs are used for transport, the information is carried in a format close to its native format (e.g., IP, 802.3, MPEG-2, etc.) with minimum additional overhead in the access network. An early deployment may support only a subset of the functions shown in *Figure 1*. An example of possible early deployment is to support data services over IP and 802.3 using VL PDUs. The backbone connectivity is through IP routers, while the connectivity between the CM and information appliances (IAs) is over 802.3 LAN (a true LAN or a point-to-point connection using LAN protocols). Another early deployment scenario may involve video retrieval with MPEG-2 packets carried as native data with tighter jitter requirements, or over ATM cells with an appropriate service class. MAC protocols are being designed to allow a smooth migration from an early deployment of a limited set of services to a full service access network with connectivity to multiple backbones and multiple services.[1–9, 14–16]

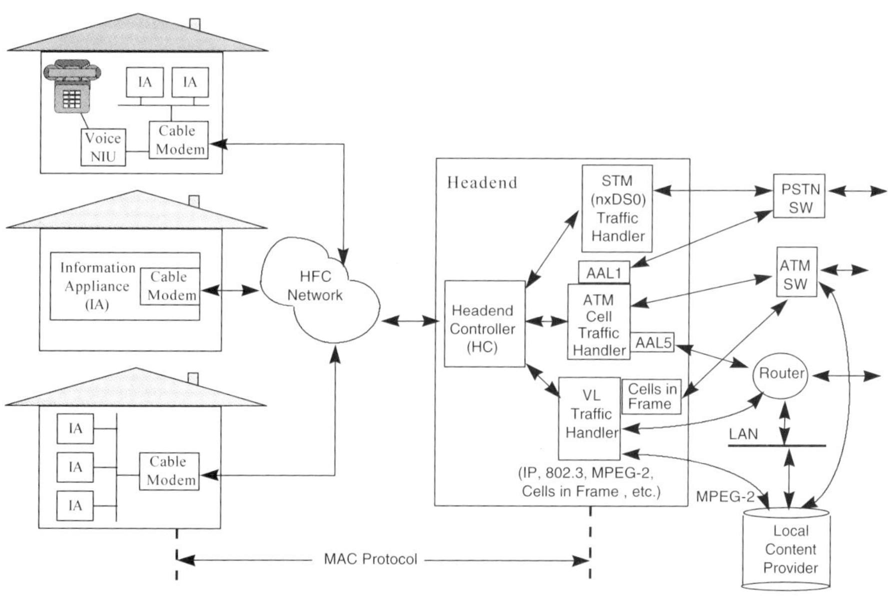

Figure 1: *The Role of a MAC Protocol in HFC Networks*

In this paper, we present a brief overview of some key features that enable superior performance and efficiency in MAC protocols for HFC and wireless access networks. These features include the following: 1) minislots for bandwidth allocation granularity, 2) flexible transport capability for fixed-length PDUs (ATM cells) and VL PDUs (e.g., IP, 802.3 frames, IPv6), 3) means for low-delay transport of STM voice information, 4) dynamically variable minislots for contention request resolution, and 5) piggybacking for additional transmission requests. We also include in this paper a comparison of some alternative MAC protocol features. The objective of the paper is to identify a set of MAC protocol features that results in significant performance and capacity gain, and (in conjunction with connection level bandwidth management algorithms) is capable of meeting different delay and bandwidth requirements for different traffic classes such as telephony, data, video, and multimedia. The resulting MAC protocol would enable efficient integration of various applications in a community access network employing HFC or wireless media. The IEEE 802.14 standards specification (draft) and the MCNS (a cable industry consortium) specification on cable modems have embraced several of the key features discussed here as part of their MAC protocols.[14–16]

Principles of MAC Protocol Operation

We consider here MAC protocols in which an HC manages the upstream and downstream components of the MAC protocol. The protocol operates in one-to-many broadcast mode in the downstream, and many-to-one multiple access mode in the upstream. The HC and CMs have a common notion of upstream and downstream frame structures based on initial synchronization and periodic control messages from HC to CMs regarding possible restructuring of the upstream and downstream frames. Because the upstream bandwidth is scarce compared to the downstream bandwidth, multiple upstream RF channels are usually associated with one downstream RF channel.[2,3]

Upstream Frame Structure and Protocol Operation

In the upstream component of the MAC protocols, minislots are used to request permission to transmit packets of information (voice, data, video, or multimedia) in the upstream channels, and the information is subsequently transmitted in packet time-slots allocated by a central controller.[3,4] The requests transmitted in request minislots are subject to collisions, and an appropriate collision resolution algorithm (CRA) is used to resolve collisions. The CRA specifies the rules by which the CMs involved in a collision can retransmit to resolve the collision. A brief overview of different types of CRAs is given in a later section.[4] The HC transmits status (feedback) messages in a downstream channel to inform the CMs about the collision status (collision or success) of requests in minislots. For the requests received successfully, the status message also includes grants of upstream packet slots in which the actual information may be transmitted.

An example upstream frame structure is shown in *Figure 2*. It includes an STM region, a contention request region, and a reservation region for ATM and VL PDUs. Synchronous traffic such as DS0 and n*DS0 calls are allocated fixed-size time-slots in the STM region. The frame duration is denoted by *T* ms. The frame duration, *T*, in *Figure 2* is shown to be 2 ms, as an example. The roundtrip propagation delay accounts for about 1 ms, and about 1 ms is budgeted for transmission and processing times at the HC and CM. Each time-slot in the STM region contains the information bytes for one STM connection collected over *T* ms plus physical layer (PHY) and MAC overheads. These overheads include guard time, preamble, and forward error correction (FEC). The contention region consists of multiple minislots. As mentioned previously, requests for transmission are made in the contention request minislots. When the request successfully reaches the HC, a reservation is granted to the requesting station for transmission of its PDU in the reservation region of a future frame. The reservation region (see *Figure 2*) also consists of minislots of the same size as the request minislots. The reservation for PDU transmission is allocated in units of minislots. The size of a minislot is selected so that an ATM PDU burst equals an integral number of minislots. A variable number of minislots are allocated for VL PDUs, depending on the VL payload size.

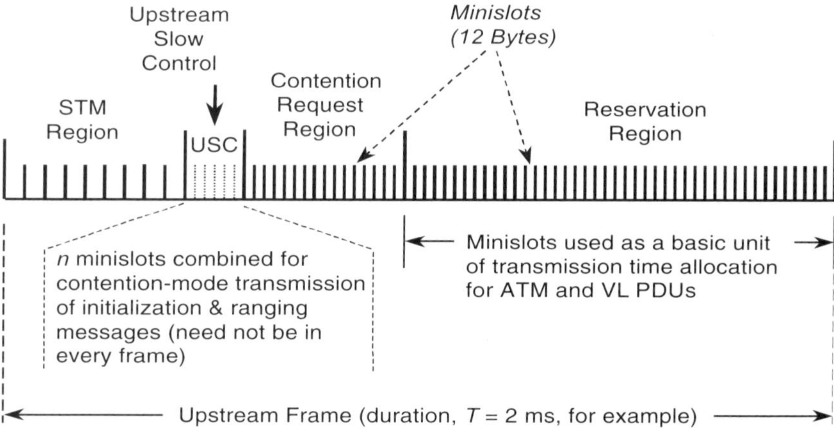

Figure 2: *An Example Upstream Frame Structure Showing STM Slots, Contention Request Minislots, and a PDU Reservation Region with a Minislot as the Basic Unit of Granularity*

Upstream bursts of different types are illustrated in *Figure 3* based on the example of the adaptive digital access protocol (ADAPt+) described in the references.[3] In this example, it is assumed that quaternary phase shift keying (QPSK) modulation is used in the upstream at a 1.28 M symbols/s rate in approximately 1.8 MHz RF channels, resulting in a 2.56 Mbs data rate per RF channel. The guard time and preamble are of the same size for all types of bursts. The size of the FEC is a function of the payload size, and is different for different size bursts. The payload of the STM burst in *Figure 3* is 16 bytes, corresponding to frame (packetization) duration, $T = 2$ ms. The Sig field in the STM burst carries A-H signaling bits of the DS0 signaling scheme. The size of the STM burst in this example is 27 bytes. The minislot request burst in *Figure 3* is 12 bytes. The CM address is 12 bits long. The traffic type field specifies if the request is for an ATM or VL PDU. The bandwidth request field specifies the length of the PDU for which request is being made. The unit of bandwidth request is an ATM cell when the traffic type is ATM, and the unit is a minislot when the traffic type is VL. An ATM burst (see *Figure 3*) is 72 bytes long, including the PHY/MAC overheads, and is equal to 6 minislots. The VL PDU exemplified in *Figure 3* equals 13 minislots in length. It carries a VL payload of 128 bytes (with 7 bytes padded to make the total burst length an integer multiple of minislots). The 2-bit BOP/COP/EOP field in the VL PDU is to indicate beginning (00), continuation (01), and end (10) of the packet in case of segmentation of long VL payloads into smaller MAC layer VL PDUs. When the 2-bit field is set to a value of eleven, it indicates to the higher layer VL data element that no segmentation occurred. For several additional details related to upstream frame and PDU structures, the reader is referred to the references.[3,14,16]

Piggybacking of Requests

A request made for an additional PDU transmission time-slot, while transmitting a presently allocated PDU burst, is called a piggybacked request. In *Figure 3*, the ATM PDU and the VL PDU bursts include piggybacked requests. The piggybacked request field is identical to that in a request minislot. The advantage of piggybacking is that a request for an additional PDU waiting in queue (if any) can be made by a CM without going through further contention. Thus, piggybacking by CMs that currently have PDU allocations helps reduce the traffic and probability of collisions in contention minislots, and thereby benefits other CMs that are transmitting requests in a contention mode.

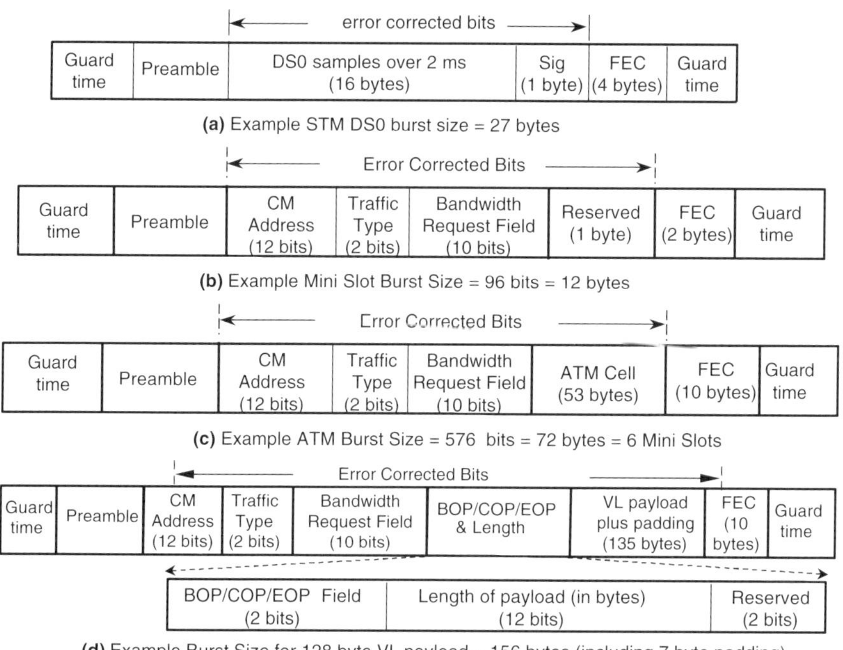

Figure 3: *Example Formats of Upstream Bursts for a) STM, b) Minislot Request, c) ATM PDU, and d) VL PDU*

Propagation Delay Effects and Upstream Frame Duration

Due to the physical distances between the HC and the CMs, the status message is subject to a roundtrip propagation delay. The roundtrip propagation delay together with the HC and CM processing delays for processing and notification of collision status information give rise to an upstream frame structure in at least two possible ways (see *Figure 4* and *Figure 5*). In *Figure 4*, the minislots are all clustered adjacently at the beginning of each upstream frame, while in *Figure 5*, they are distributed over the upstream frame interval. The frame duration, T_F , for the two cases is lower-bounded as follows:

$$T_F \geq t_R + 2\ t_p + t_{HC} + t_{CM} \quad \text{(for minislots clustered at the beginning of the frame)}$$

$$T_F \geq 2\ t_p + t_{HC} + t_{CM} \quad \text{(for minislots distributed over the frame interval)}$$

where t_R is the total time duration of the request minislots placed at the beginning of a frame; $2{*}t_p$ is the round-trip propagation delay; t_{HC} and t_{CM} are the HC and CM processing times.

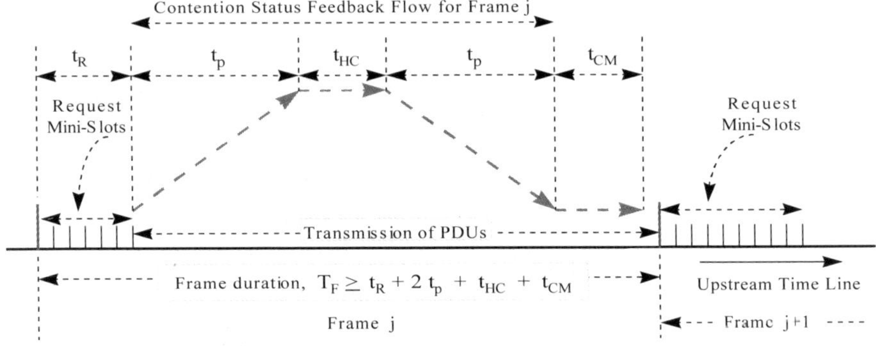

Figure 4: *Upstream Frame Structure with Contention Request Minislots Grouped Together at the Beginning of Each Framing Interval*

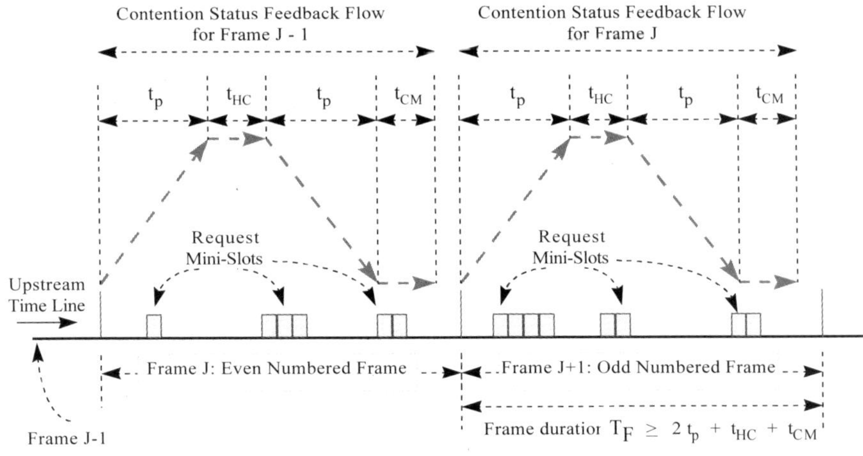

Figure 5: *Upstream Frame Structure with Contention Request Minislots Distributed over the Frame Interval*

The upstream frame structure can be chosen to be either of the two illustrated in *Figures 4* and *5*. For the case of MAC protocol operation in *Figure 4*, the status message corresponding to a cluster of minislots is received prior to the beginning of the next frame. For the case of MAC protocol operation in *Figure 5*, the status message corresponding to minislots in a frame is received during the next frame. The second case (*Figure 5*) allows the flexibility of placing minislots anywhere in the frame, and hence it would require less frequent regrouping of idle time-slots in a frame. It also requires two instances (or engines) of the CRA to be run in parallel (in an interleaved manner)—one over odd numbered frames and another over even numbered frames. However, this requirement is not expected to add any significant implementation complexity. The protocol features and results of performance analysis presented in this paper are equally applicable to either of the two cases of upstream MAC operation depicted in *Figure 4* and *Figure 5*.

Principles of Collision-Resolution Algorithms

A collision occurs in a minislot of an upstream frame when two or more CMs transmit request messages in the same minislot. Collisions are detected at the HC, and status messages are transmitted back to the CMs. A CRA is employed to resolve collisions. CRAs can be broadly classified into three categories as briefly described in this section. Although we assume the use of minislots, the descriptions of CRAs provided here are valid generally in the context of any contention-oriented packet transmission.

Random Binary Exponential Back-Off Algorithm: The CM generates a random number, i, uniformly distributed between $[0, (2^j - 1)]$, where j is the number of collisions that the CM has experienced for the request message it is attempting to transmit. If j is larger than ten, then i is selected from a uniform distribution in the range $[0, (2^{10} - 1)]$. The CM skips i minislot transmission opportunities and retransmits the request in the $(i + 1)th$ minislot (these minislots may span over one or more upstream frames).

Adaptive p-Persistence Algorithm: In the adaptive p-persistence algorithm, a parameter, say q, is communicated from the HC to the CMs as part of the status information. Upon receiving the status information, each CM generates a uniform random number, i, such that $1 \leq i \leq q$. If $i \leq k$, where k is the number of minislots available for request transmissions in the forthcoming frame, the CM transmits its request in the i *th* minislot, or else it waits for the next status message and repeats the operation by generating a new random number. The HC can vary the value of q from one frame to the next as a function of the number of collisions that are observed in a frame.[9,13] The value of q may be increased or decreased when the number of collisions observed at the HC increases or decreases, respectively. In effect, the algorithm adaptively decreases or increases the probability of retransmission by a CM as a function of the existing traffic load in the system. When the traffic load is heavier, resulting in more frequent collisions, then the value of q is increased. When the traffic load is lighter, resulting in less frequent collisions, then the value of q is decreased.

Tree Algorithm: A tree algorithm may use a binary, ternary, or n-ary tree to help spread retransmissions and thereby resolve collisions at each consecutive step in the contention resolution process. Although a binary tree is commonly used, it has been shown that the ternary tree algorithm provides higher performance.[10,11] We briefly describe here the operation of a ternary tree algorithm.

The principle of operation of a ternary tree CRA is as follows.[10–12] It is assumed that clock time increases from left to right within an upstream frame (as depicted in *Figure 4* and *Figure 5*). Each CM listens to the HC status message for the preceding frame and determines the number of minislots, say *k*, with collisions that were located to the left of the minislot in which it unsuccessfully transmitted its request. The CM skips the next 3**k* minislots (which may be contained in one or more forthcoming frames), and then it retransmits its request by randomly selecting one of the next three minislots. While one group of transmission requests is being resolved for collisions, new requests are held in buffers at the CMs and transmitted only after the current group has been resolved completely. From one frame to the next, the span of minislots over which retransmissions occur spreads wider by three times the number of minislots with collisions. Therefore, over successive frames, the probability of further collisions rapidly decreases within a group of contention requests that are being resolved. When this progressive process of contention resolution finally results in a frame with one or more successful request messages and no collisions in any of the minislots, then the next frame is opened for the waiting new requests to be transmitted. A fresh set of collisions may occur; the operation of the CRA repeats, and a new cycle of collision resolutions occurs.

The first two CRA algorithms described above have the same upper bound on the throughput per minislot, which is the same as that for slotted Aloha (about 37%). The corresponding throughput of a ternary tree CRA can be somewhat higher (up to 50%). The ternary tree and adaptive p-persistence are deemed more desirable for minimizing delay variance and maximizing stability of the system from a traffic viewpoint. A performance comparison in terms of delay, delay variance, stability, etc., between ternary tree and adaptive p-persistence algorithms is provided in the references.[9] The IEEE 802.14 specification (draft) includes a ternary tree CRA as part of its MAC protocol.[14] The MCNS document specifies the use of a random binary exponential back-off algorithm as part of its MAC protocol requirement.[15,16] Because different implementations may use different CRAs, for the purpose of an analytical study of different possible ways to use minislots, we conservatively assume that the throughput efficiency of minislots is 37% (which is achievable by use of any of the CRAs described earlier).

Downstream Frame Structure and Protocol Operation

Each downstream RF channel (e.g., 6 MHz wide) is suitably modulated (e.g., using 64 or 256 quadrature amplitude modification [QAM]) to provide a digital bit stream (e.g., 30 or 40 Mbs). Within each RF channel a frame structure, as shown in *Figure 6*, may be used to organize and schedule the transmission of STM, ATM, and VL information. For the purpose of supporting low delay STM, a 125 μs subframe may be used, and eight successive subframes would constitute a 2 ms downstream frame to match with a 2 ms upstream frame.[3] The framing bits (8 bits) help the CM to synchronize with the downstream frames. If telephony traffic were not supported in a strictly STM mode of transport in the downstream, the downstream framing as shown in *Figure 6* would not be necessary. The downstream synchronization may be derived by a header error check (HEC) sequence in the ATM and VL PDU headers. Additional details with regard to frame organization and upstream/downstream synchronization are provided in the references.[3,14,16]

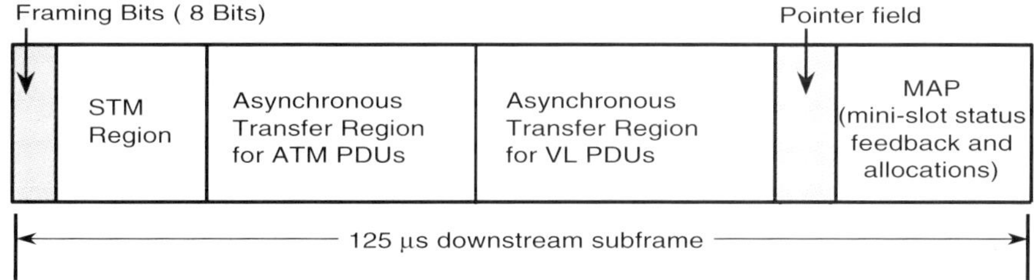

Figure 6: *An Example Downstream Frame Structure*

Example formats of ATM and VL PDUs are shown in *Figure 7* and *Figure 8*. The CM address provided in the ATM and VL PDU headers is used by CMs to pick out and process the PDUs that are intended for them. The asynchronous transfer region for ATM and VL PDUs may be separated into two byte-streams within the larger pipe (of 30 Mb/s), with a boundary as shown in *Figure 6*. This arrangement facilitates easier quality of service (QoS) guarantees to ATM connections without interference due to possibly long VL PDUs. On the other hand, the ATM and VL PDUs may be transmitted within a common asynchronous region or byte-stream, provided that the longer VL PDUs are segmented. The purpose of segmenting the longer VL PDUs would be for avoiding excessive delay jitter for ATM PDUs, which would in turn help meet ATM QoS requirements. The pointer field (PF) in *Figure 6* points to the beginning of the first full VL PDU in the next subframe. This helps as an additional means for synchronization in the downstream within the VL PDU region.

The MAP field in *Figure 6* provides a map of upstream minislots, including the collision status in the prior upstream frame, the number and locations of contention request minislots in the next upstream frame, and information about allocation of upstream minislots to CMs for ATM and VL PDU transmissions.

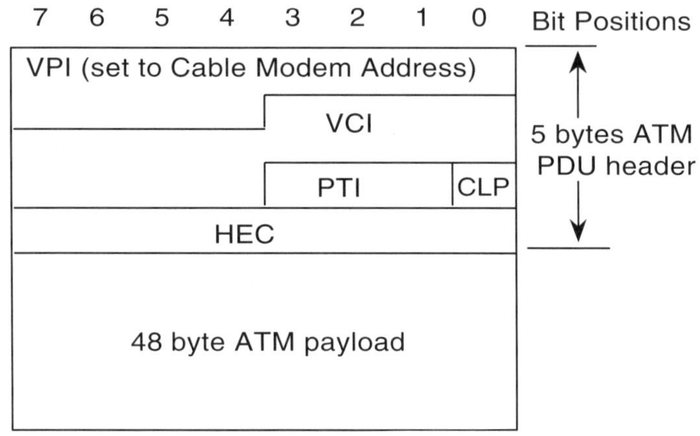

Figure 7: *An Example Downstream ATM PDU Format*

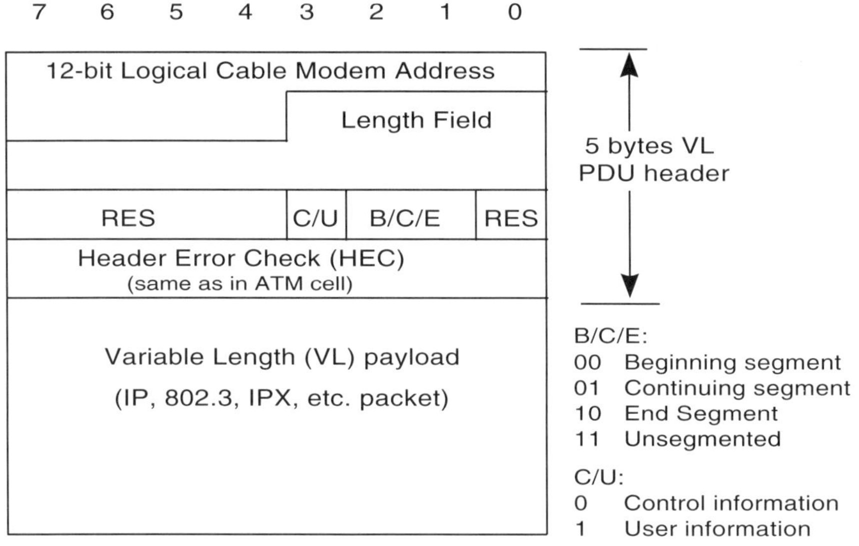

Figure 8: *An Example Downstream VL PDU Format*

Low-delay–requiring synchronous DS0 (e.g., telephony) connections may be carried in one of three possible ways:

(a) **Using an STM byte region in 125 μs subframes**:
One byte per DS0 is transmitted in an STM region within 125 μs subframes (as shown in *Figure 6*). This arrangement results in the lowest possible overhead for STM traffic while providing strictly synchronous DS0 (or n*DS0) service capability. One signaling byte overhead (per 125 μs) is allowed to accommodate the A-H signaling bits for up to 16 DS0 connections.

(b) **Scheduling composite ATM cells every 125 μs**:
One or more ATM cells are scheduled at 125 μs intervals (approximately), and each DS0 connection is allocated one byte in an ATM cell payload. Roughly, a one signaling-byte overhead is required for each group of 16 DS0 bytes to accommodate the A-H signaling bits. Hence, three bytes per ATM cell payload are allowed for signaling. Each DS0 connection is allocated a byte in a fixed position within an ATM cell payload of 48 bytes. Thus, up to 45 DS0 connections can be position–multiplexed in each ATM cell (scheduled at 125 μs intervals). ATM cell scheduling can be done so that the delay jitter does not exceed 125 μs, and the downstream STM delay is bounded to 250 μs.

(c) **Scheduling composite ATM cells every 1 ms**:
One or more ATM cells are scheduled at 1 ms intervals (approximately), and each DS0 connection is allocated eight bytes in an ATM cell payload. Roughly, a one signaling-byte overhead is required for each group of 16 DS0 bytes to accommodate the A-H signaling bits. Hence, three bytes per ATM cell payload are allowed for signaling. Each DS0 connection is allocated eight bytes in a fixed position within an ATM cell payload of 48 bytes. Thus, up to five DS0 connections can be position multiplexed in each ATM cell (scheduled at 1 ms intervals). ATM cell scheduling can be done so that the delay jitter does not exceed 125 μs, and the downstream STM delay is bounded to 1.25 ms.

The next section describes data traffic characteristics and the assumptions used in our analysis and numerical results. In the following two sections, we develop a performance model to compare message delays and bandwidth efficiencies for carrying native protocol data units (IP, 802.3 packets) over ATM vs. VL PDUs in the upstream and downstream channels, respectively. The subsequent section describes, analyses, and numerically compares MAC protocols with no minislots, with a fixed number of minislots, and with a dynamically variable number of minislots. In the final section, we quantitatively compare the bandwidth efficiencies of the three different modes of carrying synchronous DS0 traffic described above.

Traffic Characteristics

For the purpose of this study, the traffic supported by the system is considered to be primarily of two types: 1) reservation-oriented long messages (e.g., file transfers, constant bit-rate [CBR] connections) that require a fixed bandwidth allocation for a relatively long period of time, and 2) random bursty messages that arrive as batches (or bursts) of packets at random intervals. The former typically generate one contention message at the beginning of each transaction to request bandwidth for the entire duration of the transaction. The latter type of traffic sources generate frequent and random contention messages, one for each burst of data packets. In this paper, we refer to the former as reservation-oriented traffic and the latter as contention-oriented bursty traffic.

The contention-oriented bursty traffic is assumed to have different message lengths (in number of packets). For example, when Ethernet frames of sizes 64, 128, 256, 512, 1024, and 1516 bytes are converted to fixed-size ATM cells using AAL-5 adaptation, the messages would be of lengths 2, 3, 6, 11, 22, and 32 ATM cells, respectively. From available traffic measurement studies,[17] it is known that the frequency of Ethernet frames are approximately 0.6, 0.06, 0.04, 0.02, 0.25, and 0.03, for the respective sizes of 64, 128, 256, 512, 1024, and 1516 bytes. The average message length in that case computes to 8.33 packets (ATM cells) per message. The message arrival process is assumed to be Poisson. Thus, the contention-oriented bursty traffic is characterized by a Poisson batch arrival process; the distribution of batch size (in number of packets) is derived based on the MAC PDU payload size information and the message length distribution stated above. The messages may be carried using ATM or VL PDUs. For ATM PDUs, the MAC PDU payload size is 48 bytes. For VL PDUs, the MAC PDU payload is equal to the message size whenever it is smaller than 256 bytes (the maximum PDU payload size permissible for VL PDUs is assumed to be 255 bytes based on upstream FEC limitation). When the message length is larger than 255 bytes, it is segmented and transmitted in multiple VL PDUs.

Upstream Bandwidth Efficiency Comparison for ATM versus VL PDUs

In this section, we develop an analytical model for the upstream message delay. The goal is to compare the performance and throughput efficiency of using ATM PDUs versus. VL PDUs for message transmission over the upstream channel. The MAC protocol based on minislot granularity is flexible so that the messages (higher layer data units, e.g., IP, 802.3 packets) can be carried in either VL PDUs or in ATM PDUs. In the latter case, the messages are segmented using AAL-5. The queuing model developed here is generic and applicable to either case. The comparison between ATM and VL PDU usage is best highlighted if the traffic is assumed to be

100% random bursty messages, i.e., all sources (CMs) produce contention-oriented burst traffic. This would generally be true for the upstream traffic in Internet applications.

The queuing model developed here is based on a combination of delay analysis of contention access for the request messages and an *M/G/1* approximation for message queuing for transmission over an upstream channel.[2,19,20] The upstream message access delay, W_U, is defined as the time from arrival of the first byte of a message at a CM to the time the last byte is received at the HC. This includes the initial contention request delay, the HC processing and queuing delay for issue of the grants, the roundtrip propagation delay, and the message-transmission delay. The HC is assumed to provide grants to the CMs on the basis of receiving successful requests in the minislots on a first-in-first-out (FIFO) basis. Since the CMs are assumed to produce data in the same manner (i.e., Poisson batch arrivals at a certain rate), this analysis also holds for the round-robin service discipline.[20,21]

W_U can be determined as follows by adding the various contributing components:

$$W_U = D_c + a_s * (Q_U + T_U) + \tau_p \qquad (4\text{-}1)$$

where

D_C = request contention access delay (ms).
a_S = average number of PDUs per message (computed from message size distribution and segmentation information).
Q_U = upstream queuing delay at PDU level due to HC scheduling process.
T_U = average PDU transmission time in the upstream.
τ_p = one-way propagation time (ms).

The request contention access delay, D_C, is the delay for a request message to reach the HC successfully. We assume that a random binary exponential algorithm is used for collision resolutions. For a conservative approximation for W_U, we use a high percentile value of D_C, such as the mean plus five times the standard deviation. We use the results presented in an earlier paper to obtain the following approximation for D_C:[2]

$$D_c = 2 * F * \left(1 + \frac{3 * \tilde{n}}{2}\right) \qquad (4\text{-}2)$$

where

F = upstream frame duration (2 ms).

ρ = upstream channel utilization due to all sources.

In deriving (4-2), we use data from the references corresponding to a message size equal to 1, and knowledge that the x-axis there should be scaled up by at least three times to convert from throughput in minislots to data utilization, ρ.[2] For a more detailed understanding of the relationship between minislots throughput and data message throughput, please see the references.[4]

We have not included the effect of piggybacking requests in the model developed here. We assume that each message generates one request individually. Piggybacking requests would help improve the performance for ATM PDUs as well as VL PDUs by lowering the contention access delay component, D_C.

Let us define the following parameters for use in deriving an equation for Q_U:

R = source rate (kb/s).
P = average message size without overhead (bytes).
P_S = average PDU size including overhead (bytes).
P_X = average PDU size without overhead (bytes).
$a_S = P/P_X$ = average number of PDUs per message.
λ = average message rate per source (messages/ms).
S_U = average PDU service time in the upstream.
ρ = upstream channel utilization per RF channel.
C = upstream aggregate channel bandwidth over all RF channels available for data (kb/s).
C_1 = upstream channel bandwidth per RF channel (kb/s).
C_m = upstream channel bandwidth used per RF channel for provisioning minislots (kb/s).
$g_m = C_m / C_1$ = fraction of upstream channel bandwidth used for provisioning minislots.
β = coefficient of variation of PDU service time (derived from PDU size distribution).
σ_U= standard deviation of upstream message delay (ms).

The following equations hold:

$$\lambda = \frac{R}{8 * P} \text{ (packets per ms)} \tag{4-3}$$

$$S_U = \frac{8 * P_s}{C * (1 - g_m)} \text{ (ms)} \tag{4-4}$$

$$\rho = \lambda * S_U * N = \frac{R * P_s * N}{P * C * (1 - g_m)} \tag{4-5}$$

$$T_U = \frac{8 * P_s}{C_1 * (1 - g_m)} \text{ (ms)} \tag{4-6}$$

Now using an *M/G/1* queuing model at the PDU level,[19,20] we obtain the following set of equations for upstream performance measures of interest (i.e., Q_U, W_U, and σ_U):

$$Q_U = \frac{\tilde{n} * S_U * (1 + \hat{a}^2)}{2 * (1 - \tilde{n})} \tag{4-7}$$

$$W_U = 2 * F * \left(1 + \frac{3 * \tilde{n}}{2}\right) + a_s * \left(\frac{\tilde{n} * S_U * (1 + \hat{a}^2)}{2 * (1 - \tilde{n})} + \frac{8 * P_s}{C_1 * (1 - g_m)}\right) + \tau_p \text{ (ms)} \tag{4-8}$$

$$\sigma_U \approx a_s * Q_U * \sqrt{\frac{4 - \tilde{n}}{3 * \tilde{n}}} \text{ (ms)} \tag{4-9}$$

Numerical Examples for Comparison of ATM vs. VL PDUs in Upstream

Messages belonging to random bursty traffic may originate from home computers with an interface such as a 10baseT (802.3) to the CMs. These messages can be transported using the VL PDUs or ATM PDUs (with AAL-5 adaptation in the latter case). When these packets are segmented into smaller ATM PDUs, the upstream bandwidth efficiency suffers because of a) partially filled ATM cells, b) ATM header overhead, and c) repeated PHY & MAC layer overheads for each segment. The VL mode would also require segmentation for payloads over 256 bytes because of FEC considerations, but this level of segmentation is much less compared to the segmentation required by the ATM mode. For these reasons, the overheads for carrying the messages in upstream channels are much lower for VL PDUs than for ATM PDUs (see *Figure 9* and *Table 1*). The upstream bandwidth efficiencies for ATM and VL PDUs as a function of message sizes are compared in *Figure 10*. The average bandwidth efficiencies for ATM and VL PDUs are also shown in *Figure 10* (see the rightmost pair of bars).

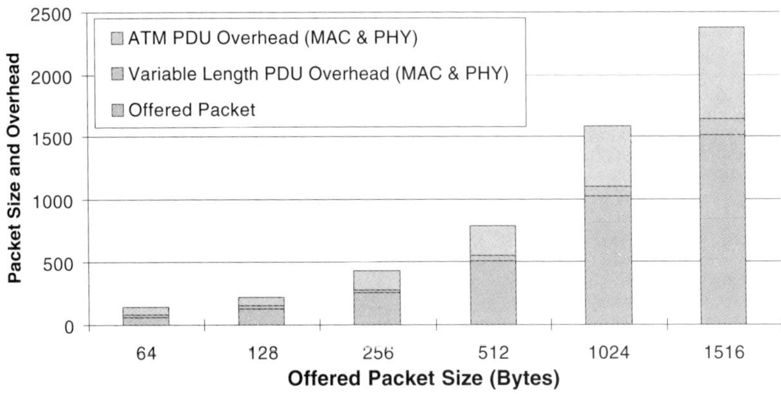

Figure 9: *Comparison of Upstream Overheads for Carrying 802.3 Packets in VL PDUs vs. ATM PDUs*

- ATM cell PDUs: Payload = 48 bytes; OH (PHY + MAC + ATM) = 24 bytes
- Variable Length PDUs: Maximum payload = 256 bytes; Total OH = 20 bytes

Offered Packet (Bytes)	Variable PDU segments including MAC&PHY OH and minislot quantization OH (bytes)	ATM PDU segments including MAC&PHY OH and AAL-5 segmentation OH (bytes)
64	84 (7 minislots)	144 (2*6 = 12 minislots)
128	156 (13 minislots)	216 (3*6 = 18 minislots)
256	276 (23 minislots)	432 (6*6 = 36 minislots)
512	2*276 (2*23 = 46 minislots)	792 (11*6 = 66 minislots)
1024	4*276 (4*23 = 92 minislots)	1584 (22*6 = 132 minislots)
1516	5*276 + 264 (5 *23+22 = 137 minislots)	2304 (32*6=192 minislots)

Table 1: *Packet Segmentation and Overhead Comparisons for ATM and VL PDUs (Upstream)*

The upstream message access delay is defined as the time from arrival of the first byte of a message (or packet) at a CM to the time the last byte is received at the HC. This includes the initial contention request delay, the HC processing and queuing delay for issue of the grants, the roundtrip propagation delay, and the packet transmission delay. The delay plotted in *Figure 11* is the mean plus five times the standard

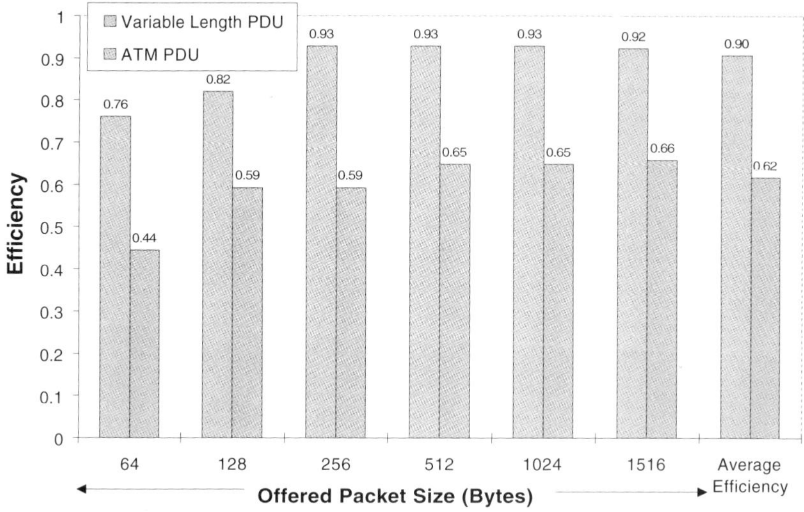

Figure 10: *Comparison of Upstream Message Access Delays for VL vs. ATM PDUs*

deviation of message access delay. The total upstream bandwidth is assumed to be 9 Mbs (three RF channels each providing 3 Mbs bandwidth). Clearly, VL PDUs facilitate lower delays as well as higher throughput. If the delay requirement were 20 ms, then from *Figure 11* it can be seen that VL PDUs allow a per user throughput of 120 kbs, whereas ATM PDUs would allow only 80 kbs (for 50 users sharing 9 Mbs). For the upstream channels, VL PDUs provide about 50% higher throughput and capacity than ATM PDUs for the random bursty traffic considered here.

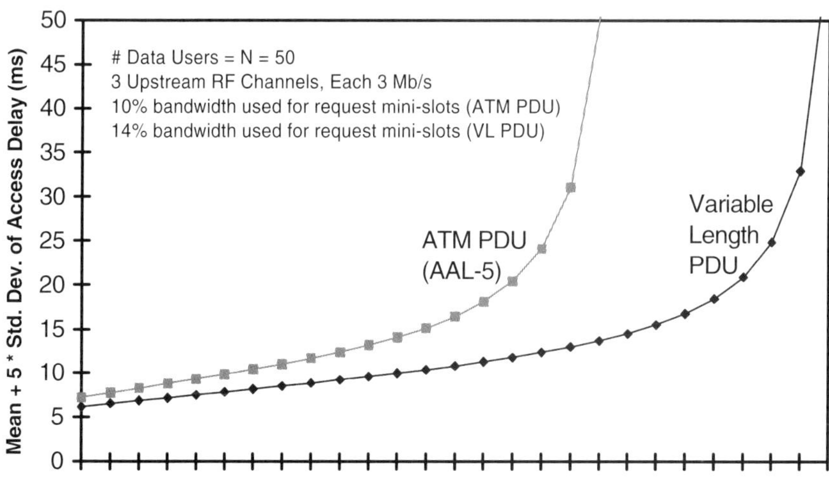

Figure 11: *Comparison of Upstream Message Access Delays for VL vs. ATM PDUs*

Downstream Bandwidth Efficiency Comparison for ATM versus VL PDUs

In this section, we develop an analytical model for the downstream message delay. The goal is to compare the performance and throughput efficiency of using ATM PDUs versus VL PDUs for message transmission over the downstream channel. The MAC protocol is flexible so that the messages (higher layer data units, e.g., IP, 802.3 packets) can be carried in the downstream either in VL PDUs or in ATM PDUs. In the latter case, the messages are segmented using AAL-5. The queuing model developed here is generic and applicable to either case. The comparison between ATM and VL PDU usage is best highlighted if the traffic is assumed to be 100% random bursty traffic. In other words, all CMs receive bursty message traffic (e.g., randomly generated IP type of messages as in Internet applications). The messages destined for the CMs are queued, and the HC forwards them to the CMs using a FIFO service discipline.

The queuing model developed here is based on an *M/G/1* approximation for message queuing for transmission over a downstream channel. The downstream message access delay, W_D, is defined as the time from arrival of the first byte of a message at the HC to the time the last byte is received at the destination CM.

W_D can be determined as follows by adding the various contributing components:

$$W_D = a_S * (Q_D + T_D) + \tau_p \qquad (5\text{-}1)$$

where

a_S = average number of PDUs per message (computed from message size distribution and segmentation information).
Q_D = downstream queuing delay at PDU level at the HC.
T_D = average PDU transmission time in the downstream.
τ_p = one-way propagation time (ms).

Let us define the following parameters for use in derivation of an equation for Q_D:

R = average source rate (same for each receiving CM) (kb/s).
P = average message size without overhead (bytes).
P_S = average PDU size including overhead (bytes).
P_X = average PDU size without overhead (bytes).

$a_S = P/P_X$ = average number of PDUs per message.
λ = average message rate per source (messages/ms).
S_D = average PDU service time.
ρ = downstream channel utilization per RF channel.
C = downstream aggregate channel bandwidth over all RF channels available for data (kbs).
C_1 = downstream channel bandwidth per RF channel (kbs).
f_e = fraction of bandwidth used for downstream error protection.
f_c = fraction of bandwidth used for downstream control and management.
β = coefficient of variation of PDU service time (derived from PDU size distribution).
σ_D = standard deviation of message delay (ms).

The following equations hold:

$$\lambda = \frac{R}{8 * P} \quad \text{(packets per ms)} \tag{5-2}$$

$$S_D = \frac{8 * P_s}{C * (1 - f_e - f_c)} \quad \text{(ms)} \tag{5-3}$$

$$\rho = \lambda * S_D * N = \frac{R * P_s * N}{P * C * (1 - f_e - f_c)} \tag{5-4}$$

$$T_D = \frac{8 * P_s}{C_1 * (1 - f_e - f_c)} \quad \text{(ms)} \tag{5-5}$$

Using an *M/G/1* queuing model at the PDU level,[19,20] we obtain the following set of equations for performance measures of interest, i.e., Q_D, W_D, and σ_D:

$$Q_D = \frac{\rho * S_D * (1 + \beta^2)}{2 * (1 - \rho)} \quad \text{(ms)} \tag{5-6}$$

$$W_D = a_s * \left(\frac{\rho * S_D * (1 + \beta^2)}{2 * (1 - \rho)} + \frac{8 * P_s}{C_1 * (1 - f_e - f_c)} \right) + \tau_p \quad \text{(ms)} \tag{5-7}$$

$$\sigma_D \approx a_s * Q_D * \sqrt{\frac{4 - \rho}{3 * \rho}} \quad \text{(ms)} \tag{5-8}$$

Numerical Examples for Comparison of ATM vs. VL PDUs in Downstream
Messages belonging to random bursty traffic may also originate from an Internet content provider. These messages require transmission in the downstream direction to reach their destination CMs. The messages can be transported downstream on a CM using the VL PDUs or ATM PDUs (with AAL-5 adaptation in the latter case). When data message traffic is subjected to segmentation into smaller ATM PDUs (AAL-5 adaptation), the downstream bandwidth efficiency suffers because of a) partially filled ATM cells, and b) ATM header overhead. The VL mode may also require some segmentation for meeting the QoS for other (e.g., CBR) traffic that may also be present in the downstream. However, the effect of VL PDU segmentation on the total packet overhead is small compared to that of partially filled ATM cells. For these reasons, the overheads for carrying data messages in downstream channels are much lower with VL PDUs than with ATM PDUs (see *Figure 12* and *Table 2*). The width of the VL PDU overhead in *Figure 12* is almost invisible because it is very small compared to the packet size. The downstream bandwidth efficiencies for ATM and VL PDUs as a function of packet sizes are compared in *Figure 13*. The average bandwidth efficiencies for ATM and VL PDUs (averaged over the packet size distribution described earlier) are also shown in *Figure 13* (see the rightmost pair of bars).

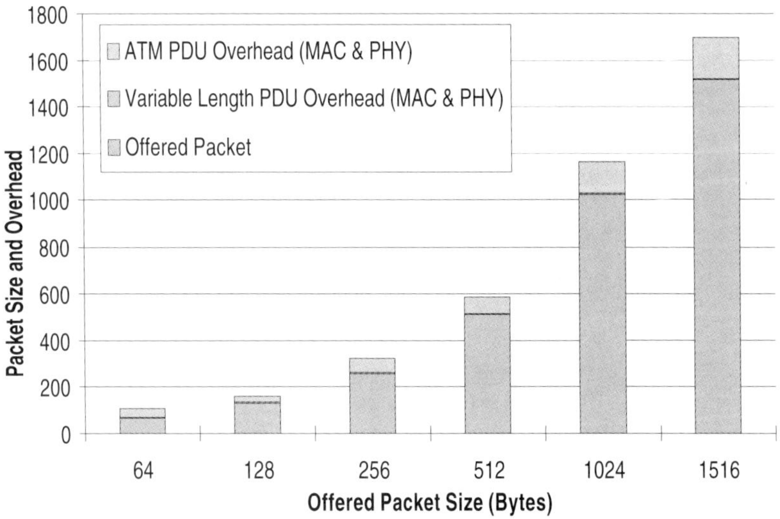

Figure 12: *Comparison of Downstream Overheads (in Bytes) for Carrying Messages in VL vs. ATM PDUs*

- ATM cell PDUs: Payload = 48 bytes; OH (MAC + ATM) = 5 bytes
- Variable Length PDUs: No limit on maximum payload; MAC OH = 5 bytes

Offered Packet (bytes)	Variable Length PDUs (bytes)	ATM PDUs (AAL-5) (bytes)
64	69	2 * 53 = 106
128	133	3 * 53 = 159
256	261	6 * 53 = 318
512	571	11 * 53 = 583
1024	1029	21 * 53 = 1113
1516	1521	32 * 53 = 1696

Table 2: *Packet Segmentation and Overhead Comparisions for ATM and VL PDUs (Downstream)*

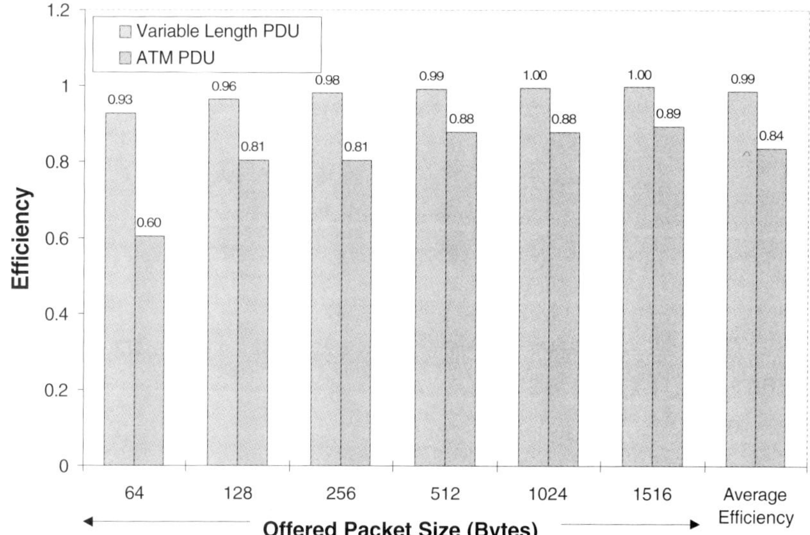

Figure 13: *Bandwidth Efficiencies as a Function of Packet Sizes for ATM and VL PDUs (Downstream)*

The downstream message access delay is defined as the time from arrival of the first byte of a message (or packet) at the HC to the time the last byte (of the same message) is received at the CM. This includes the HC processing and queuing delay, the one-way downstream propagation delay, and the packet transmission delay. The delay plotted in *Figure 14* is the mean plus five times the standard deviation of message access delay. The downstream bandwidth is assumed to be 10 Mb/s (e.g., using QPSK in one 6 MHz RF channel), and the FEC and control/management overheads are assumed to be 10%. Hence, the ($fe + fc$) term in equation (5-7) is equal to 0.1. So the available bandwidth for data traffic is 9 Mbs. As shown in *Figure 14*, VL PDUs facilitate lower delays and higher throughput. With the use of VL PDUs for the bursty message traffic (described earlier in this paper), the system saturates at a per-user-throughput value that is approximately 15% higher compared to that for ATM PDUs (see *Figure 14*).

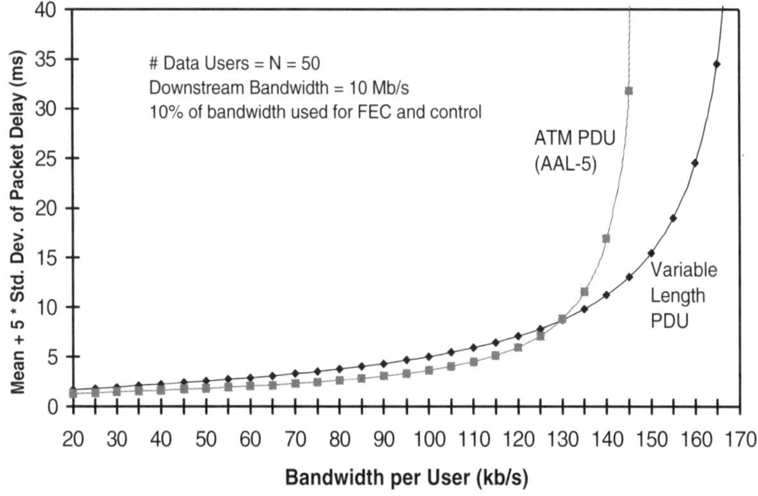

Figure 14: *Comparison of Downstream Message Access Delays for VL vs. ATM PDUs*

Upstream Capacity Enhancement Due to Dynamic Request Minislots

In this section, we describe the operation of MAC protocols for three cases regarding the use of minislots in the upstream channel. It is assumed that the information packets are all of a fixed size (e.g., ATM cells) for the purpose of this study. For the remainder of this paper, we use the term "data packet" instead of "information packet" for ease, with the understanding that "data" may mean voice, computer data, or video information.

MAC Protocol Without Minislots

Figure 15 shows an upstream frame for a MAC protocol that does not use minislots. In this scheme, when a batch or burst of data packets (or ATM cells) arrives at a CM, the first packet of the burst is transmitted in a contention mode using a contention time-slot (denoted by *C*). A request for bandwidth for the remainder of the packets of that burst is transmitted along with the first packet (also known as piggybacking requests). When the request is granted by the HC, the subsequent packets of the burst are transmitted in a reservation mode, in time-slots denoted by *R* in *Figure 15*, as per allocation by the HC. In the case of a reservation-oriented transaction, a connection set-up request is made by sending a request message in one of the *C* slots. Subsequently, the call is admitted, depending on bandwidth availability, and reserved *(R)* slots are periodically allocated to the call for its duration. The frame duration in *Figure 15* is shown to be 2 ms, as an example. The roundtrip propagation delay accounts for about 1 ms, and about 1 ms is budgeted for transmission and HC/CM processing times. At an upstream channel rate of 2.56 Mbs, typically there would be about 10 data (ATM cell) time-slots in a 2 ms frame duration. As shown in *Figure 15*, at least one time-slot in each frame is designated for contention access so that new call requests and other signaling messages may be transmitted without undue delay.

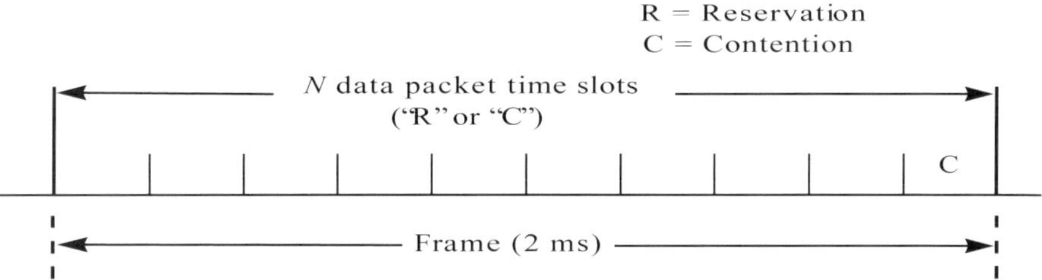

Figure 15: *Upstream Frame Without Minislots*

MAC Protocol with Fixed Number of Minislots

Figure 16 illustrates an upstream frame for a MAC protocol that operates with a fixed number of minislots. Let N represent the total number of data (packet) time-slots per frame. The value of N is a function of the frame duration, the physical link bandwidth, and the packet size. Out of a total of N data time-slots in a frame, the duration (or span) of j_f data slots is designated by the HC for use as minislots. Thus the number of minislots is $k * j_f$ per frame. It is assumed here that a data time-slot is the same size (i.e., duration) as k contention request minislots ($k = 4$ in the example of *Figure 16*). The value of k depends on the PHY layer overheads (guardtime, preamble, FEC, etc.) and the length of the request message transmitted in a minislot. For an upstream transmission rate of 2.56 Mbs, typically four to six minislots can be accommodated in the time duration of one data (ATM cell) time-slot. In *Figure 16*, n_1 is the number of time-slots used by the reservation-oriented traffic, and hence typically the value of n_1 remains constant for a long duration. The number of time-slots available for contention-oriented bursty data packets is n_2. The value of j_f is determined as a function of N and k only and is not adjusted when n_1 changes.[18] Hence, we refer to this scheme as a MAC protocol with fixed number of minislots.

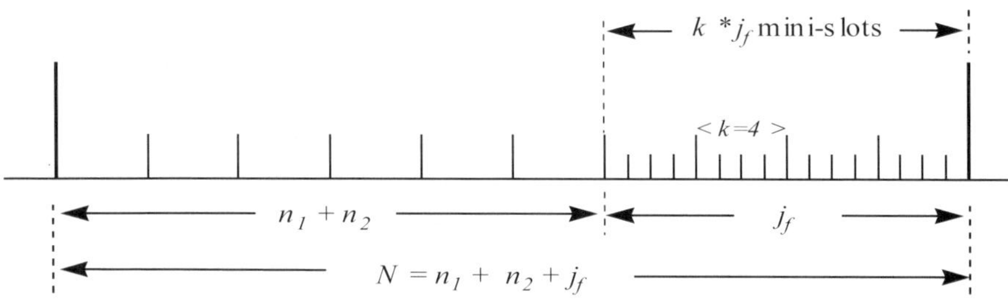

Figure 16: *Upstream Frame with a Fixed Number of Minislots*

MAC Protocol with Variable Number of Minislots

Figure 17 illustrates an upstream frame for a MAC protocol that uses a variable number, j_v , of minislots. In this scheme, the number of request minislots , j_v , is a function of N , k , and n_1, and hence varies with the number of packet time-slots, n_1 , used by reservation-oriented traffic. Hence, j_v may dynamically vary from one frame to the next as a function of the bandwidth currently in use by the reservation-oriented traffic. The potential benefits of this scheme may be qualitatively explained as follows (the quantitative comparisons will be provided in subsequent sections). When a greater number of packet time-slots is being used in a reservation mode for a relatively long period of time, fewer minislots are needed for contention requests. Just enough minislots should be created so that the average throughput (per frame) of the contention request minislots closely matches the number of available data time-slots (n_2 per frame) for the contention-oriented bursty messages. An algorithm for varying the number of minislots that is based on maintaining this dynamic balance will be described later in this paper.

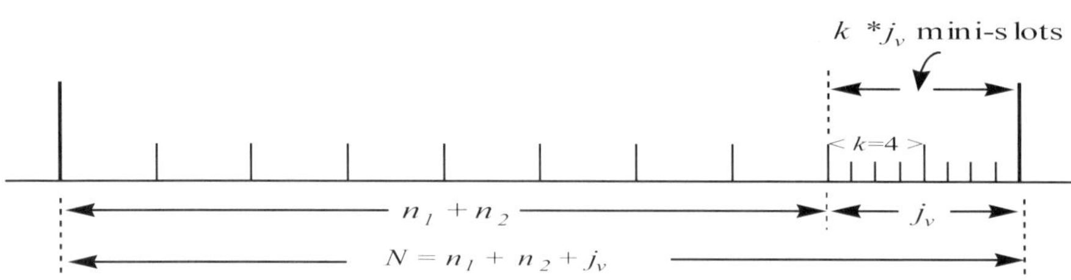

Figure 17: *Upstream Frame with a Variable Number of Minislots*

Throughput Efficiency Analysis

Let us define the following parameters to assist in the analytical derivations:

α = PHY efficiency (including effects of guardtime, preamble, FEC, etc.).

N = total number of packet time-slots per frame (including those used for creation of minislots).

k = number of minislots that can be accommodated within the duration of a packet time-slot (k depends on α and quantization effects thereof).

n_1 = number of packet time-slots in use by reservation-oriented traffic (long holding time).

n_2 = number of packet time-slots available for bursty messages.

$$= \begin{cases} N - n_1 & \text{for protocol with no minislots} \\ N - n_1 - j_f & \text{for protocol with fixed number of minislots} \\ N - n_1 - j_v & \text{for protocol with variable number of minislots} \end{cases}$$

E_a = throughput efficiency of all traffic for the MAC protocol without minislots.

E_f = throughput efficiency of all traffic for the MAC protocol with fixed number of minislots.

E_v = throughput efficiency of all traffic for the MAC protocol with variable number of minislots.

S_a = throughput efficiency of bursty messages for the MAC protocol without minislots.

S_f = throughput efficiency of bursty messages for the MAC protocol with fixed number of minislots.

S_v = throughput efficiency of bursty messages for the MAC protocol with variable number of minislots.

u_1 = throughput of reservation-oriented traffic. (%) $= \dfrac{100 * n_1 * \alpha}{N}$

B = average bursty message length including segmentation and PHY/MAC overheads (bytes).

B_a = size of a packet (one data PDU) including PHY/MAC overheads (bytes),

ℓ = average number of packets per bursty message = $\dfrac{B}{B_a}$

Determination of Number of Minislots

In this section, we present the methodology for determining the number of request minislots. The request messages are subject to collisions. A suitable CRA facilitates successful retransmission in the event of collision(s). With the use of a simple CRA, such as the random binary exponential back-off algorithm, the theoretical throughput efficiency of a minislot is 37%. For practical reasons, primarily based on delay considerations, the efficiency of each minislot would be closer to 0.9 * 37% = 33.3%.[2,9] Three request minislots are needed to achieve 100% throughput efficiency in an associated data time-slot (for random bursty messages of size one packet each).

MAC with Variable Number of Request Minislots: Based on the preceding observations, the ratio of the number of minislots to the number of data slots should be at most three to one. This ratio can be smaller when the bursty messages are multi-packet messages. When a bursty message consists of multiple packets, only the first packet generates a contention request, and when that request successfully reaches the HC, the rest of the packets of the multi-packet message are transmitted based on reservations allocated by the HC. Given that the average bursty message length is ℓ, on average ℓ data packet time-slots are utilized for each request message that successfully makes it through the minislot contention process. However, the bursty messages vary randomly in size and are often smaller than ℓ. Hence, we introduce a multiplication factor, f_m ($f_m < 1$), for conservative use of the benefit of ℓ. We compute an effective message length, ℓ_e, (in number of packets or MAC PDUs) as follows:

$$\ell_e = max\left\{\lfloor \ell * f_m \rfloor, 1\right\} \qquad (6\text{-}1)$$

We set $f_m = 0.5$ for the numerical examples so that when $\ell \geq 4$, ℓ_e helps to reduce the number of minislots needed per frame.

Further, in computing the number of minislots, it is recognized that even one minislot per frame is sufficient for the request messages originating from reservation-oriented traffic. Such requests are very sparse compared to those originating from bursty messages because one reservation is made per reservation–based transaction, which lasts a long time. At least one data time-slot is always designated by the HC for use as k minislots. Hence, it is safe not to require any additional minislots to be created for the reservation-oriented traffic.

Algorithm for Computing the Variable Number of Minislots (j_v): Based on the insights presented so far into the nature of the relationship between minislot throughput and data slot throughput, the following equations represent a procedure for computing the number, j_v, of packet time-slots that are designated by the HC to be used as k*

j_v minislots for the variable minislot case. These equations help optimize the throughput for the bursty message traffic as a function of n_1, N, k, and ℓ_e:

$$\text{Let } j_x = max\left\{\left\lceil \frac{3(N-n_1)}{3+k*\ell_e} \right\rceil, 1\right\}, \quad [A]$$

$$\text{and } P = Logical\{j_x \geq 2 \, and \, 0.33*(j_x-1)*k*\ell_e \geq N-n_1-j_x\}, \quad [B] \qquad (6\text{-}2)$$

then,

$$j_v = \begin{cases} j_x & \text{if P = FALSE,} \\ j_x - 1 & \text{if P = TRUE} \end{cases} \quad [C]$$

In equation 6-2A, an initial value for j_x is computed based on the observations made earlier. This equation is further refined in equations 6-2B and 6-2C, by noting that if j_x is at least two and the throughput (in equivalent packet units) of requests per frame for bursty messages via $(j_x - 1) * k$ contention minislots exceeds the number of data time-slots, $N - n_1 - j_x$, available for bursty messages (for the case of $j_x * k$ minislots), then the value computed in equation 6-2A is recognized to be a little too high. Accordingly, an adjustment is made in 6-2C by decrementing the value obtained in 6-2A by one to arrive at the final value for j_v. If this adjustment step were not followed, then it is possible that the request throughput per frame would frequently exceed the number of available data time-slots for the bursty messages. Thus the data queue for bursty messages could back up and overflow into the next several upstream frames.

Figure 18 illustrates that for $n_1 = 5$, $N = 10$, $k = 4$ and $\ell_e = 1$, we would initially compute a value 3 for j_x following equation 6-2A. However, it is recognized that for the value of $j_v = 3$, the minislot request throughput is 3.96 while the maximum available bursty message packet throughput is smaller at $n_2 = 2$ packets per frame. However, for $j_v = 2$, the maximum available bursty message packet throughput is three while the request throughput is 2.64. The actual packet throughput for the bursty message traffic would be the smaller of these two quantities. Thus, with $j_v = 2$, the throughput in consideration is minimum (3, 2.64) = 2.64 packets per frame (see the lower part of *Figure 18*), which is better than that with $j_v = 3$, since minimum (2, 3.96) = 2 (see the upper part of *Figure 18*).

It may also be noted that if the distribution of bursty message sizes is not known *a priori* but estimated by traffic measurement during system operation, then the algorithm of equation (6-2) can still be effectively used by incorporating the most recent measurement estimate of ℓ_e. In doing so, the system performance can be dynamically tuned to the changing nature of bursty message traffic characteristics over time.

MAC with Fixed Number of Request Minislots: The following somewhat simplistic rationale may be followed to approximately estimate the number of minislots per frame for a MAC with a fixed number of minislots. Three request minislots per data

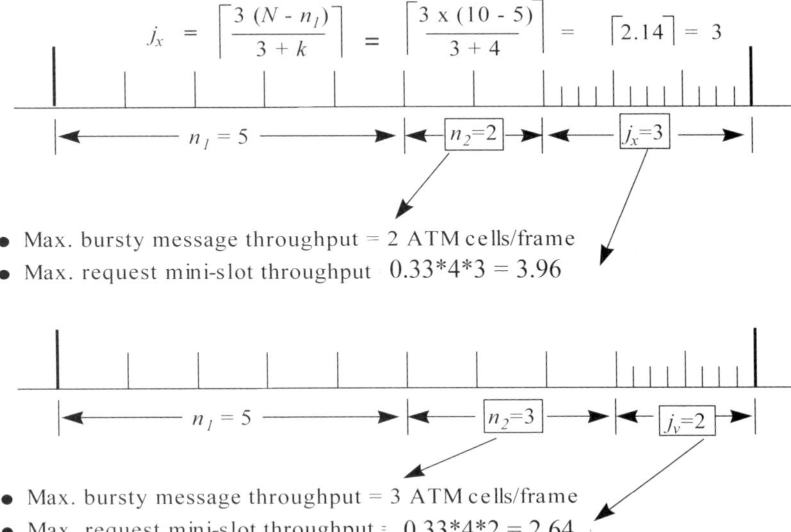

Figure 18: *How the Refinement in the Value of j_v Using Equations 6-2B and 6-2C Helps in Fine-Tuning the Throughput Efficiency (for n_1=5, N=10, k=4, and ℓ_e =1)*

time-slot would be required if the traffic were 100% purely sporadic traffic with single packet messages ($\ell = 1$). However, some of the traffic is reservation–oriented with long transaction time, and some of it is bursty messages with a mix of message lengths. Therefore, it may be reasoned that assigning two minislots per data slot would be sufficient.[18] By doing so, the overhead due to the presence of minislots is better compared to the case of the three-to-one ratio. Hence, the following equation ensues for the number, j_f, of packet time-slots that are designated for use as $k*j_f$ minislots (per frame):

$$j_f = max \left\{ \left\lceil \frac{2N}{2+k} \right\rceil , 1 \right\} . \qquad (6\text{-}3)$$

Throughput Efficiency Results

MAC with No Request Minislots: In this case, the MAC operates essentially like CSMA/CD regarding throughput. Due to the presence of multi-packet bursty messages, the throughput efficiency is better than that of slotted Aloha (with random binary exponential back-off algorithm for collision resolution). The average efficiency, β, for the contention time-slots allocated to bursty messages is as follows:

$$\beta = \sum_i p_i * \eta_i , \tag{6-4}$$

where p_i are the probabilities describing the frequency of various message lengths and i are the contention throughput efficiencies (similar to those for the CSMA/CD scheme) corresponding to different message lengths.[2,22–24] The message lengths shown earlier are 2, 3, 6, 11, 22, and 32 packets (ATM cells) with probability distribution given by 0.6, 0.06, 0.04, 0.02, 0.25, and 0.03. The corresponding values of i for these message lengths obtained from the analysis in the references are approximately 0.4, 0.45, 0.55, 0.65, 0.7, and 0.8.[2] The value of β now computes to about 0.5, or 50%. Taking the PHY efficiency, α, into account, the throughput efficiency, S_a, of bursty messages for the MAC protocol without minislots is as follows:

$$S_a = \frac{(N - n_1) * \beta * \alpha}{N} \tag{6-5}$$

Further, the throughput efficiency, E_a , of the combined (reservation-oriented plus random bursty message) traffic for the MAC protocol without minislots is as follows:

$$E_a = \frac{n_1 * \alpha}{N} + S_a = \frac{[n_1 + (N - n_1) * \beta]\alpha}{N} \tag{6-6}$$

MAC with Fixed Number of Request Minislots: The throughput efficiency, S_f , of the random bursty messages for the MAC protocol with a fixed number of mini-slots is as follows:

$$S_f = \frac{min\{(N - n_1 - j_f), 0.33 * k * j_f * \ell_e\}\alpha}{N} \tag{6-7}$$

The throughput efficiency, E_f , of the combined (reservation-oriented plus random bursty message) traffic for the MAC protocol with fixed number of minislots is as follows:

$$E_f = \frac{n_1 * \alpha}{N} + S_f = \frac{[n_1 + min\{(N - n_1 - j_f), 0.33 * k * j_f * \ell_e\}]\alpha}{N} \tag{6-8}$$

MAC With Variable Number of Request Minislots: The throughput efficiency, S_v, of the random bursty messages for the MAC protocol with variable number of mini-slots is as follows:

$$S_v = \frac{min\{(N - n_1 - j_v), 0.33 * k * j_v * \ell_e\}\alpha}{N} \qquad (6\text{-}9)$$

The throughput efficiency, E_v, of the combined (reservation-oriented plus random bursty message) traffic for the MAC protocol with variable number of minislots is as follows:

$$E_v = \frac{n_1 * \alpha}{N} + S_v = \frac{[n_1 + min\{(N - n_1 - j_v), 0.33 * k * j_v * \ell_e\}]\alpha}{N} \qquad (6\text{-}10)$$

Numerical Examples

The parameter values used for generating numerical results that are shown in *Figure 19*, *Figure 20*, and *Figure 21* are as follows: $N = 10$, $\alpha = 0.8$, $k = 5$, $\beta = 0.5$, $\ell = 8$, $f_m = 0.5$, and $\ell_e = f_m * \ell = 4$. The numbers shown in *Figure 19*, namely j_f and j_v, should be multiplied by k to get the number of minislots used per frame for the cases of MAC with fixed number and MAC with variable number of minislots, respectively. The values of j_v and j_f shown in *Figure 19* are computed in accordance with equations 6-2 and 6-3, respectively. As the reservation-oriented traffic increases, the MAC with variable minislots dynamically varies the number of minislots so as to maximize the throughput for the bursty message traffic in the time-slots that are available for the latter traffic.

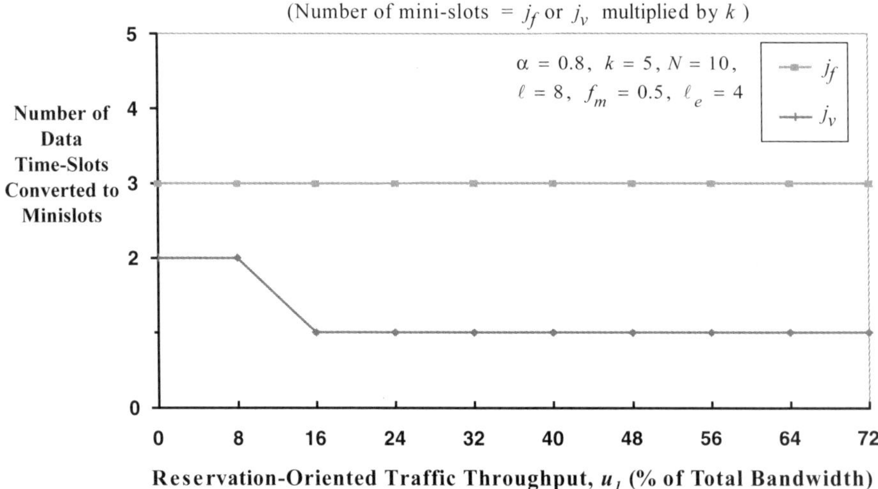

Figure 19: *The Number of Packet Time-Slots Converted to Request Minislots as a Function of the Throughput Used by Reservation-Oriented Traffic*

Figure 20 and *Figure 21* show the throughput efficiencies for the bursty messages and combined traffic, respectively. Clearly, the MAC protocol with variable minislots provides the best throughput efficiency because it dynamically adjusts the number ($j_v * k$) and hence the overhead due to the presence of minislots as a function of the given mix of traffic. The plots for the case of MAC protocol with fixed minislots (denoted by S_f in *Figure 20* and E_f in *Figure 21*) do not extend all the way to the right because three time-slots out of ten are always given over to creating minislots. Accordingly, the number of time-slots that can be allocated to reservation-oriented traffic is limited to seven out of ten. In the case of a MAC protocol with a variable number of minislots, it is possible to lower the value of j_v all the way to one when there is a high amount of reservation-oriented traffic in the system. Hence, the plots for the case of variable minislots (denoted by S_v in *Figure 20* and E_v in *Figure 21*) extend farther out to higher values of reservation-oriented traffic throughput.

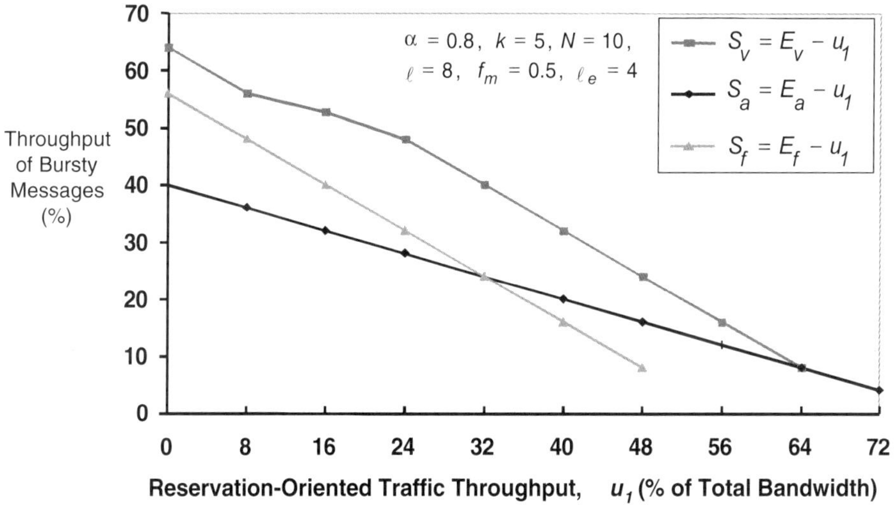

Figure 20: *Throughput Efficiency of Random Bursty Traffic as a Function of the Throughput Used by Reservation-Oriented Traffic*

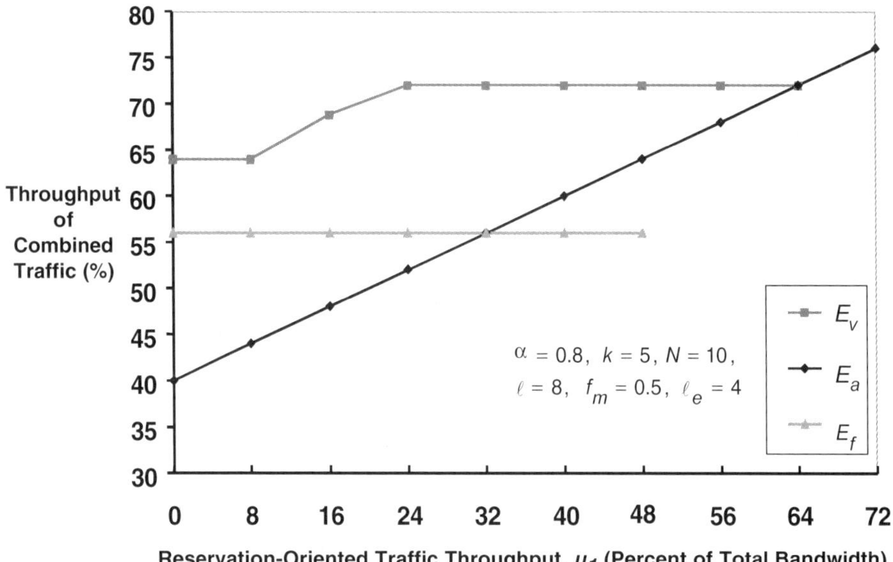

Figure 21: *Throughput Efficiency of the Combined (Reservation-Oriented Plus Random Bursty) Traffic as a Function of the Throughput Used by Reservation-Oriented Traffic*

Delay and Efficiency Comparisons for Alternative Methods of Supporting STM in Downstream

Earlier in the paper, three methods were described for supporting DS0 traffic in the downstream for telephony applications. Let n denote the number of voice calls (64 kb/s DS0 connections) that are currently in progress. Let V_a, V_b, and V_c denote the downstream bandwidths for supporting n voice calls for the three methods (a), (b), and (c). Including the effects of signaling and ATM overheads as described earlier, the following equations hold for V_a, V_b, and V_c:

(a) Using an STM byte region in 125 µs subframes:

$$V_a = \left(\left\lceil \frac{n}{16} \right\rceil + n \right) * 64kbps \tag{7-1}$$

This method is the usual strict STM approach. The fixed downstream voice delay for this case is 125 µs.

(b) Scheduling composite ATM cells every 125 μs:

$$V_b = \left\lceil \frac{n}{45} \right\rceil * \left(\frac{53*8}{0.125} \right) kbps \qquad (7\text{-}2)$$

The voice delay for this case is 250 μs, including the receiver buffer play-out delay for smoothing over the limited delay jitter that may happen, due somewhat to the asynchronous nature of ATM CBR service.

(c) Scheduling composite ATM cells every 1 ms:

$$V_c = \left\lceil \frac{n}{5} \right\rceil * (53*8)\ kbps \qquad (7\text{-}3)$$

The voice delay for this case is 1.25 ms, including the receiver buffer play-out delay for smoothing over the limited delay jitter that may happen, due somewhat to the asynchronous nature of ATM CBR service.

Numerical Results on Bandwidth Efficiencies for Telephony in Downstream

The comparison of bandwidth needed for the three approaches of supporting telephony in the downstream is shown in *Figure 22*. Method (a) has the least delay and most bandwidth efficiency. However, one may use composite ATM cells as in methods (b) and (c) to support DS0 connections while avoiding a synchronous 125 μs framing structure in the downstream. If such is the case, it is evident from *Figure 22* that method (b) is very bandwidth inefficient when the voice call volume is low. If about 2 ms (1.25 ms due to buffering plus the rest for propagation delay) of the available access delay budget for voice can be allocated for the downstream channel, then method (c) would be desirable over (b). If echo cancellors are used in the access loop, then the delay budget of 2 ms can be easily allocated to downstream.

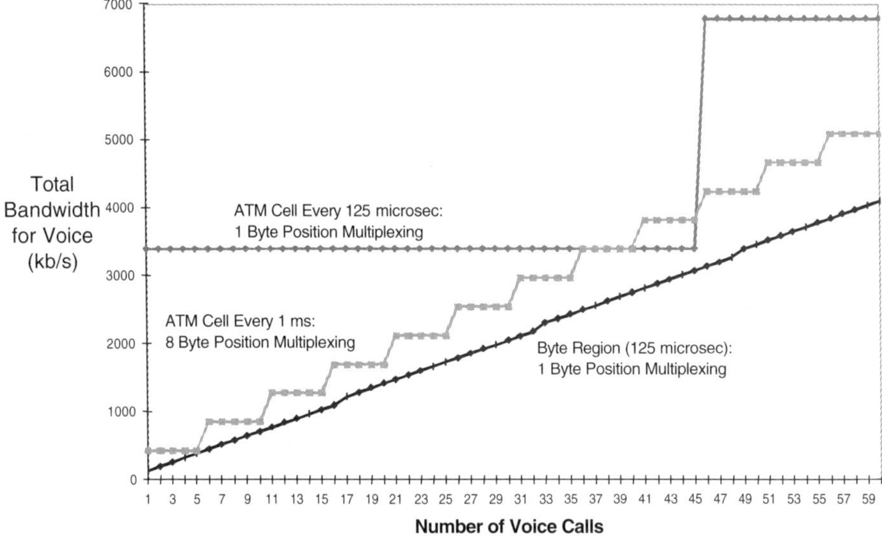

Figure 22: *Comparison of Bandwidth Efficiencies for Alternative Approaches for Supporting DS0 Connections for Telephony Applications in the Downstream*

Conclusions

Due to the wide-ranging traffic characteristics and performance requirements of applications supported over broadband HFC and wireless access networks, it is important to design MAC protocols with several key features for flexible and efficient operation. In this paper, we described several key features that enable superior performance and efficiency in MAC protocols. These features include: 1) minislots for bandwidth allocation granularity, 2) flexible transport capability for fixed-length (ATM) PDUs and VL PDUs (e.g., IP, 802.3 frames, IPv6), 3) means for low delay transport of STM voice information, 4) dynamically variable minislots for contention request resolution, and 5) piggybacking for additional transmission requests. We also included in this study a comparison of some alternative MAC protocol features.

For upstream channels, the use of VL PDUs for carrying data traffic results in about 50% higher bandwidth efficiency compared to the use of ATM PDUs (with adaptation based on AAL-5). The corresponding number for the increase in bandwidth efficiency for downstream channels is 10% to 15%. Accommodation of VL PDUs as well as ATM PDUs is best accomplished in the upstream by means of the minislot granularity. The support of VL PDUs in the downstream may be accomplished by creating two separate regions (or byte-streams) for data transmission, and for carrying ATM and VL PDUs, respectively. The boundary between these two data regions should be dynamic and specified by the HC. A system designed on the above principles would initially carry mostly VL PDUs to efficiently support 10baseT interfaces at the customer premises, and it can evolve gracefully to an ATM PDU–based transport in the access network as the information appliances and backbone networks become increasingly ATM capable. We also discussed and compared alternatives for supporting low-delay–requiring DS0 (and n*DS0) connections, with the telephony application in mind.

The paper compared MAC protocols for three cases with regard to request minislots: 1) with no minislots (in this case, the first of a batch of data packets from a station is transmitted in a full data slot, using contention-mode access and carrying a reservation request for the remainder of packets in that batch), 2) with a fixed number of minislots per frame, and 3) with a dynamically variable number of minislots per frame. We also described an algorithm for a dynamically varying number of minislots as a function of the traffic mix. The results show that a MAC protocol with dynamically variable minislots has the highest throughput efficiency among the different alternatives discussed.

The IEEE 802.14 standards specification (draft) and the MCNS (a cable industry consortium) specification on cable modems include several of the key features discussed in this paper as part of their MAC protocols.[14–16]

Acknowledgments

The author would like to thank Chia-Chang Li and James Dail of AT&T Laboratories, and Bharat Doshi, Subrahmanyam Dravida, Peter Magill, Miles Murdocca, and Curtis Siller, Jr., of Lucent Technologies–Bell Laboratories for comments and suggestions during the course of this work.

References

1 Sriram, K., Li, C.-C., Magill, P. D., Whitaker, N. A., Dail, J. E., Dajer, M. A., and Siller, C. A., Jr., "An Adaptive MAC-Layer Protocol for Multi-Service Digital Access via Tree and Branch Communication Networks," *SPIE Proceedings: Hybrid Fiber-Coax Systems*, vol. 2609, pp. 10–21, October 1995.

2 Dail, J. E., Dajer, M. A., Li, C.-C., Magill, P. D., Siller, C. A., Sriram, K., and Whitaker, N. A., "Adaptive Digital Access Protocol: A MAC Protocol for Multiservice Broadband Access Networks," *IEEE Communications Magazine*, pp. 104–112, March 1996.

3 Doshi, B. T., Dravida, S., Magill, P. D., Siller, C. A., Jr., and Sriram, K., "A Broadband Multiple Access Protocol for STM, ATM, and Variable Length Data Services on Hybrid Fiber-Coax Networks," *Bell Labs Technical Journal*, vol. 1, no. 1, pp. 36–65, summer 1996.
4 Sriram, K. and Magill, P. D., "Enhanced Throughput Efficiency by Use of Dynamically Variable Request Minislots in MAC Protocols for HFC and Wireless Access Networks," *Telecommunications Systems* (special issue on Multimedia), vol. 9, nos. 3 & 4, 1998, pp. 315–333.
5 Xie, H., Narasimhan, P., Yuan, R., and Raychaudhuri, D., "Data Link Control for Wireless ATM Access Channels," *Proceedings of the IEEE ICUPC (1995)*, pp. 753–757.
6 Karol, M., Liu, Z., and Pancha, P., "Implications of Physical Layer Overhead on the Design of Multiaccess Protocols, " *IEE Electronics Letters*, vol. 32, pp. 2062–2063, October 24, 1996.
7 Karol, M., Liu, Z., and Pancha, P., "Performance Implications of Physical Layer Overhead on Demand-Assignment MAC Protocols, " *Proc. of the Eleventh Annual IEEE Workshop on Computer Communications*, Reston, VA, September 1996.
8 Sala, D., Limb, J. O., and Khaunte, S., "Adaptive MAC Protocol for a Cable Modem," Georgia Inst. Tech. Technical Report, GIT-CC-97/14, May 1997.
9 Golmie, N., Masson, S., Pieris, G., and Su, D. H., "A MAC Protocol for HFC Networks: Design Issues and Performance Evaluation," *Computer Communications,* vol. 20, no. 12, pp. 1042–1050, November 1997.
10 Merakos L. F. and Bisdikian, C., "Delay Analysis of the n-ary Stack Random Access Algorithm," *IEEE Trans. on Information Theory*, pp. 931–942, September 1988.
11 Bisdikian, C., McNeil, B., Norman, R., "ms-START: A Random Access Algorithm for the IEEE 802.14 HFC Network," *Computer Communications* (special issue on Recent Advances in Networking Technology), pp. 876–887, September 1996.
12 Bisdikian, C., "The n-ary Stack Algorithm for the Wireless Random Access Channel," *Mobile Networks and Applications*, vol. 2, pp. 89–99, 1997.
13 Citta, R., Xia, J., and Lin, D., "A Tree-Based Algorithm with Soft Blocking," IEEE Project 802.14, Doc. No. IEEE 802.14/96-244, November 4, 1996.
14 IEEE 802.14 Working Group, Web page at http://www.walkingdog.com
15 MCNS Holdings, Inc., Web page at http://www.cablemodem.com
16 CableLabs, Boulder, Colorado, Web page at http://www.cablelabs.com/cable-spec.html
17 Gusella, R., "A Measurement Study of Diskless Workstation Traffic on an Ethernet," *IEEE Transactions on Communications*, vol. 38, no. 9, pp. 1557–1568, September 1990.
18 Wu, C.-T. and Campbell, G. "CBR Channels on DQRAP-based HFC Network," *SPIE Proceedings: Hybrid Fiber-Coax Systems*, vol. 2609, pp. 154–167, October 1995.

19 Sriram, K., and Whitt, W., "Characterizing Superposition Arrival Processes in Packet Multiplexers for Voice and Data," *IEEE Journal on Selected Areas in Commun.*, vol. SAC-4, no. 6, pp. 833–846, September 1986.
20 Sriram, K., "Methodologies for Bandwidth Allocation, Transmission Scheduling, and Congestion Avoidance in Broadband ATM Networks," *Proc. of Globecom'92*, Orlando, pp. 1545–1551, December 1992. Also, in the *Computer Networks and ISDN Systems* (special issue on Traffic Issues in ATM Networks), vol. 26, pp. 43–59, September 1993.
21 Dail, J. E., Li, C.-C., Magill, P. D., Sriram, K., and Whitaker, N. A., "Method and Apparatus Enabling Synchronous Transfer Mode and Packet Mode Access for Multiple Services on a Broadband Communication Network," U.S. Patent No. 5,570,355, October 29, 1996.
22 "Carrier Sense Multiple Access with Collision Detection (CSMA/CD) Access Method and Physical Layer Specifications," ANSI/IEEE Std. 802.3-1985.
23 Roberts, L. G., "ALOHA Packet System With and Without Slots and Capture," *Computer Communications Review,* vol. 5, no. 2, pp.28–42, April 1975.
24 Lam, S. S., "Packet Broadcast Networks—A Performance Analysis of the R-ALOHA Protocol," *IEEE Transactions on Computers*, vol. C- 29, no. 7, pp. 596–603, July 1980.

A Review of Contention Resolution Algorithms for IEEE 802.14 Networks

Nada Golmie, Yves Saintillan, and David Su
High Speed Networks Technologies Group, National Institute of Standards and Technology

Abstract

Bidirectional cable TV networks using hybrid fiber/coaxial (HFC) systems are good examples of broadcast environments in which a contention resolution algorithm allocates the multi-access medium (in this case, the upstream link) among the various nodes. The IEEE 802.14 standards group defining the physical (PHY) and medium access control (MAC) layer protocols for HFC cable networks focused on the study and evaluation of several contention resolution solutions for inclusion in the MAC protocol specifications.

This paper reviews several contention resolution algorithms considered by the IEEE 802.14. The paper also presents different implementations for several well-known contention resolution algorithms, such as tree-based and p-persistence. Their performance is evaluated in the HFC context with respect to upstream channel allocation, roundtrip delay, various traffic types, and the number of stations in the network. Simulation results for configurations and scenarios of interest are also presented.

Introduction

Emerging residential cable networks with burst speeds up to 40 Mbps—more than 2000 times faster than an ordinary dial-up modem—will most likely provide the next generation of data communication services, including Internet access to homes and small businesses.

This giant leap forward with higher bandwidth and better quality of TV signals is the direct result of introducing the HFC technology in the cable plant. Cable

An earlier version of this paper appears in the IEEE Communications Surveys, First Quarter 1999.
http://www.comsoc.org/pubs/surveys/

television (CATV) operators have already replaced long chains of copper wiring and amplifiers with fiber cable and electro-optical converters and multiplexors called fiber nodes. Today, most of the major multiple systems operators are either upgrading their cable plants or rolling out a new broadband infrastructure.

Cable modem standards have been pursued by several industry groups, including the Society of Cable Telecommunications Engineers (SCTE),[1] the IEEE 802.14 standard specifications,[2] the International Telecommunications Union (ITU),[3] and Multimedia Cable Network Systems (MCNS).[4] Of these, the MCNS standard is actively supported by both industry vendors and cable service providers. The next generation of cable modems will conform to the data-over-cable services interface specifications (DOCSIS) requirement set forth by MCNS.

In 1994, the IEEE 802.14 Working Group was formed to define a PHY layer as well as a MAC layer for bidirectional data transfer over an HFC-CATV network. The draft specifications (as of the writing of this paper) support mainly asynchronous transfer mode (ATM) cell transfers, thus providing high quality of service (QoS) and accommodating current and future CATV applications and television broadcast, namely, video on demand, interactive computer games, and video telephony.

The 802.14 Working Group spent considerable time on the selection of contention resolution algorithms. The two major contenders were the tree-based and the p-persistence algorithms. After nine months of debate, the group selected the tree-based algorithm, mainly because it generates the lower access delay variations critical to support QoS. During the course of the debate, proponents of each camp produced enhancements, many of which were incorporated into the selected tree algorithm. As an unbiased third party, the National Institute of Standards and Technology (NIST) conducted extensive evaluations of the various proposals. The results of these evaluations are presented here.

This paper will reveal the process the IEEE 802.14 committee underwent to select a specific contention resolution algorithm suited for HFC networks. The paper presents a wide range of algorithms and features, and guides the reader through the selection process by presenting the advantages and disadvantages of each as discovered via extensive simulations. This paper's main contribution is to inform researchers of the state-of-the-art in this area and open the window for further improvements in future implementations.

The rest of the paper is structured as follows. Section two describes the HFC environment and gives details about the MAC operation. Section three focuses on the main elements that constitute a contention resolution mechanism and discusses their relative performance. Section four studies a number of factors such as bandwidth allocation, roundtrip delays, traffic types, and number of stations that may have a significant impact on performance. A conclusion is offered in section five.

HFC MAC Protocol Overview

CATV networks are characterized by a tree-and-branch topology. At the root of the tree, a base station or head end controls the traffic. The bandwidth is divided into several channels, some dedicated to downstream communication (from the head end to the stations) while others are for upstream transmission (from the stations to the head end).

The upstream data streams are segmented into fixed size slots, or minislots. Several minislots can be concatenated to form a data slot, while one contention slot maps into one minislot. Data slots (DS) containing subscriber data packets are assigned to stations by reservation. The IEEE 802.14 draft specifications assume that data traffic is carried via ATM cells. The MAC layer prepends an octet to each ATM cell to form a MAC packet data unit (PDU) that is carried in a data slot. The contention slots (CS) carry the stations' request for bandwidth in "shared" or contention mode. Since more than one station can transmit a request at the same time and cause a collision, a contention resolution algorithm must be implemented as part of the MAC protocol.

The downstream data streams are segmented into a fixed-slot size of MAC PDUs.

The MAC reservation cycle, or request/feedback cycle, is defined as the time elapsed between request transmission and feedback reception at the station that is farthest from the head end. This ensures that all stations have the same opportunity of transmission in any given slot and prevents unfairness due to the relative location of the stations with respect to the head end (see *Table 1*).

A bandwidth allocation algorithm running at the head end controls the number of CS and DS contained in a MAC reservation cycle. The head end also decides on the distribution and location of the CS and DS on the upstream channel. CS can either be grouped together in clusters or distributed over the request/feedback interval. Unless otherwise specified, we assume that the minislots are grouped at the beginning of each request/feedback interval. The information about the contention and data slot location is sent to the stations on the downstream channel.

The basic MAC operation is as follows. Upon the arrival of the first data packet, a station sends a request packet in a contention slot that conforms to the first transmission

Simulation Parameter	Values
Number of active stations	200
Distance from furthest station to head end	80 km
Downstream data transmission rate	30 Mbits/s
Upstream data transmission rates channels	3 Mbits/s \\
Propagation delay	5 : /km for coax and fiber
Length of simulation run	30 sec
Length of run prior to gathering statistics	3 sec
Guardband and pre-amble between stations transmissions	Duration of 5 bytes
Data Slot size	64 bytes
Contention Slot size	16 bytes
DS/CS size ratio	4:1
MAC Reservation Cycle	1.536 ms (36 minislots)
Maximum request size per Contention Slot	32
Bandwidth Allocation Scheme	fixed ratio (12 CS, 6 DS)
Collision Resolution Scheme	Ternary-tree, T_{bound} Access
Head end processing delay	0 ms

Table 1: *MAC Default Parameters*

rule (FTR). (FTR, which governs the access of newcomer stations, is discussed in greater detail later in this paper.) Then, the station waits for the request feedback from the head end. If more than one station sends a request in the same contention slot, a collision results. Requests are then retransmitted according to a collision resolution algorithm. In the next section, different algorithms are presented and evaluated.

It must be stressed, however, that the characteristics of HFC networks impose a number of constraints. Stations cannot monitor collisions because modem receivers and transmitters are tuned to different frequencies for the downstream and upstream channels. Feedback about collisions must be provided by the head end, either explicitly or implicitly. The algorithm must also take into account the delay before a station receives the feedback information to ensure the best utilization of the network resources. In the case of a successful request transmission, the station waits for a data grant in order to send its data. At this point, if the station has additional requests for bandwidth, it may choose to bypass the contention process and use the extended bandwidth request field available in the MAC PDU. This is known as "piggybacking."

In order to allocate DS, the head end uses schemes such as first-come-first-served (FCFS) or round robin (RR) (we use RR in our simulations). It is assumed that the head end can deal with priority traffic for both contention resolution and bandwidth allocation. Similarly, it is able to distinguish between different connections and provide QoS.

In this study, a generic MAC model conforming to the above description is implemented to evaluate the various contention resolution algorithms. The default parameters used in the simulations are set according to *Table 1* and, unless otherwise mentioned, the cluster mode ternary-tree algorithm is used to resolve collisions.

The traffic type used, unless specified otherwise, is based on 48-byte packets generated according to a Poisson distribution with a mean arrival rate of λ (λ is a function of the offered load and the total number of stations).

Contention Resolution

Contention resolution algorithms were subject to much interest in the early 1970s for usage in packet radio transmission, especially during the development of the ALOHANET project.[3] Much research since then has been devoted to devising efficient contention resolution mechanisms for multi-access media for local-area networks (LANs), metropolitan area networks (MANs), satellite networks, and radio networks.

The many strategies that have been developed to solve the generic problem of having two transmitters send packets simultaneously can be divided into two basic paradigms. One is the "free-for-all," similar to the one used for the ALOHANET, in which nodes attempt to retransmit collided messages and hope for no interference from other nodes.[4] A variation of this technique adapted to HFC networks, p-persistence, associates a transmission probability, p, with each slot, usually controlled by the head end.[5] Hence, collided requests are retransmitted with a probability p. This process is repeated until a request is successfully received at the head end.

The other paradigm consists of splitting collided nodes into a tree structure. In this tree-based mechanism,[6] all nodes involved in a collision split into a number of subsets. The first subset transmits first, followed by the second subset, then the third, etc. The chances of future collisions are reduced by forcing stations that collided in the same slot (assuming a slotted frame structure) to retransmit their requests in different slots in the future (thus distributing the number of contending stations over several slots).

These two basic schemes have evolved to adapt to various network environments and constraints. Unlike carrier sense, multi-access techniques are employed when the ratio of propagation delay to packet transmission time is relatively small (less than one), and the stations can monitor the transmission channel reservation techniques (request/grant) that are used for the HFC networks. Roundtrip delays typically range between 0.8 ms to 2ms, including the propagation time for distances up to 80 km.

Also, to further increase the throughput of the upstream channel where data slot intervals are long (about 170 s for a 3 Mbps channel), short packets reserve long noncontending slots for sending data. Thus, only short slots are wasted by idles or collisions, which leads to better overall channel efficiency.

The key functions provided by a contention resolution algorithm consist of the following: 1) controlling the transmission of new requests by means of the FTR; 2) giving collision feedback; and 3) managing retransmissions. The FTR regulates when a newcomer station is allowed to send its first request on the upstream link. A feedback message informing the station about the status of its reception at the head end is associated with each contention slot. The status of a contention slot can be empty, successful, or collided, depending on whether there is zero, one, or more than one station transmitting in it. Finally, a mechanism is needed to resolve collisions and control the retransmission of request packets. This process is best visualized by means of a stack and is designated as stack management.

In this paper, several solutions for FTR and stack management are presented and compared. Some were considered during the IEEE 802.14 standardization process. Others appeared in the literature. Also, simulation results are presented. Each of the following test scenarios uses different assumptions to stress the impact of each solution on the overall system performance.

First Transmission Rule

This section presents different strategies to control the admission of newcomer stations. These strategies, also called first transmission rules (FTRs), define in which minislot a station with a new packet is allowed to send a request. The FTRs can be classified as follows:

Free Access: The first transmissions of requests are allowed to take place on the same minislots used to retransmit collided requests. Newcomer requests are mixed with "old," or retransmitted, requests.

Blocked Access: New requests are not allowed on the minislots that resolve current collisions. This is illustrated by a contention interval split into two regions. One is reserved for the ongoing collision resolution, and another, denoted as the newcomer minislot region, is open for newcomer requests (see *Figure 1*).

Furthermore, different strategies can be applied to each mode. For example, the free access mode can be either persistent or non-persistent. In this section, we consider only the non-persistent free access strategy, but several blocked access mechanisms considered by the IEEE 802.14 are reviewed.

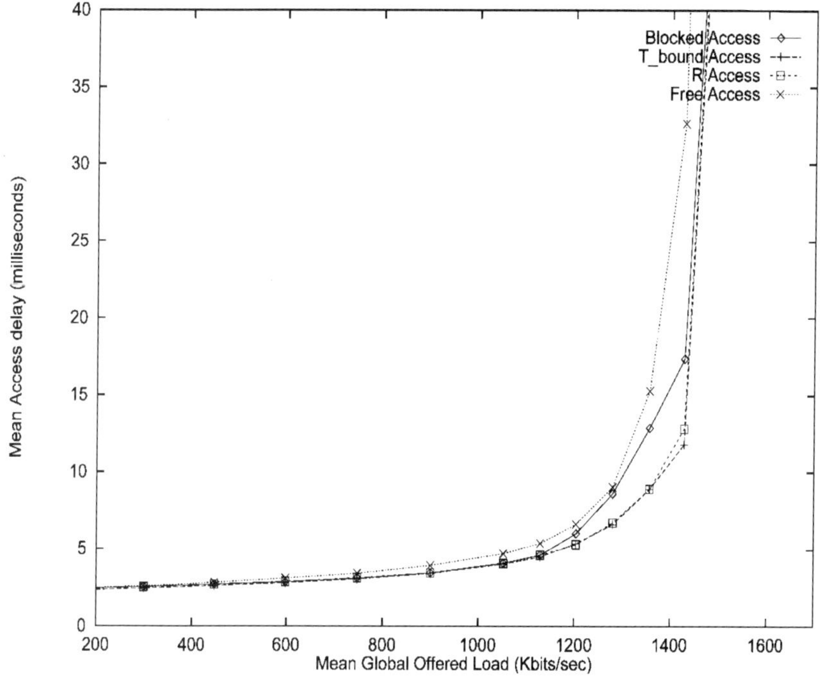

Figure 1: *FTR: Mean Access Delay vs. Offered Load*

In the non-persistent blocked access (simply called blocked access) all newcomer stations are allowed in the current newcomer region. In case of an opening, even if only one contention minislot is available for newcomer contention, all backlogged stations with new requests transmit in that minislot.

In the blocked access using persistence R, only a portion *MS*(j+1)/*R* of newcomer stations are allowed to transmit in the newcomer region of the current *j+1* cycle, where *MS(j)* is the number of contention minislots in the region and *R* is a range parameter computed at the head end. *R* determines the range over which a station can randomly select a slot of the newcomer region. If the random number chosen is greater than *MS*, the station repeats the process with the next cycle.

This so-called "soft blocking" or "R unblocking" strategy was voted on in November 1996.[7,8] In this paper, it is referred to as *R Access,* and *R* is computed according to the following:

$$R(j+1) = max\{min\left[n, R(j) - MS(j) + col(j) * \frac{(e-1)}{(e-2)} + \frac{MS(j)}{e}\right], MS(j+1)\} \quad (1)$$

where e=2.718, n is the number of stations, and $col(j)$ is the number of collided minislots in the jth cycle. This computation of R is based on the approximation of the traffic load used for the stable ALOHA (also called adaptive p-persistent).

On the other hand, the blocked access using T_{bound} was proposed by Bisdikian and was adopted by the committee in November 1997.[9] It is based on a time moving window and emulates a first-in-first-out (FIFO) access ordering known to be optimal.[4] Every cycle, the head end sets a value T_{bound}. Only if its message is generated prior to T_{bound} is a station allowed to transmit its request in the newcomer region (it selects a minislot randomly). Thus, the newcomer admittance becomes deterministic and is no longer a purely random process. T_{bound} can be computed from the R used previously according to the following:

$$T_{bound}(j+1) = T_{bound}(j) + \frac{MS(j)}{(R+1)} * (T_{current} - T_{bound}(j)) \qquad (2)$$

where $T_{current}$ is the time at which R is computed. This method, called T_{bound} access, significantly reduces the delay variability. To compare the relative performance of the non-persistent free access with the variants of the blocked access defined above, we look at the mean access delay and collision multiplicity for different offered loads, and the access delay versus time for a particular load. The mean access delay is defined as the average transmission time for a packet from the moment it is generated until it is received at the head end. This includes all queuing delays at the MAC layer, the time to transmit a request, and resolve collisions. Collision multiplicity is another significant performance measure of the FTR. It is defined as the number of stations that collide in one slot. The offered load represents the total packets generated by all stations in kbps.

Figure 1 plots the mean access delay versus the offered load for all four admittance policies described earlier. Starting from 20% of the channel capacity (600 kbps), the access delay for the free access is slightly above the access delay for the other mechanisms. At 45% capacity (1440 kbps), the difference in mean access delay is 15 ms between the free access and the blocked access and 20 ms between the free access and the T_{bound} access. Since requests from newcomer stations can mix with collided stations in free access, the collision resolution is less effective than in blocked access, which results in higher mean access delays at higher loads. The blocked access method was mainly introduced in order to resolve collisions more efficiently.

Blocked access starts to exhibit problems starting from 40% (1200 kbps), where its mean access delay is higher than the delays for R access and T_{bound} access (5 ms difference at 45%). This is mainly due to a high collision multiplicity factor. With blocked access, all backlogged stations are released into the system at once. As the load is higher, the number of CS reserved for newcomer access is relatively small, which may cause a large number of stations (a maximum of thirty-three at 50% from *Table 2*) to collide in one slot. The maximum values for the blocked access collision multiplicity are much higher than for all other methods: we measure 13, 33, and 38 at 40%, 50%, and 60% capacity, respectively, compared to maximums of 6, 8, and 8 at the same loads for free access (see *Table 2*). While controlling the station's entry, R access and T_{bound} access have a relatively low collision multiplicity since in both cases the newcomers are let into the system gradually. The maximums, means, and standard deviations of collision multiplicity are similar to those for free access. Although the mean access delays for R access and T_{bound} access are almost identical, there is a significant difference in terms of access delay between the two methods (*Figure 2*). The FIFO ordering of T_{bound} access reduces the difference between the minimum and maximum access delays or the delay variation. A relatively low access delay variability generally translates into a low delay jitter for higher layer applications. This may be critical for certain ATM applications with stringent constraints on cell delay variations.

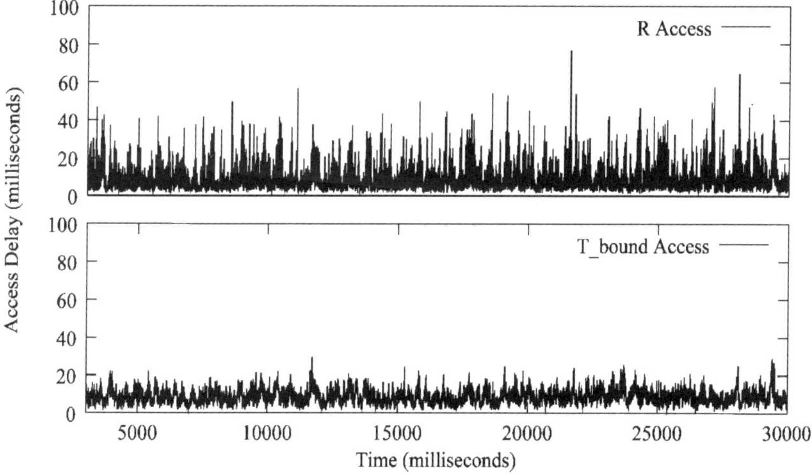

Figure 2: *FTR: Access Delay vs. Time*

Stack Management

Often in the literature, contention resolution algorithms are defined in terms of operations performed on a stack.[10] The idea is based on managing a stack of collided and newcomer stations where each stack level corresponds to a different group of stations ranked from top to bottom. Note that the stack in this case is only a visualization and need not be implemented at the station or the head end.[11]

Algorithm	Load	Max	Mean	F
T_{bound} Access	40	7	2.211	1.695
	50	8	2.277	1.651
	60	8	2.275	1.649
Blocked Access	40	13	2.541	1.973
	50	33	3.589	3.049
	60	38	4.000	3.664
R Access	40	8	2.217	1.704
	50	7	2.283	1.657
	60	8	2.277	1.649
Free Access	40	6	2.230	1.643
	50	8	2.343	1.530
	60	8	2.345	1.526

Table 2: *FTR: Collision Multiplicity*

In this section, we study the various stack implementations available for contention resolution algorithms. We consider the ternary-tree LIFO, ternary-tree FIFO, and p-persistent algorithms with respect to the relative performance of their stack management. For ternary-tree algorithms, colliding requests are split into three groups and each group is transmitted in a different minislot. The process repeats until all collisions are resolved. For this analysis we consider that each group of requests that results in a split occupies a level of the stack. We will look at the various ways for a group to enter and leave the stack.

Assuming that the next group of stations allowed to transmit occupies the bottom level, collided stations entering the stack may either go on top or at the bottom of the stack, thus achieving a FIFO or a LIFO ordering, respectively. In *Figure 3*, a stack visualization for the ternary-tree algorithm is shown. *Figure 3(a)* illustrates a LIFO ordering. Stations at level zero transmit their request at the next contention opportunity. In the case of collision feedback, the group of stations splits into three sub-groups at levels zero, one, and two. Stations previously at level one are pushed two levels up. On the other hand, in the case of a successful feedback, all stations are pushed down one level. Newcomer stations entering the system for the first time can either be placed at the top or bottom level, depending on the FTR. For blocked access mode, stations are put at the highest level of the stack, while stations start at level zero for free access mode. *Figure 3(b)* depicts a FIFO stack ordering. In the case of successful feedback, the stack dynamics are similar to the LIFO ordering (*Figure 3(a)*). In the case of collision feedback, the stations at level zero go on top, while stations previously at level one go to level zero, thus achieving a time-based sorting.

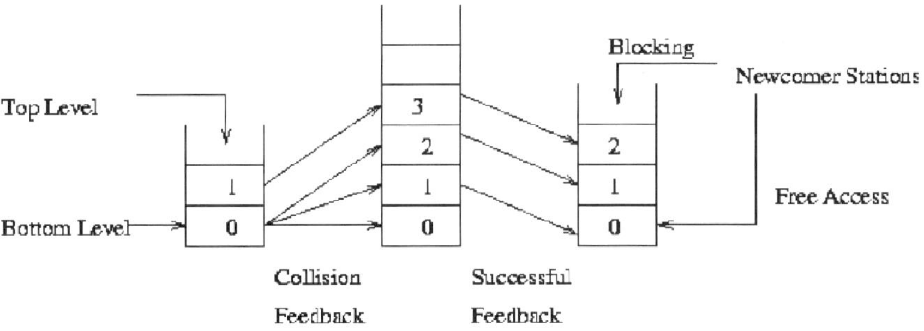

(a) LIFO Stack Ordering

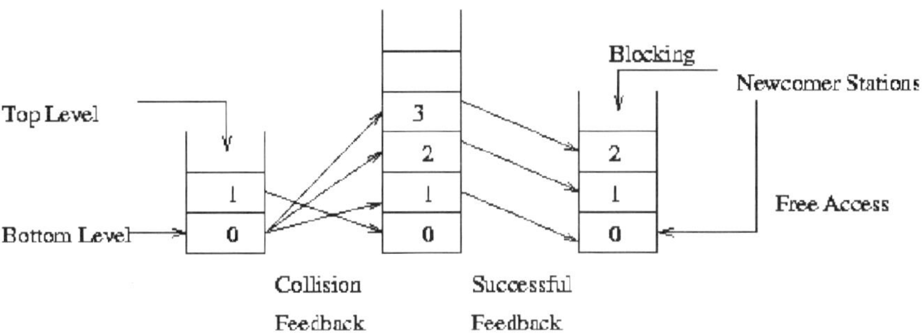

(b) FIFO Stack Ordering

Figure 3: *Ternary-tree Stack Visualization*

In *Figure 4*, LIFO ordering has higher access delays than FIFO, starting from 40% of capacity (15 ms and 30 ms for FIFO and LIFO, respectively, at 47% capacity). Note that this is not due to differences in the mean collision resolution delay, which is shown to be the same, but to high access delay variations of the LIFO ordering (*Figure 4*).

Although the stack is a good visual aid for understanding the dynamics of splitting algorithms and is usually associated with tree-based contention resolution mechanisms, the idea can be applied to p-persistence schemes as well.[10] The result is a two-level stack, as shown in *Figure 5*. Based on a collision feedback, stations at levels zero

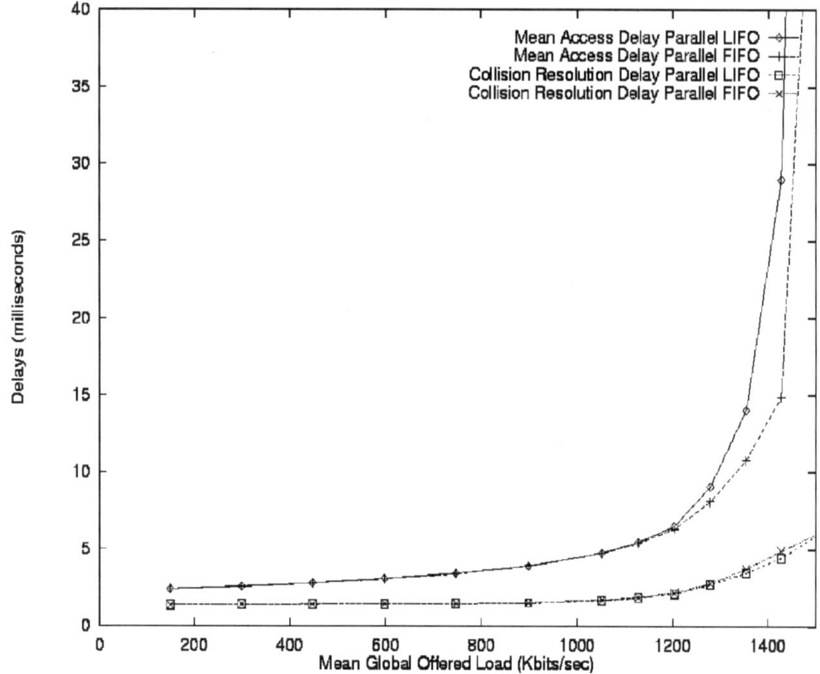

Figure 4: *Tree Stack: Mean Access/Collision Resolution Delay vs. Offered Load*

and one remain at the same level with a probability of p and $1-p$, respectively. On the other hand, stations at levels zero and one change levels with probability $1-p$ and p, respectively. After a successful feedback, stations at level one either go to level zero with probability p or remain at level one with probability $1-p$. *Figure 6* compares the mean access and collision resolution delays of the p-persistence to the ternary-tree implementation.

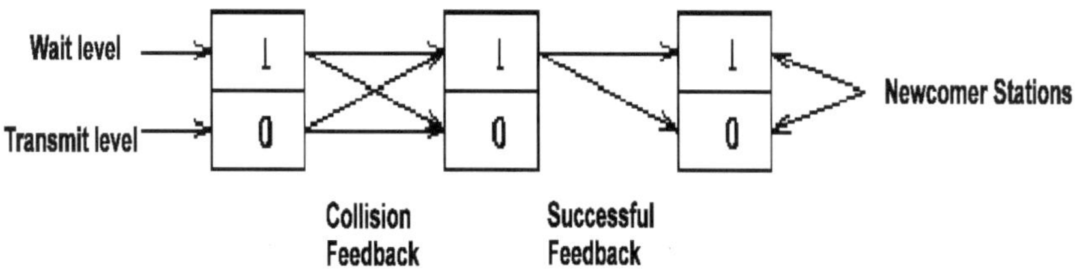

Figure 5: *p-Persistence Stack Visualization*

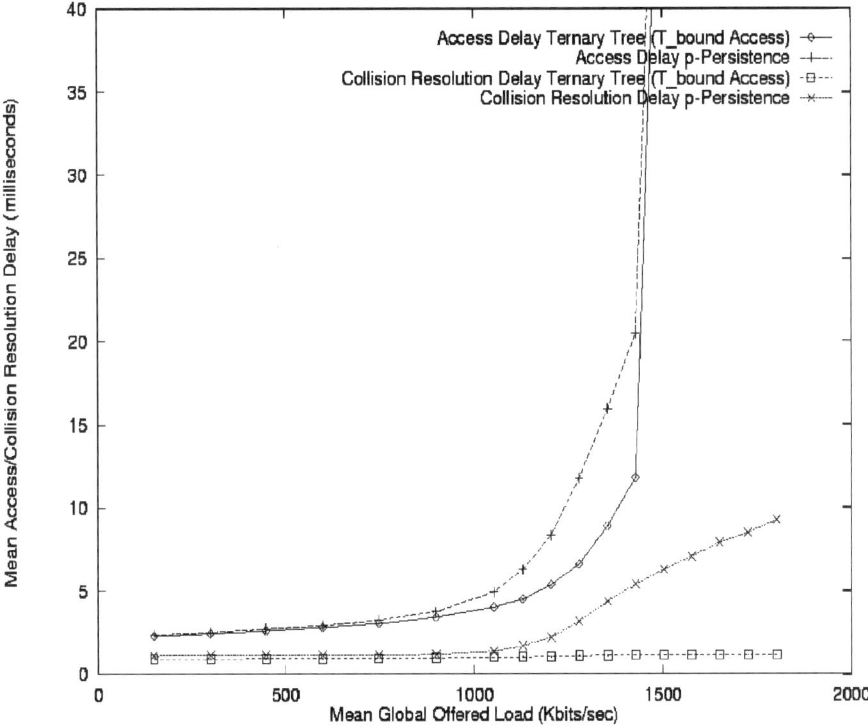

Figure 6: *Ternary-tree vs. p-Persistence: Mean Access/Collision Resolution Delay vs. Offered Load*

For the ternary-tree, we use T_{bound} access for newcomer stations. As the offered load increases, the mean collision resolution for the ternary-tree is kept constant, while it is always increasing for the p-persistence. This is a direct result of keeping the transmission of collided requests and newcomer requests in separate contention intervals for the ternary-tree T_{bound} access. Starting at 30% of capacity, the mean access delay for the p-persistence is higher than for the ternary-tree; however, the difference is almost 10 ms at 45% of capacity. Both mean access delay curves converge after 45% of capacity, since the maximum throughput is 50%. *Figure 7* shows that the percentage of collisions (relative to the total number of requests) is almost 30% higher with p-persistence than with ternary-tree, starting from 40% of capacity.

Figure 8 plots the cumulative density function for the tree-based and p-persistence with a 50% applied load. For the tree-based probability, the access delay is less than 20 ms, almost 100%, while it is close to 75% for the p-persistence. This result is expected, considering the random nature of the p-persistence and the blocking

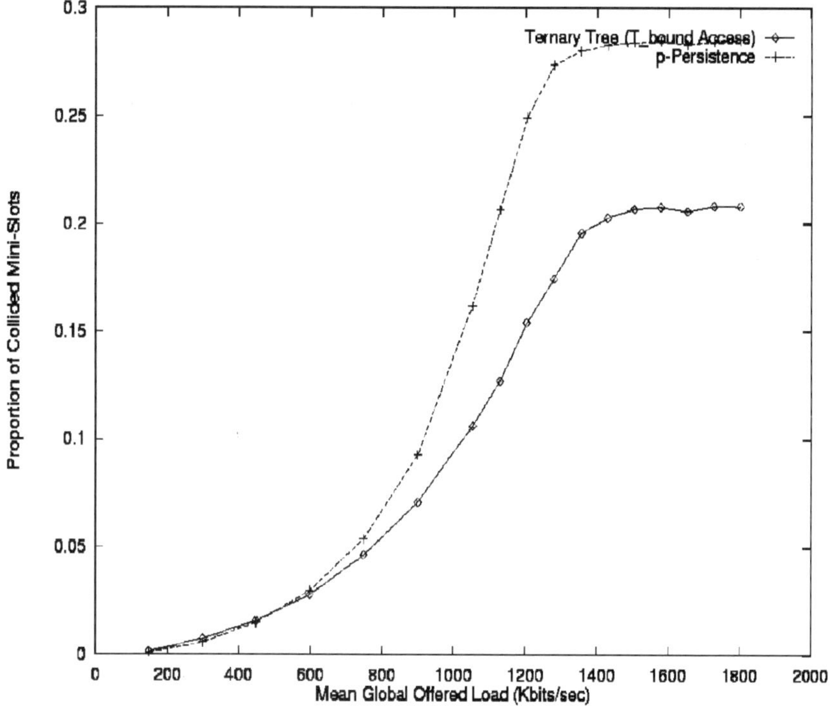

Figure 7: *Tree vs. p-Persistent: Percentage of Collisions vs. Offered Load*

feature with the FIFO ordering used in the tree-based T_{bound} access. This result was crucial in the decision of the IEEE 802.14 group to adopt the ternary-tree (LIFO stack) algorithm for contention resolution.

Ternary-Tree Implementations

This section focuses on the ternary-tree contention algorithms and presents three different implementations for use with a LIFO or a FIFO stack management scheme.

After the station sends its request, it waits for feedback from the head end. In case of collision feedback, the station enters the contention resolution cycle and activates its resolution state machine. The complexity of the state machine depends on the algorithm considered. Splitting algorithms make use of the type and frequency of the feedback in order to develop more intelligent (but also more complex) collision resolution state machines. In *Figure 9,* the dynamics of three different implementations of the ternary-tree are presented over an interval of twelve CS. Note that DS are omitted from the picture for a better understanding of the feedback dynamics.

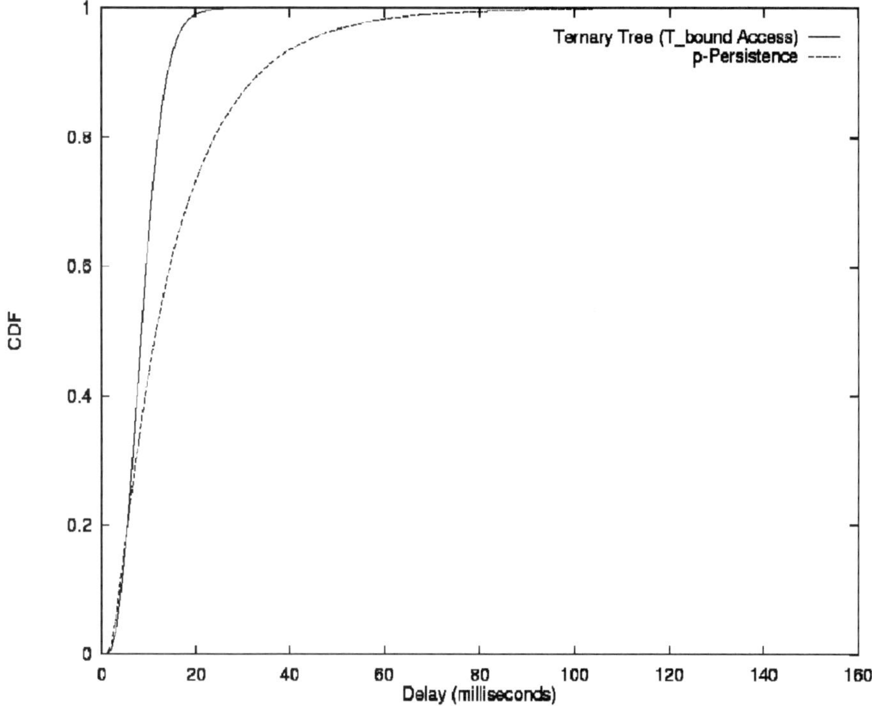

Figure 8: *Tree vs. p-Persistent: Probability that Mean Access Delay < **x** ms, Offered Load = 50% of Capacity*

Both the sequential and parallel implementations (*Figure 9(a)* and *(b)* respectively) use the feedback information given on each slot. However, they differ in the number of feedback messages a station must monitor after a collision has occurred. In the sequential implementation,[12] the station monitors the feedback message every interval equivalent to a roundtrip delay to the head end. When the station counter reaches zero, the station is allowed to transmit its request. In case of a collision *and* if the station counter is equal to zero, the station selects a random value for its counter between zero and two. Otherwise, the counter is incremented by two for every collision feedback. The station decrements its collision counter by one for an empty or successful feedback. In the parallel implementation,[13,14] the station's counter dynamics are similar to the sequential case, but the counter is updated for every slot rather than every roundtrip delay. Thus, the station needs to monitor the channel for all feedback messages following a collision until the station can retransmit its request.

In *Figure 9 (a)* and *(b)*, stations A, B, and C collide in slot zero. The feedback is received at the end of slot two, and station C retransmits in slot three in both the sequential and parallel implementations. However, since stations A and B update

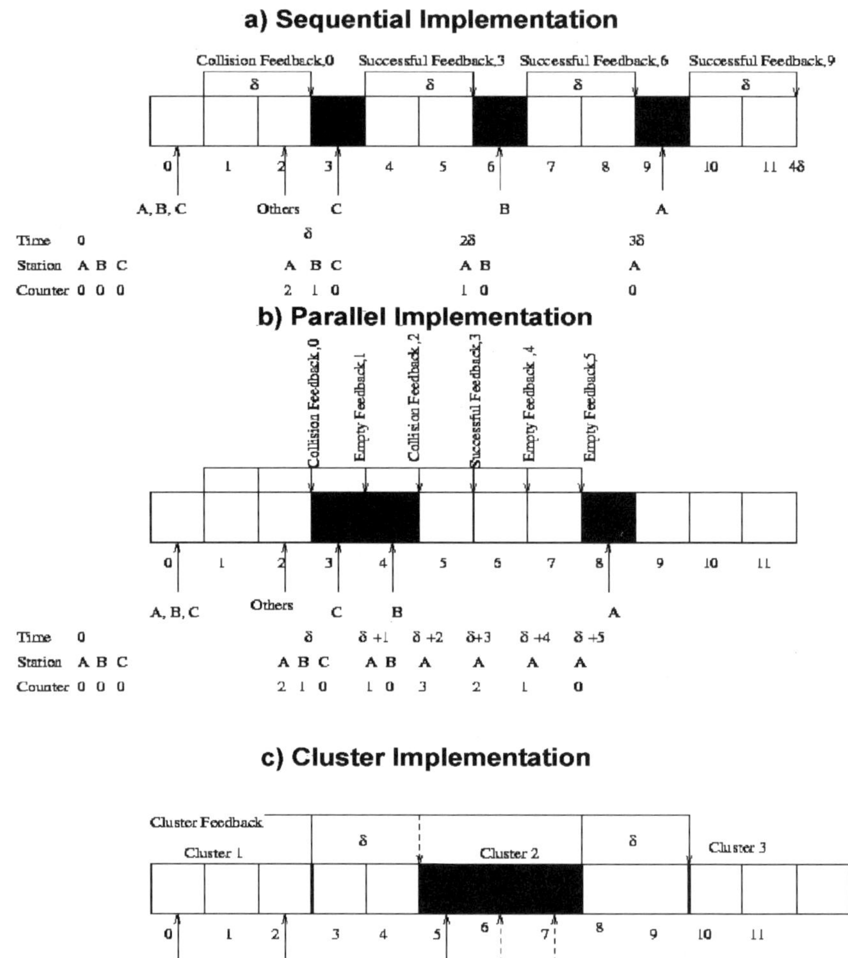

Figure 9: *Tree Implementation*

their counter every slot in (b), an empty feedback message at the end of slot three causes station B's counter to go to zero and retransmit in slot four. In (a), station B's counter is only updated every δ. In this case, at 2δ, B retransmits in slot six. As illustrated in this example, the parallel implementation may reduce the request delay by allowing stations to retransmit their collided request earlier than with the sequential implementation. The average request delay in (a) is equal to six CS ([3+6+9]/3) while the average request delay in (b) is equal to five slots ([3+4+8]/3). But one has to keep in mind that less request delay in the parallel implementation case requires extra processing of the feedback messages at the station. In addition, the parallel implementation requires a feedback message for each contention slot, regardless of whether it is used or not. This constitutes an extra bandwidth overhead in the downstream in order to carry control information for the upstream channel.

The cluster implementation chosen by the IEEE 802.14 is shown in *Figure 9(c)*. It is analogous to the sequential implementation, since the station needs to tune in for the feedback every roundtrip delay and not every contention slot as in the parallel implementation *(b)*. There is even "bonus" information contained in the feedback message received at the beginning of each cluster. Note that the feedback delay in this case accounts for the roundtrip propagation time plus the transmission time for all of the CS contained in the cluster. This guarantees that feedback information from transmissions in CS in a cluster is available to the stations prior to the occurrence of the CS in the next cluster. The head end can then use its knowledge of the number of collisions that occurred in the cluster to distribute the collided stations in the next cluster. For example, consider cluster one in *Figure 9(c)* containing collided minislots zero and two. All stations involved in a collided minislot (zero and three in this case) split into three subsets (for the ternary-tree), each choosing a random number between zero and two. The stations involved in the first collided slot, zero, are dispatched in the first three slots of cluster two (slots five, six, and seven). Those involved in the second collided slot my use the next three slots, and, more generally, the stations involved in the ith collided slot select a subset between ($i * n$ + 1) and ((i + 1) * n) slots. If cluster j contains p CS, then the first p subsets are able to retransmit, the ith subset transmitting in the ith slot. The other subsets wait in the stack. In (c), stations collided in slot two wait until cluster three. If c_j new collisions occur in cluster two, the waiting subsets must be shifted by $3 * c_j - p$ positions in the stack to make room for the new subsets.

Collisions are resolved sequentially but within each group or cluster of CS. If it is a LIFO ordering of collided slots, more recent collisions preempt older ones. The average request delay for the cluster implementation (see *Figure 9(c)*) is equal to (5+6+7)/3 = 6, and in this case it is the same as the sequential implementation.

Figure 10 plots the mean access delay and the mean collision resolution delay for the sequential, parallel, and cluster implementations discussed earlier. A free access policy is used for newcomer stations. We define the mean collision resolution display as the average time required to resolve a collision, and that is the time elapsed between the reception of collision feedback at the station and the reception of a successful feedback for a given request.

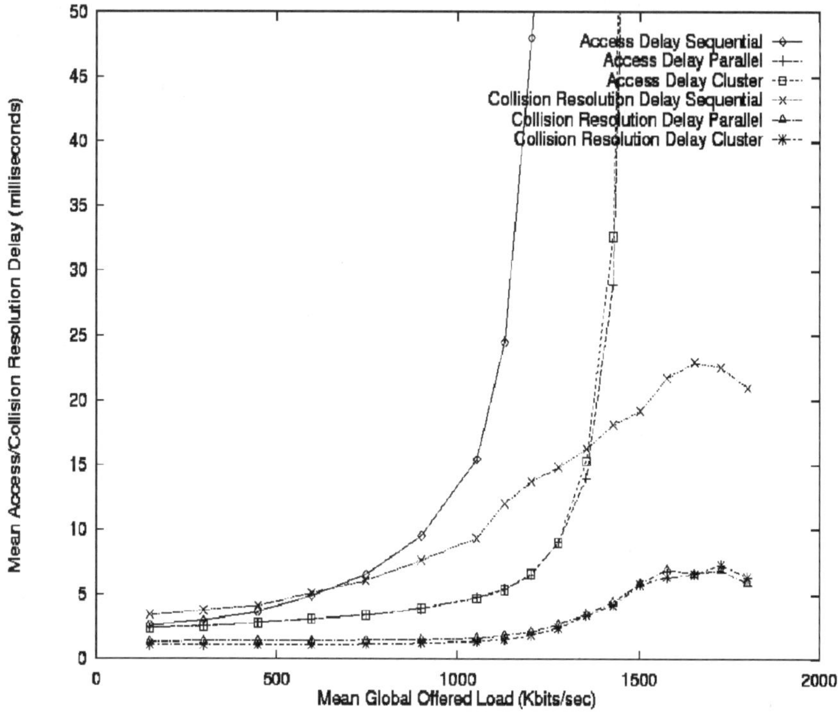

Figure 10: *Ternary-tree Implementation: Mean Access/ Collision Resolution Delay vs. Offered Load*

First, note that the mean collision resolution delay curves for all three implementations at high loads (greater than 50% capacity) are subject to a slight reduction due to the accumulation of packets in the stations' buffer and an increase in the request size. Thus stations are able to request for multiple packets. This keeps the number of collisions from increasing, and it stabilizes in *Figure 11* when the channel completely saturates at 50% of its capacity (which is also the theoretical throughput limit for a fixed allocation of 12 CS and 6 DS). It seems rather counterintuitive that the mean collision resolution delay for the sequential implementation is slightly higher than the mean access delay at low loads (between 5% and 10%), but this is due to the addition of time to make a successful request in one roundtrip to the mean collision resolution delay in order to obtain a mean request delay. Note that the mean access delay is the sum of the mean request delay plus the packet transmission time (depending on the bandwidth allocation). As expected, the mean access delay and the mean collision resolution delay for the sequential implementation are much higher at higher loads than the two other methods: the collision resolution delay is 15 ms higher at 50%, and the mean access delay is 40 ms higher at 40%

capacity (*Figure 10*). The percentage of collisions (defined as the number of collided slots divided by the total number of minislots) incurred by the sequential implementation, however, is less than the percentage of collisions for the cluster and parallel methods as seen in *Figure 11*. In this case, reducing the number of collisions increases the delays. For both the parallel and cluster implementations, the results are almost identical in terms of mean access delay, collision resolution delay, and the number of collisions (*Figurse 10 and 11*).

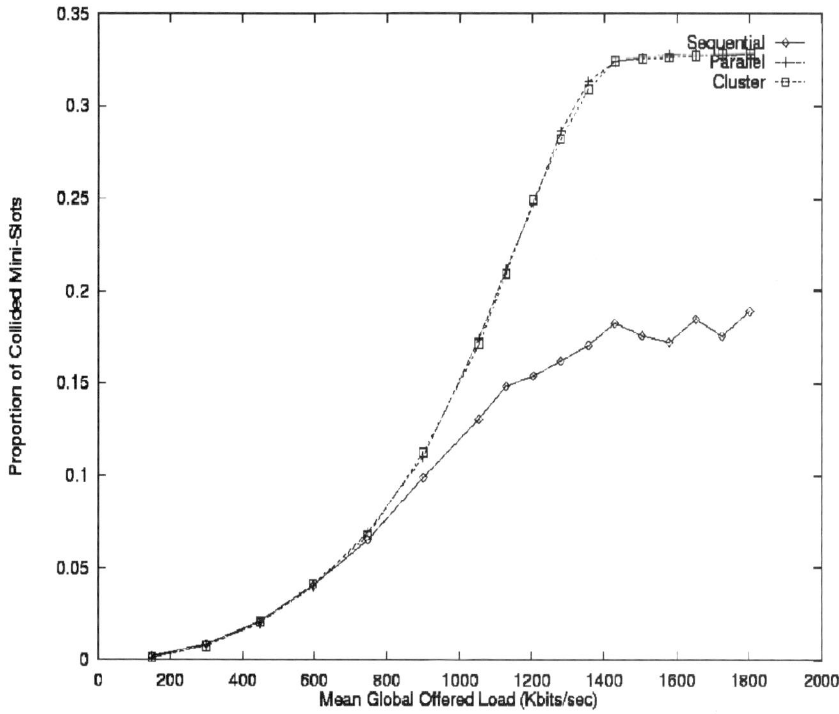

Figure 11: *Ternary-tree Implementation: Percentage of Collisions vs. Offered Load*

IEEE 802.14 Solution

After comparing the performance of the algorithms as described, the rationale for the IEEE 802.14 solution adopted becomes clear. As of the writing of this paper, the IEEE 802.14 draft specifications include T_{bound} access for FTR, a LIFO stack ordering, and a cluster implementation for the tree-based collision resolution algorithm.[2]

Related Factors and Impact on Performance

This paper has compared various contention resolution features under the same network constraints to stress the differences among them. However, the overall system performance depends on a number of design choices and specific network conditions. This section examines some of those factors and their impact on performance. The experiments are conducted using the tree-based cluster mode with T_{bound} access for contention resolution.

Bandwidth Allocation

This section focuses on the bandwidth allocation element of the MAC protocol and studies its impact on performance. We compare three different strategies to improve bandwidth, namely, varying the CS/DS ratio over the request/feedback cycle, piggybacking, and increasing the request size.

Variable Slot Allocation: We first will compare the fixed slot allocation we have been using so far in the experiments (2:1 CS/DS ratio) to a variable slot allocation strategy that we shall define shortly. In the fixed scheme, the number of contention minislots and reservation DS in each cycle remains the same. In the variable scheme, the head end varies the ratio of CS to reservation DS from cycle to cycle based on the traffic in the system. This mechanism is similar to what is presented in the references with some slight modifications.[16–18] The number of CS, N_{CS}, contained in each upstream cluster is dynamically adjusted as the head end converts the number of DS, N_{DS}, into CS, according to the following expression:

$$N_{DS} = \frac{2 * MAX\_DATA}{(2 + m * k)}$$

where MAX_DATA is the maximum number of DS in a frame, m is the number of minislots that a data slot occupies, and k is the average number of DSs that can be requested at a time. As a result, N_{CS} can be determined by:

$$N_{CS} = \begin{cases} 0 & \text{if } DQ \geq \alpha * (MAX\_DATA - N_{DS}) \\ m * N_{DS} & \text{else} \end{cases}$$

where DQ is the length of the request queue at the head end and α is a design parameter set to 2.5. In addition, any unused DS resulting from the previous allocation are then converted into CS, as suggested in the references.[15] *Figures 12* and *13* compare the mean access delay and the throughput, respectively, for the fixed and variable allocations. As expected, variable allocation performs better than the fixed ratio policy in terms of lower delays at higher loads and maximum throughput. At higher loads, keeping the number of CS fixed constitutes a significant bandwidth overhead. Stations may take advantage of requesting more than one packet at a time, thus reducing the need for CS. At low loads, the frequency and location of CS could reduce the delays as shown in the references.[19]

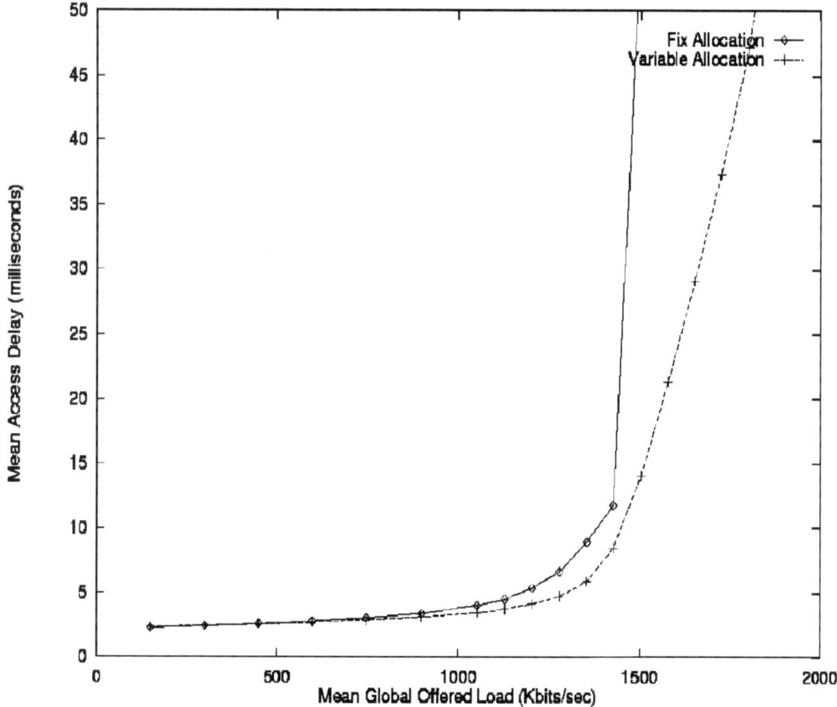

Figure 12: *Fixed Allocation: Mean Access Delay vs. Offered Load*

Piggybacking: Piggybacking consists of sending requests for additional data transmission with a data packet without having to go through contention. Note that the first request for data transmission is sent in contention. *Figure 14* compares the mean access delay with and without the piggybacking feature. For both cases, we use the variable allocation algorithm described earlier. We note 2 ms to 3 ms lower delays when using piggybacking, starting from 50% capacity.

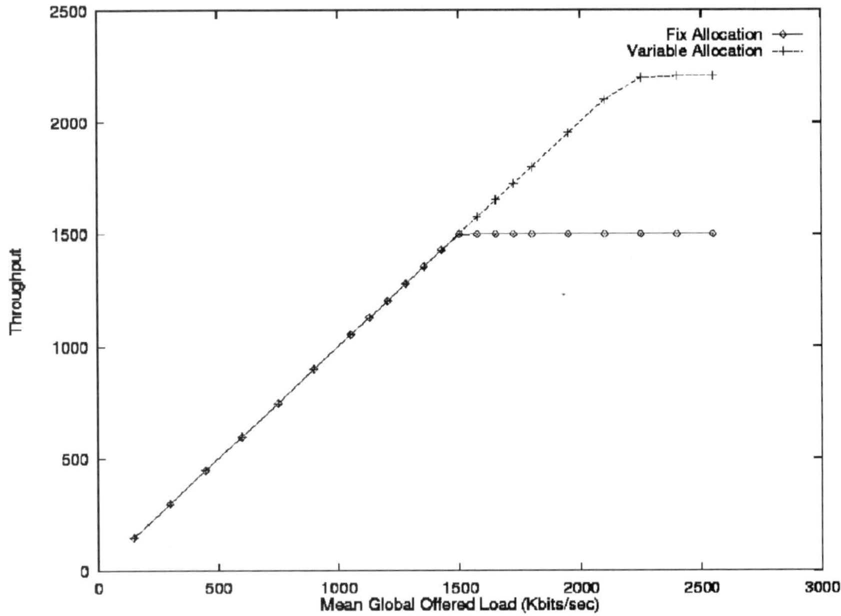

Figure 13: *Variable Allocation: Throughput vs. Offered Load*

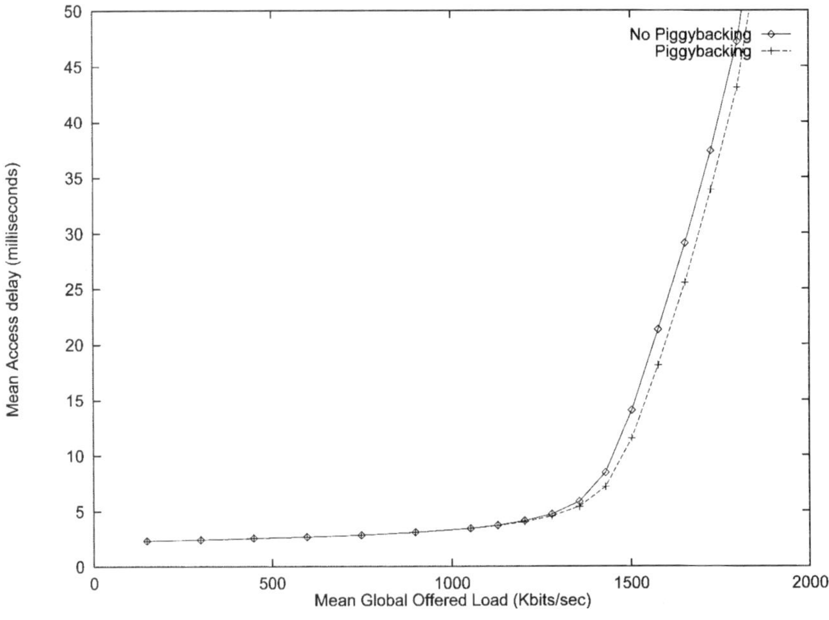

Figure 14: *Piggybacking: Mean Access Delay vs. Offered Load*

Request Size: As the offered load is increased, short packets tend to accumulate in the station's buffer. This is true even at lower loads, depending on the traffic type. Large packets for some applications may generate several ATM cells at once when segmented into 53-byte chunks (assuming ATM cell transfer). Thus, to avoid sending several requests in contention and to reduce the access delay, the station can send a request for all the packets in its buffer at once. For the IEEE 802.14 MAC, the request size is limited to 32 packets (the request field is 8 bits long). *Figure 15* compares the mean access delay for different request limits of one, two, and thirty-two with the variable allocation scheme. Since short packets are used in the experiment, the differences in the three curves become significant only at high loads, starting at 45% of capacity. For a request size of one, delays are 3 seconds higher than for a request size of two at 75% capacity. Delays for the request size of thirty-two are still in the order of milliseconds (~ 100 ms) at the same load.

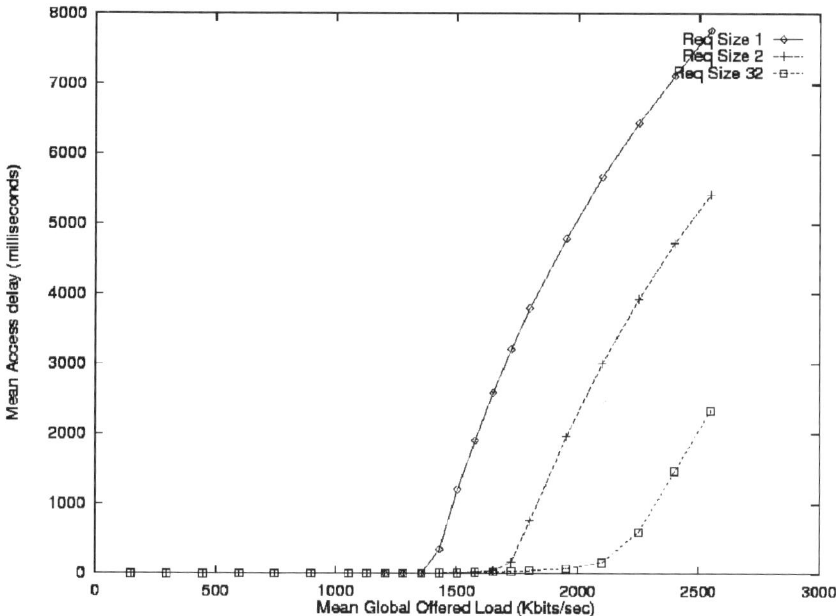

Figure 15: *Request Size: Mean Access Delay vs. Offered Load*

Number of Stations

An interesting issue for cable operators and subscribers is the number of stations that can share the upstream channel. Should there be a limit on how many stations share the same channel? It is definitely a design issue and a trade-off between cost and performance. Adding more stations to the HFC network not only increases the overall load but also increases the number of contenders. The previous experiments demonstrated how to better utilize resources by using a variable allocation scheme, a larger request size, and piggybacking when the number of stations remained constant. Although these strategies are scalable to various numbers of stations, there is still a price to pay for increasing the number of subscribers. It may be more economical to the cable operator, but it increases the access delay and lowers the throughput. In *Figure 16,* the mean access delay at 47% of capacity is close to 7 ms for 50 stations, while it is 11 ms, 23 ms, and 123 ms for 200, 500, and 2000 stations, respectively.

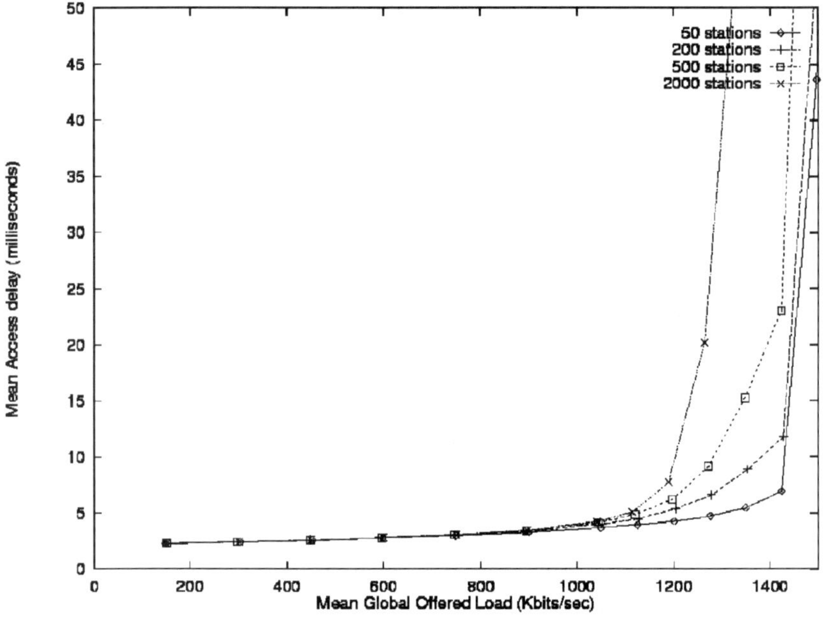

Figure 16: *Varying Number of Stations: Mean Access Delay vs. Offered Load*

Feedback Delays

Another design parameter of HFC networks is the total length of the cable or the maximum distance between the head end and the farthest station in the network. Increasing this distance will require additional amplification of the signal. In addition, it will increase the roundtrip delay time for the entire distribution network, since the feedback delay is set to the roundtrip delay between the head end and the farthest station. Note that this requirement is necessary in order to ensure fairness of access among all stations. For example, in the case of a collision feedback, all stations have an equal opportunity for retransmission. It is interesting to note that regardless of the distance used, the maximum throughput is the same. In *Figure 17*, all delay curves converge around 50% of capacity (maximum throughput with fixed allocation). Between 5% and 50% capacity, the differences between the delay curves for 40 km and 80 km and 80 km and 200 km are constant and equal to 1 ms and 2 ms, respectively.

As previously discussed, in the ternary-tree cluster implementation (refer to the Stack Management section), the length of the cluster is usually equal to or greater than the roundtrip delay plus the transmission delay of the CS in the cluster. Thus, if a station used a contention slot in cluster *i*, it gets the feedback before cluster *i*+1 in case it needs to retransmit its request. Another way to fully utilize the channel and get feedback to the stations at the beginning of each cluster is to split the cycle time into smaller clusters (up to sixteen) and thus have several collision resolution engines operate in parallel on each cluster. This method is known as interleaving. A station with a request to send can randomly select an interleave. Collisions are resolved for each interleaf separately. *Figure 17* uses an interleave of two for a distance of 200 km. The mean access delay with two interleaves for 200 km is slightly less than with one interleave (~1 ms).

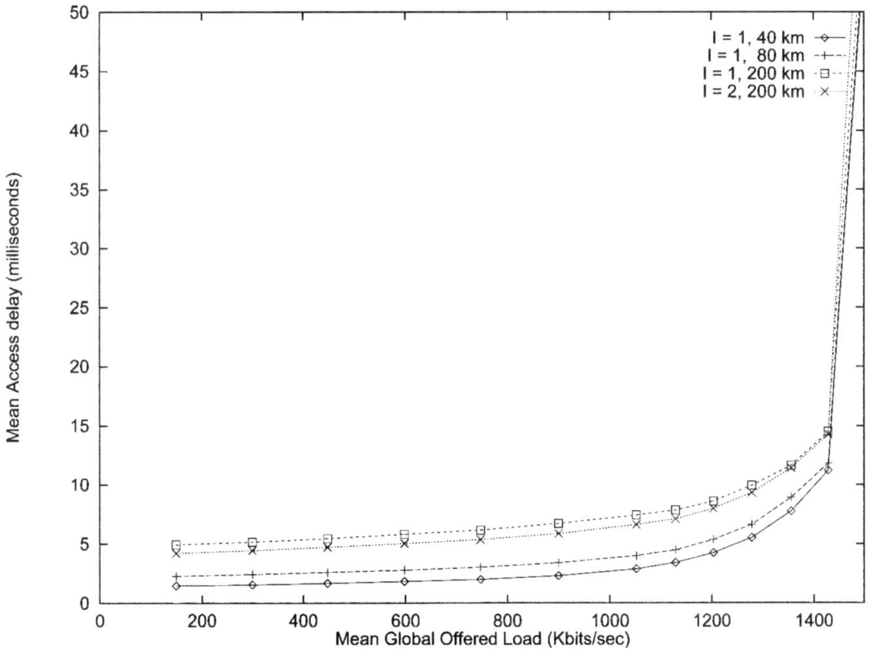

Figure 17: *Varying Feedback Delays: Mean Access Delay vs. Offered Load*

Traffic Types

The source generating long packets in the experimental results of *Figure 18* is a bursty source with a batch Poisson arrival model. The message size distribution is equal to 64, 128, 256, 512, 1024, and 1518 bytes with a probability of 0.6, 0.06, 0.04, 0.02, 0.25, and 0.03, respectively. The message interarrival time is exponentially distributed with mean T= 1/λ, where λ varies according to the load. The source generating short packets is the one used in all the other experiments so far (48-byte packets with an arrival rate of λ).

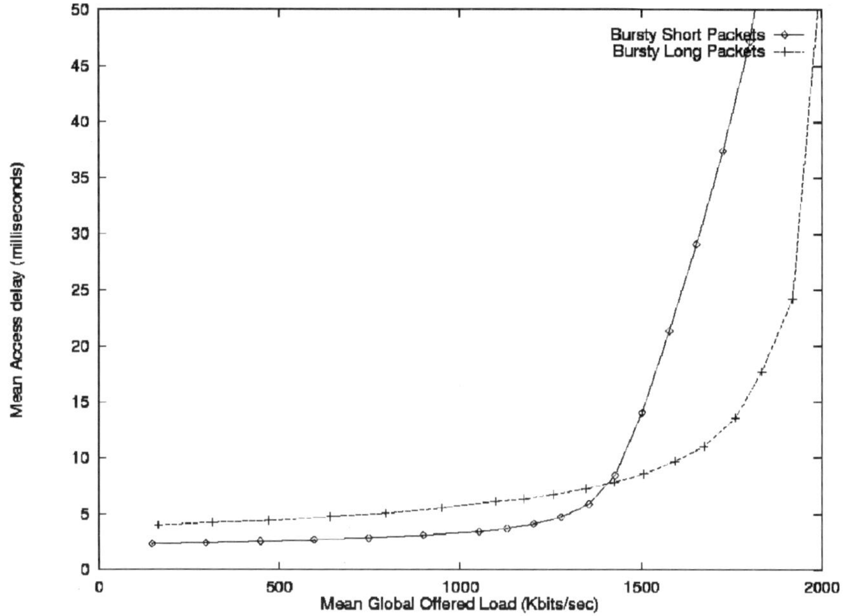

Figure 18: *Traffic Types: Mean Access Delay vs. Offered Load*

For loads of less than 40% of capacity, delays for longer packet bursts are higher (2 ms) than shorter packets. This is mainly due to the larger transmission delay incurred by long packets (*Figure 18*). At higher loads, shorter packets cause more requests to be sent in contention, resulting in more collisions and less throughput. With large packet sizes, the delays are lower for higher loads (knee of the curve is pushed farther to the right). The large average packet size results in larger request sizes, so that less contention minislots are needed and more DS can be allocated.

Conclusion

This paper provides an overview of contention resolution design considerations for HFC networks and paves the way for further enhancements in protocol development. We study important elements of contention resolution algorithms, such as the first transmission rule and stack management, and show that mean access delays for blocked access with a FIFO ordering of newcomers are lower than those for free access. In addition, blocked access significantly reduces the delay variation while maintaining a low collision multiplicity. We also note that the p-persistence scheme exhibits higher delays and a larger probability distribution tail than the tree mechanism.

References

1 MCNS Holdings L.P., Data-Over-Cable Service Interface Specifications, Radio Frequency Interface Specification, SP-RFI-I02-9710008, October 1997.
2 IEEE 802.14 Working Group, Media Access Control, IEEE Draft Std. 802.14, Draft 3 R3, October 1998.
3 Abramson, N., ``Development of the ALOHANET," *IEEE Transactions on Information Theory*, vol. IT-31, no. 2, March 1985.
4 Bertsekas, D. and Gallager, R., *Data Networks*, 2nd ed., Prentice Hall, 1992.
5 R. Citta, C.C. Lee, and D. Lin, ``Phase 2 Simulation Results for Adaptive Random Access Protocol," contr. no. IEEE802.14-96/114, *IEEE 802.14 working group (WG) meeting*, May 1996.
6 Capetanakis, J., "Tree Algorithms for Packet Broadcast Channels," *IEEE Transactions on Information Theory*, vol. IT-25, no. 5, September 1979.
7 Citta, R., Xia, J., and Lin, D., "The Tree-based Algorithm with Soft Blocking," contr. no. IEEE 802.14-96/244, IEEE 802.14 working group (WG) meeting, November 1996.
8 Golmie, N. and Su, D., "First Transmission Rule: Unblocking Techniques," contr. no. IEEE 802.14-96/245, IEEE 802.14 working group (WG) meeting, November 1996.
9 Bisdikian, C., "Enhancing the Collision Resolution Algorithm: Comments to d2R2," contr. no. IEEE 802.14-97/121, IEEE 802.14 working group (WG) meeting, November 1997.
10 Bisdikian, C., "Performance Analysis of the Multi-slot N-ary Stack Random Access Algorithm (msSTART)," contr. no. IEEE 802.14-96/117, IEEE 802.14 working group (WG) meeting, May 1996.
11 Bisdikian, C., "A Review of Random Access Algorithms," Proceedings of the Int'l. Workshop on Mobile Communications, pp. 123–127, Thessaloniki, Greece, September 1996.

12 Bisdikian, C., McNeil, B., Norman, R., and Zeisz, R., "MLAP: A MAC Level Access Protocol for HFC 802.14 Network," *IEEE Communication Magazine*, vol. 34, no. 3, pp. 114–121, March 1996.

13 Jacquet, P., Muehlethaler, P., and Robert, P., "Performance Implementations of Tree Collision Resolution on CATV Network," contr. no. IEEE 802.14-96/, IEEE 802.14 working group (WG) meeting, April 1996.

14 Sala, D., Limb, J., and Khaunte, S., "Adaptive Control Mechanism for Cable Modems MAC Protocols," Proceedings of INFOCOM '98, San Fransisco, March 1998.

15 Bisdikian, C., McNeil, B., and Norman, R., "msSTART: A Random Access Algorithm for the IEEE 802.14 HFC Network," *Computer Communications*, vol. 19, no. 11, pp. 876–887, September 1996.

16 Sriram, K., and Magill, P., "Enhanced Throughput Efficiency by Use of Dynamically Variable Request Minislots in MAC Protocols for HFC and Wireless Access Networks," *Telecommunications Systems: Special Issue on Multimedia*, vol. 9, nos. 3, 4, 1998.

17 Sriram, K., "Performance of MAC protocols for Broadband HFC and Wireless Networks," *Advances in Performance Analysis*, vol. 1, no. 1, pp. 1–37, march 1998.

18 Doshi, B., Dravida, S., Magill, P., Siller, C., and Sriram, K., "A Broadband Multiple Access Protocol for STM, ATM, and Variable Length Data Services on Hybrid Fiver-Coax Networks," *Bell Labs Technical Journal (BLTJ)*, vol. 1, no. 1, pp.36–65, summer 1996.

19 Golmie, N., Masson, S., Pieris, G., and Su, D., "A MAC Protocol for HFC Networks: Design Issues and Performance Evaluation," *Computer Communications*, vol. 20, pp. 1042–1050, 1997.

Multimedia Cable-Network System Media Access–Control Protocol Performance Simulation

David Hartman
Senior Development Engineer, Residential Broadband Business Unit
Broadcom Corporation

Thomas J. Quigley
Director, Residential Broadband Business Unit
Broadcom Corporation

Introduction

Much has been written and discussed about how the Internet is an unparalleled agent of change in our society. However, the World Wide Web, e-mail, chat groups, town hall meetings, and virtual communities barely scratch the surface of the possibilities that a ubiquitous IP–based network promises when coupled with the access speeds of cable modems. A high bandwidth pipe into the home, covering the "last mile," is a service that can expand beyond best effort–type IP applications to incorporate a burgeoning array of services, including basic and video telephony, multiuser games, and new heights of Web interaction. High-speed access is not enough, however. Without the ability to provide classes of service with committed and guaranteed bit rates and latencies, providing consistent, reliable service is very difficult. The data over cable service interface specifications (DOCSIS) cable modem specification [DOCSIS1] provides the basis of a high-bandwidth, multi-vendor interoperable service over HFC plants. In its first iteration, it did not provide enough tools to deliver all the services anticipated. This paper describes a number of mechanisms needed to provide multiple classes of service including best effort, committed minimum bit rate, and constant bit rate, which should allow a dynamic mix of services that include the low jitter constant bit rate demands. The paper describes a bandwidth allocation mechanism utilizing basic DOCSIS MAP messages and constructs to provide low latency access to the upstream medium. Simulation results of the concepts applied to a DOCSIS environment are shown, starting with committed information rates and moving to mixed quality of service (QoS) situations that are likely to be found in the "last mile" of the Internet age.

Background

The DOCSIS MAC provides both the basic tools for mapping cable modem packets to upstream time intervals as well as mechanisms for bandwidth and latency control for various classes of service. The latter are furnished in a manner consistent with the Internet Engineering Task Force (IETF) integrated services architecture (ISA) [RFC 1633].

DOCSIS MAC Basics

For the purposes of the work described here we focus on the features of the MAC that manage the upstream bandwidth on the cable plant. There are a number of interesting components of the MAC, such as baseline privacy, that we will consider only insofar as they impact throughput and latency on the upstream. The description provided here is limited in scope. For details of the protocol operation, see the DOCSIS RF interface specification [DOCSIS1]. Fundamentally, from the MAC protocol perspective, there are two key components of upstream bandwidth allocation.

1. A time base for the upstream channel: Time on the channel is divided into minislots, which are an integer number of 10.24 MHz system clock time ticks. Each modem locks to the system clock via timestamp messages transmitted periodically in the downstream. At initial ranging time, and through periodic ranging messages, the head end cable modem termination system (CMTS) is able to calculate the time offset of each modem from the source system clock and transmit the information to each cable modem. The cable modems can then adjust their notion of minislot time accordingly.

2. A method for the CMTS to communicate upstream allocation information to cable modems: Every minislot of upstream time is allocated explicitly by the CMTS in maps transmitted in the downstream. While there is more than one taxonomy that can be applied to upstream transmission opportunities, the one of interest here has only two classes: contention and non-contention opportunities. Contention regions are made available to multicast or broadcast groups of cable modems. Initial registration requests, requests for future minislots in which to transmit data uncontested, as well as packet data may be transmitted in contention regions. A non-contention opportunity is granted to a particular cable modem and may be used for transmitting a MAC-related message or a higher-layer protocol data unit (PDU).

Packets from a cable modem reach the CMTS in one of several ways.

1. A cable modem may transmit a request in a contention region for time in a future map. If an allocation, or acknowledgment of the request, is not forthcoming in a future map the cable modem will back off, in truncated binary exponential fashion, and retry the request in a future contention interval. This goes until the request is acknowledged or the maximum sixteen retries have been attempted. In the latter case the packet is discarded.

2. A packet may be transmitted in a contention region large enough to accommodate it. Explicit acknowledgment may be requested from the CMTS. If the acknowledgment fails to appear in some future downstream map, the cable modem backs off and either retries sending the packet or attempts to send a contention request.

3. If the cable modem is in the process of transmitting a packet and has one or more additional packets remaining in its queue, it can piggyback a request for the additional packets in the MAC header of the transmitted packet.

4. The CMTS may send unsolicited grants to a cable modem. The mechanism for establishing the allocation schedule is not part of the original DOCSIS MAC specification. An approach has been recommended by a DOCSIS working group [DOCSIS2], and is described later.

5. The CMTS may send unicast, or non-contention, request opportunities to a cable modem as a form of polling. If the cable modem has data in its queue, it is able to make an uncontested request for a transmit interval. As with unsolicited grants, the method for settling on the polling interval is not part of the original specification.

Figure 1 is a diagram of the DOCSIS MAC allocation mechanism. A single upstream frame is shown on the return path, illustrating the types of transmit opportunities that a CMTS may include in a frame. The sample map in the CMTS block describes a single frame, where the symbol IE refers to map information elements. Each IE allocates one or more upstream minislots for a specific purpose burst type.

The downstream path in the figure, with the sequence of timestamp, data, map, and upstream channel descriptor (UCD) frames, gives an example of the sort of MAC protocol messages and packet data that are carried to the cable modems. Each IE in the map has an associated interval usage code (IUC), which is described in a UCD burst type. Burst-type parameters include things like the type of FEC to apply to an interval.

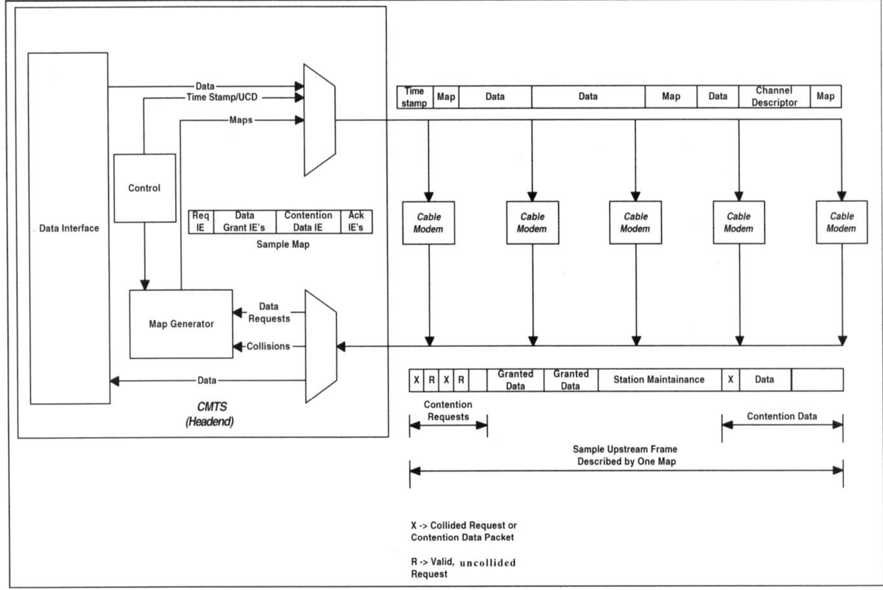

Figure 1: *DOCSIS MAC Upstream Allocation Functional Block Diagram*

An important point to note here is that the CMTS actually makes grants to a service ID (SID) rather than to a cable modem. Each cable modem may have multiple independent SIDs. For instance, a cable modem might have one SID for all of its best effort traffic and another for telephony traffic.

There are two additional features of the bandwidth request/grant mechanism of interest here. First is the ability of the cable modem to concatenate packets into larger "super-packets" in order to reduce physical layer overhead. The obvious drawback is that very large bursts can be generated, which tend to create problems when dealing with traffic streams requiring low latency. Naturally, concatenation must be considered when requesting a transmit opportunity, and may not be appropriate in all situations. The second feature, which will be present in future versions of the specification, is a way for the CMTS to make partial, or fragmented grants. This is essential when considering fixed length frames whose length may be less than the largest packet that must be carried. An approach to fragmentation has been recommended by a DOCSIS working group [DOCSIS3] and is described later in this paper.

The Internet Engineering Task Force (IETF) Integrated Services Architecture (ISA)

Elements of the original Internet protocol suite, consisting of TCP/IP and a simple set of applications such as file transfer protocol (FTP) were designed to deliver fairly distributed, best-effort service to all users of the network. The aim was to arrive at a protocol that was robust in the presence of a sometimes hostile environment. More recently, the IETF has been actively working on improving support for differentiated QoS in the Internet. The ISA is the resulting set of protocol extensions [RFC 1633].

Each ISA capable element, which includes bridges, routers, and subnets, has three basic components:

- Packet classifier: Packets arriving at a node must be routed to a particular flow, defined as a logical aggregation of packets based on some characteristic of those packets. An example would be all outgoing IP packets with a given destination address and TCP port ID. Packets arriving at a node are classified and then placed in the output queue associated with their flow.

- Packet scheduler: Once a packet is queued, it awaits service. The packet scheduler decides the service order for output queues. One example might be a strict priority scheme in which the queue with the highest priority is always serviced first.

- Admission control: In order to offer anything like a guarantee of service, a node must take steps to limit the traffic it handles. As new flows attempt to reserve bandwidth along a multielement link, each element must determine if it can meet the demand for additional bandwidth. Strict limits on capacity are not required, and in fact may not be desirable. Rather, an element can grant admission based on the statistical properties of the flow. For example, a router with capacity for an additional 64 kbps of traffic might accept a reservation for a flow with an average bandwidth of 64 kbps, even though the same flow has a peak bandwidth of 128 kbps.

The protocol tools necessary to manage these components are:

- Packet classifier table, or filter spec: The packet classifier table is a matrix with one row, or filter spec, per flow spec. The table describes the mapping of input packets to output flows.

- Flow specification: The basic tool for communicating the bandwidth and latency requirements of a flow is the flow spec. The Tspec, or token bucket specification, is described as a token input rate and bucket depth. There are other elements of the flow spec, such as the minimum policed unit (MPU) and the maximum transmission unit (MTU).

- Reservation protocol: A reservation protocol is used to communicate information about a flow to the connection admission function, and then, if a connection is established, to the packet scheduler. The most prevalent reservation protocol in the literature is the resource reservation protocol (RSVP) [RFC 2205].

The DOCSIS MAC as a Link in the ISA

Version 1.0 of the DOCSIS MAC has only rudimentary QoS support. During configuration time for a modem SID's may be assigned with a basic flow specification. However, there is no mechanism in place for setting up filter specifications, dynamically creating new flows, or fragmenting large packets into smaller ones. DOCSIS has chartered two working groups to hash out the details of adding these features to the specification.

The QoS working group [DOCSIS2] bears the responsibility for establishing the protocol mechanisms necessary for setting up flow and filter specifications as well as codifying the various service types that will be available. Protocol support for at least four service types is expected:

- Unsolicited grant service: Periodic grants of a fixed size are sent to a SID. When the flow is set up, the grant size and range of interarrival times are established.

- Real-time polling: Periodic unicast request opportunities are made available to a SID. If the cable modem has data to send on that SID, it makes a request in only these unicast request regions.

- Non-real-time polling: Like real-time polling, except the cable modem may make requests in both contention and unicast request region.

- Best-effort, or committed information rate, service: The CMTS can establish token buckets to allocate bandwidth for best-effort SIDs.

Fields will be available in the configuration message for disabling contention requesting, piggyback requesting, and request/data transmissions for a particular service queue. This will allow the cable modem to act appropriately when, for instance, a queue should be responding only to unsolicited grants by disabling the contention request and request/data mechanism.

The second new DOCSIS working group is chartered with developing fragmentation [DOCSIS3]. As attention in the DOCSIS community shifted from carrying best-effort TCP/IP traffic to carrying a wide variety of multimedia traffic, it became clear that some way of controlling the length of packets on the wire would be necessary. In order to carry toll-quality voice data, for instance, the total delay and delay variation from packet to packet must be minimized. Large Ethernet packets can cause problems. For example, a 1500-byte packet on a 2.56-Mbps link can take more than 5 ms to transmit. On a channel-carrying telephony, for which the frame interval is expected to be on the order of 5 ms, such a large Ethernet frame can have a significant impact on voice packet delay variation.

The detailed logical flow for transmitting fragmented packets is shown in *Figure 2*. Essentially, a cable modem SID with a packet to send requests adequate bandwidth to transmit the entire packet. If the CMTS can satisfy some portion of the request, it makes a grant. At the cable modem, if the grant is smaller than the packet, the largest fragment that will fit into the grant is sent. A piggyback request for the remaining data is sent along with the fragment. A special fragmentation header designed to aid the CMTS in reconstructing the received packet encapsulates the fragments.

The System Model

We simulated the DOCSIS MAC using Opnet from MIL3. The foundation for the model was the common simulation framework (CSF) model of the DOCSIS MAC protocol, codeveloped by MIL3 and Cable Labs [OPNET1]. We made a number of modifications to version ten of the basic model in support of our QoS efforts.

Model Description

The model is comprised of three types of elements: cable modems, the head end, and a statistics collection block. Packet sources, which can be configured to produce traffic with a range of arrival time and packet-size distributions, are present in both the cable modems and the CMTS. The model collects packet latency and throughput statistics as well as statistics on the contention regions. All packets are transmitted sized with the correct physical layer and MAC overhead added to the original generated packet size. A five-node system network model from Opnet is shown in *Figure 3*.

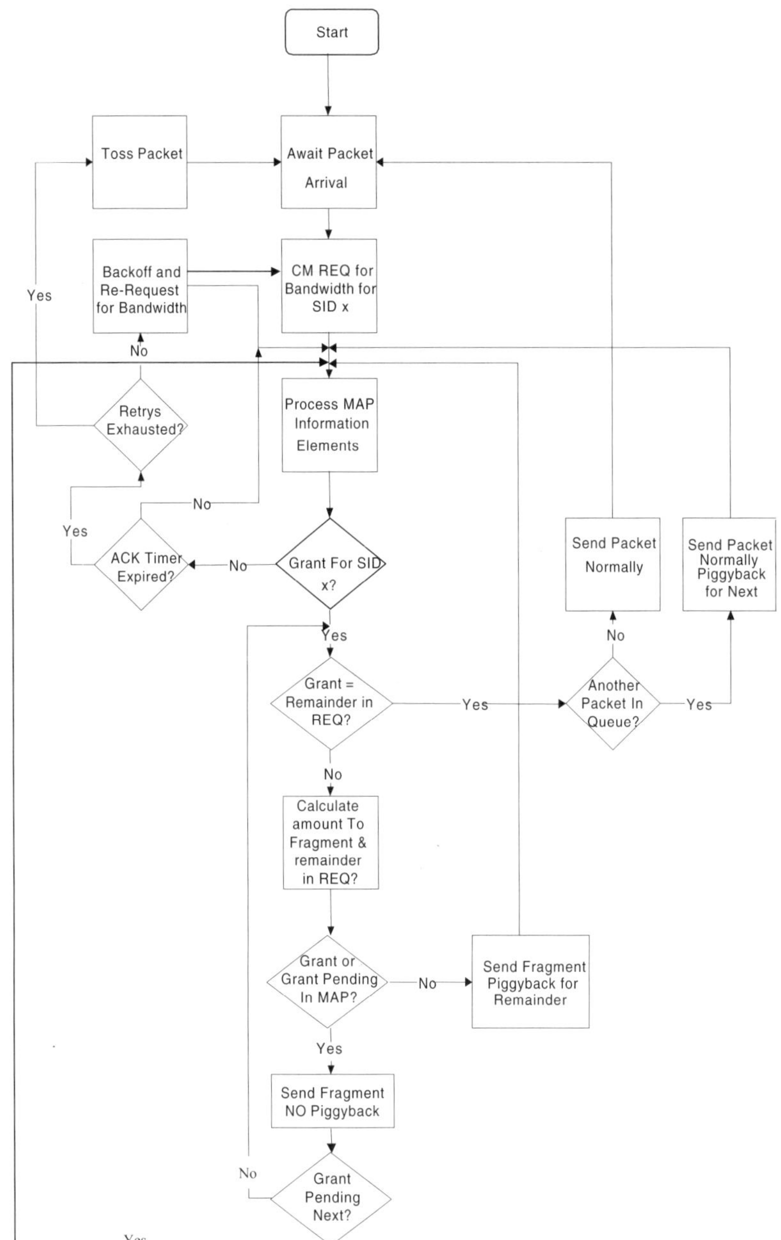

Figure 2: *DOCSIS Proposed Fragmentation Flow Diagram*

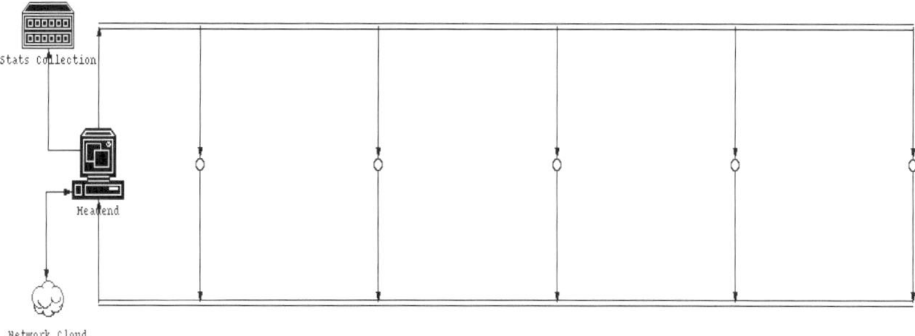

Figure 3: *Opnet Network Model*

The most significant functional modifications to the baseline model were changes to the bandwidth allocation algorithm at the CMTS and in QoS features at the cable modem. QoS parameters are assigned uniquely to each cable modem when the model is built and transmitted to the CMTS in registration packets at the start of a simulation. The CMTS stores the parameters on a per SID basis and uses the information in configuring the bandwidth allocation algorithm.

At the CMTS, we replaced the FIFO allocation algorithm with a token bucket-based one for committed information rate (CIR) flows, and unsolicited grants for constant bit rate (CBR) flows. Both the CMTS and cable modem functions were given a fragmentation capability. The cable modem maintains all fragmentation state information, requesting enough additional bandwidth via piggyback on each transmitted fragment to complete the packet in the next transmission. We also added a new grant type and associated IUC, which can be used to describe bursts with forward error correction (FEC) optimized for voice packets.

Model Assumptions

The system-wide model parameters are given in *Table 1*. Some parameters, like the modulation formats and symbol rates, were set based on assumptions about the physical plant. Others, like the initial and final backoff windows, were set through system performance analysis or trial and error.

The rationale for using twenty voice data cable modems in the simulation is described in the *Traffic Assumptions* section. We use fifty TCP/IP cable modems, which is in line with the TCI Internet Data Over Cable (IDOC) RFP [TCI1]. In our simulations we fix the number of active modems and allow the offered load from each to vary.

Parameter	**Value**				
Upstream data rate	5.12 Mbps (16QAM, 1.28 Msps)				
Downstream data rate	44.0 Mbps (256 QAM)				
Length of network	Last modem at 90Km from head end First modem at 10Km from head end CM's uniformly spaced				
Number of Active TCP/IP Data Cable Modems	50				
Number of Active Voice Data Cable Modems	20				
Packet Concatenation	Enabled				
Packet Fragmentation	Enabled (128 Byte Minimum Fragment Size)				
Request/Data Transmissions (Immediate Mode)	Enabled (8 Minislot boundaries, for 64 Byte Packets)				
Contention Backoff Windows (Power of 2)	Initial = 4 Final = 8				
Upstream FEC/Physical layer parameters	**Long**	**Short**	**Request**	**Request/Data**	**Voice**
Guard Time (Bits)	40	40	32	32	40
Preamble (Bit)	56	48	48	16	48
FEC Message Length (Bytes)	226	75	N/A	30	N/A
FEC Parity (Bytes)	16	10	0	6	0
Shortened Last Codeword Enabled	Yes	Yes	N/A	Yes	N/A
Downstream FEC/Physical layer parameters	**(No MPEG Framing)**				
FEC Message Length	1				
FEC Parity	0				
Interleaver Delay	0				

Table 1: *Model Parameters*

Because each service queue's request/grant engine is independent from all others in a cable modem, with the notable exception that two queues from the same modem can't collide in transmission, we have assumed one SID per cable modem. This results in a slightly pessimistic measure of MAC performance, as observed in some earlier simulation work [BRCM1], but a much simplified simulation model.

A DOCSIS MAC domain is defined as one or more upstream channels under the control of a single downstream channel. The model used here has a single upstream channel.

Traffic Assumptions

We considered two different types of upstream traffic. The first, with randomly distributed packet interarrival times and sizes, is meant to model TCP/IP traffic. In the system this traffic is given committed information rate or best-effort service. Modeling telephony data, constant bit rate traffic makes up the balance of the system load. Guaranteed low latency service is given to CBR packets.

TCP/IP Traffic

It is expected that most upstream TCP/IP packets will be acknowledgments of the higher bandwidth downstream packets. Hence, the majority of packets generated will be 64 bytes long. This size includes 20 bytes each of a TCP and IP header, and 24 bytes of Ethernet encapsulation overhead. The distribution of packet sizes given in *Table 2* was taken from the IEEE 802.14 simulation scenarios document [IEEE1]. Packets are generated with exponentially distributed interarrival times, with a mean value selected to produce the desired average data rate.

Packet Size (Bytes)	Probability of Occurrence	Transmitted Packet Size (Bytes)
64	0.6	96
128	0.06	176
256	0.04	320
512	0.02	592
1024	0.25	1136
1518	0.03	1664

Table 2: *Packet-Size Distribution*

CBR Traffic

Our CBR source models CBR telephony traffic. The protocol stack is described in H.323, as are list voice CODEC standards. We have settled on G.711, the 64 Kbps microLaw compounding scheme, for our simulations. We might have chosen one of the low bit rate encoders, such as the 8 Kbps G.729 CODEC, but G.711 places more stress on the system and is thus a more conservative model. The basic voice frame size, which is based on a 5-ms packet interval, is detailed in *Table 3*.

Element	Size (Bytes)
Voice Data	40
RTP Header	16
UDP Header	8
IP Header	20
Ethernet LLC Header/CRC	18
DOCSIS MAC Header	6
DOCSIS Baseline Privacy Header	5
Total Size	113

Table 3: *CBR Encapsulation without Header Compression*

From the table, we see that without header compression only 35% of the transmitted frame is voice data. There is interest in the DOCSIS vendor community in removing the requirement for LLC encapsulation of voice packets in the MAC. This would save 18 bytes, increasing the fraction of voice data in each packet to 42%. Further compression, involving the TCP/IP headers, would result in even more significant savings.

For our simulation we have assumed that the RTP/UDP/IP headers can all be suppressed and that the additional voice IUC will be available. This will require the CMTS to maintain additional state information on all active voice connections. The DOCSIS MAC header will still be transmitted with each voice packet. The efficiency, calculated from *Table 4*, is 78%. The smaller packet size, coupled with the voice interval usage code (IUC) mentioned in the system model description, results in 64-byte voice packets. These fit nicely into four 16-byte minislots.

Element	**Size (Bytes)**
Voice Data	40
RTP Header	0
UDP Header	0
IP Header	0
Ethernet LLC Header/CRC	0
DOCSIS MAC Header	6
DOCSIS Baseline Privacy Header	5
Total Size	51

Table 4: *CBR Encapsulation with Header Compression*

The Bandwidth Scheduling Algorithm

Some Requirements

The bandwidth scheduling and upstream framing algorithms must meet the latency requirements of our voice traffic without severely penalizing the data traffic. A number frequently used as a bound on the access network delay is 20 ms. This is the total loop delay, which must include all signal processing, queuing, and transmission delays in both the upstream and downstream directions. Taking this into account, and recognizing that selecting a packet size that would fit in an ATM cell would be beneficial, we settled on the 5-ms sample size and 40-byte packet described in the traffic assumptions. This in turn drives a 5-ms upstream frame interval in the MAC.

The number of voice connections carried in the simulation was based on the maximum number of voice slots required in each frame for a call blocking probability of 0.01%. Using the anticipated node size and service penetration figures from TCI [TCI1], and the 0.05 to 0.1 Erlang traffic intensity figure from Bellamy, we find that we need somewhere between sixteen and twenty-five voice slots per frame to meet the needs of 108 customers.

Scheduling the TCP/IP transmission on the upstream should allow differentiated service offerings. A cable modem should be able to contract for a specified assured data rate. If a modem exceeds its assured data rate, its requests should be granted only after the requests of compliant modems.

Finally, the algorithm should be scalable. The processing power required to maintain state information and the downstream bandwidth required to carry maps should be minimized.

Some Assumptions

For the purposes of our modeling efforts we made some simplifying assumptions. Most notably, since we are not addressing connection admission control algorithms we decide through hand calculation how many of what each type of connection to allow. This means that all connections are static throughout the duration of the simulation. For telephony connections this seems a reasonable assumption since most calls last more than a few seconds.

Framing

To meet the low latency and jitter requirements for CBR traffic, the CMTS maps fixed duration upstream frame, as illustrated in *Figure 4*. This might be wasteful at low-offered loads if it weren't for the request/data region that pads the frame out to the fixed size. With light traffic, the region tends to grow, as few requests are made. As the load increases, the CMTS grants more time in the allocated transmit region, and the request/data region shrinks. Small 64-byte packets, which dominate the upstream TCP/IP traffic, may be transmitted immediately when there are request/data slots available.

Broadcast Request Region (10 Minislots)	Unsolicited Grant Region (4*20 = 80 Minislots)	Allocated Transmit Region	Request/Data Region (64-Byte packets only)
0.25 ms (at 5.12 Mbps)	2 ms	2.75 ms	

Figure 4: *Fixed Duration Frame*

Each frame is 5 ms in duration. On a 5.12-Mbps link, with 16-byte minislots, this is equivalent to 200 minislots. Of that total, 10 are taken up for broadcast request regions and 80 for voice packets, leaving at most 110 slots for TCP/IP data.

Traffic Shaping at the CMTS (Scheduling)

The primary tool for managing bandwidth is the service description for each cable modem. At the beginning of a simulation the cable modems transmit their QoS requirements in an upstream message. At the CMTS the information is placed in a set of arrays with entries of the type shown in *Figure 5.* There is one array list for each service type, and the service types are unsolicited grant, polling, and best effort. The top group of entries in the figure are communicated from the cable modem, and the bottom group are initialized and maintained by the scheduling algorithm.

```
typedef struct {
int         SID;
int         qosPriority;
int         qosFragmentationMode;
int         qosMaxBitRate;
int         qosCommitedBitRate;
int         qosMaxPacketSize;
int         qosMaxBurstSize;
double      qosUnicastRequestInterval;
double      qosMaxUnsolicitedGrantInterval;
double      qosMinUnsolicitedGrantInterval;

/* These entries are used to manage a SID. */
int         tokenBucketLevel;
int         fragmentFlag;
double      lastBucketUpdate;
double      lastPollingUpdate;
double      usGrantSize;
}qosEntryStruct ;
```

Figure 5: *CMTS Quality of Service*

Upstream allocation proceeds in two phases. First, requests and statistics on received packets are collected during the ongoing frame interval. Each time a packet is received, the token bucket for that cable modem is updated using the algorithm shown in *Figure 6*. If a request is received, it is queued in the input request buffer.

```
/* Compute the bytes to add to the bucket. Update bucket time*/
deltaTime = op_sim_time() - qosEntry->lastBucketUpdate;

qosEntry->lastBucketUpdate = op_sim_time();
bytesToAdd = (int) (qosEntry->qosCommitedBitRate * deltaTime);
bytesToAdd = (bytesToAdd%8 == 0) ? (int) bytesToAdd/8: (int) bytesToAdd/8 + 1;

/* Limit bucket size to RT + B */
qosEntry->tokenBucketLevel += bytesToAdd;
qosEntry->tokenBucketLevel =
(qosEntry->tokenBucketLevel <= qosEntry->qosMaxPacketSize) ?qosEntry-
>tokenBucketLevel : qosEntry->qosMaxPacketSize;

/* Subtract incoming from bucket. */
bytesToSubtract = (packetSize%8 == 0) ? (int) packetSize/8: (int) packetSize/8 + 1;
qosEntry->tokenBucketLevel -=bytesToSubtract;

/* Limit the negative going bucket size to RT + B */
qosEntry->tokenBucketLevel =
(qosEntry->tokenBucketLevel > -qosEntry->qosMaxPacketSize) ?qosEntry-
>tokenBucketLevel : -qosEntry->qosMaxPacketSize;
```

Figure 6: *Token Bucket Code Segment*

The second phase is the map building phase. First, a ten-minislot region of the map is set aside for contention requests. This is followed by grants for modems with unsolicited grant service. The remainder of the frame is allocated to stations that have requested time on the channel and, if there is any room left once all the requests have been satisfied, a region of request/data slots.

Prior to mapping requests, the input request queue is reordered based on the difference between the fill level of the token bucket and the size of the request from each modem. The scheduling algorithm requires at most two accesses to the token bucket for each upstream packet and a sort operation on the request queue for each map. Requests that can't be satisfied in the current map are held in the queue, and zero length grants are made in the map.

Our assumption is that cable modems that are being held off by the CMTS will eventually run out of queuing resources, and will begin to discard packets. The dropped packets will indirectly signal the congestion condition to the TCP, which will operate to reduce the offered load from the modem. Thus, the congestion control mechanism is distributed between the cable modem's TCP kernel and the CMTS bandwidth allocation algorithm.

Results

Best Effort

The best-effort simulation models a single class of best-effort TCP/IP traffic in the presence of telephony data. For these simulations the voice traffic load was fixed at 20 connections, and the TCP/IP offered load was allowed to increment. By using a fixed size frame and unsolicited grants, we insured that delay for voice packets would be tightly grouped around the frame duration irrespective of the TCP/IP load. For these simulations, on a given run each cable modem had the same input offered load. *Figure 7* shows the aggregate TCP/IP throughput as a function of the aggregate input offered load.

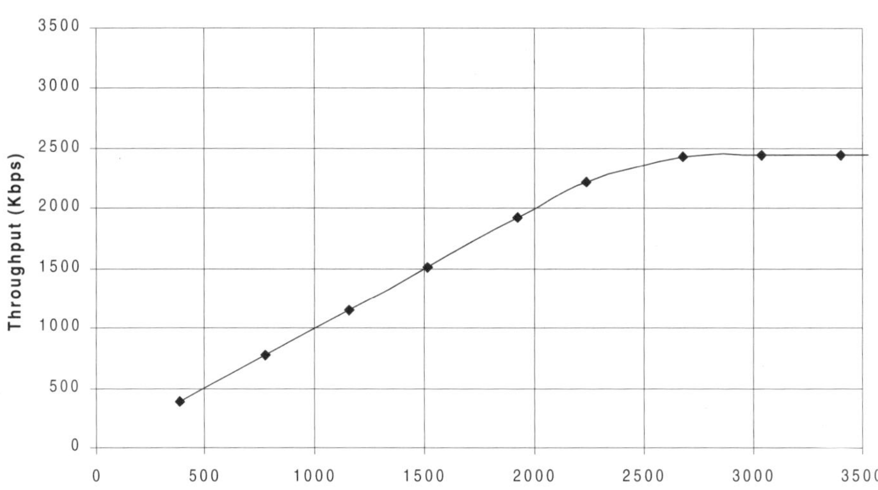

Figure 7: *Best-Effort TCP/IP Throughput vs. TCP/IP Offered Load*

Since we expect at most 110 slots of TCP/IP in each 200-slot frame, on a 5.12-Mbps channel the resulting maximum available bandwidth is 2.82 Mbps. The figure shows a maximum throughput of 2.4 Mbps. The difference is a result of MAC framing and physical layer overhead. The peak throughput from each individual modems was 48 kbps.

The average delay experienced by a TCP/IP packet is shown in *Figure 8*. For input loads below saturation the delay is less than 35 ms, or 7 frame intervals. Given that a packet can arrive any time during an ongoing frame and be scheduled for

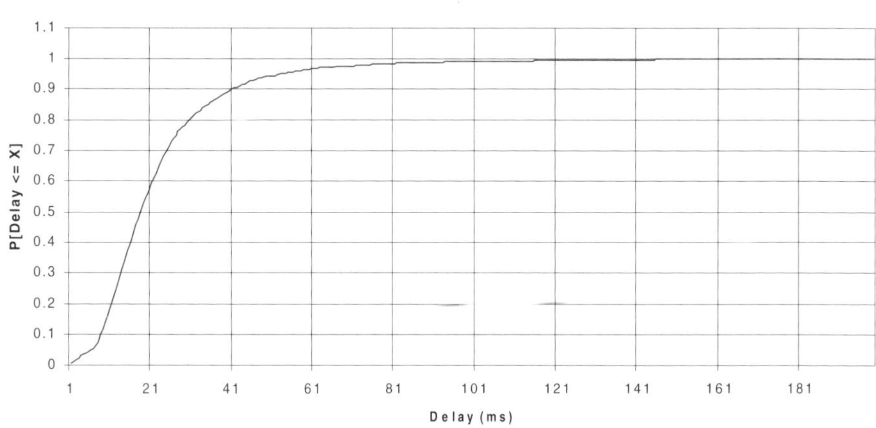

Figure 8: *TCP/IP Mean Access Delay vs. Offered Load*

transmission any time in the TCP/IP region of some following frame, it takes an average of six frame intervals for a modem to gain access to the channel.

Perhaps more interesting, from a TCP connection point of view, are the statistics of delay near saturation. The CDF of *Figure 9*, taken at the edge of saturation for the network, shows that 90% of all TCP/IP packets experience less than 41 ms of delay traversing the network.

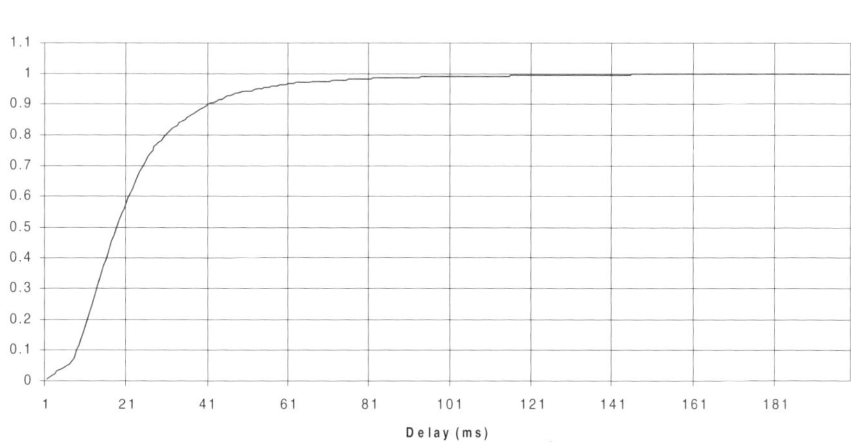

Figure 9: *TCI/IP Access Delay CDF for an Aggregate Offered Load of 2.0 Mbps*

Figure 10 shows contention throughput performance beside the theoretical slotted aloha performance. Note that for all offered traffic loads, the contention offered load never exceeds more than one third of the available contention opportunities. This is a result of the piggybacking mechanism taking the load off the contention region when the channel becomes congested.

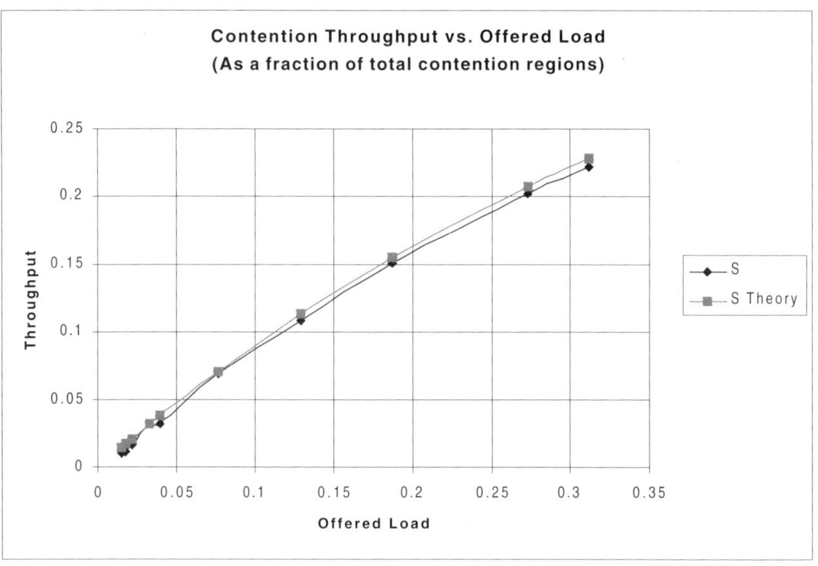

Figure 10: *Measured and Theoretical Contention Region Performance*

With Differentiated QoS

Map bandwidth is one measure of the load placed on the system in managing upstream bandwidth. For the differentiated QoS simulation, we measured the required map bandwidth when managing two different levels of upstream TCP/IP CIR service in the presence of CBR traffic. As with the best-effort simulation, twenty voice connections were carried and each TCP/IP cable modem offered the same input load. In this case, however, half of the 50 cable modems established a CIR of 64 kbps and the other half one of 28.8 kbps. The resulting throughput from one modem from each set is shown in *Figure 11*.

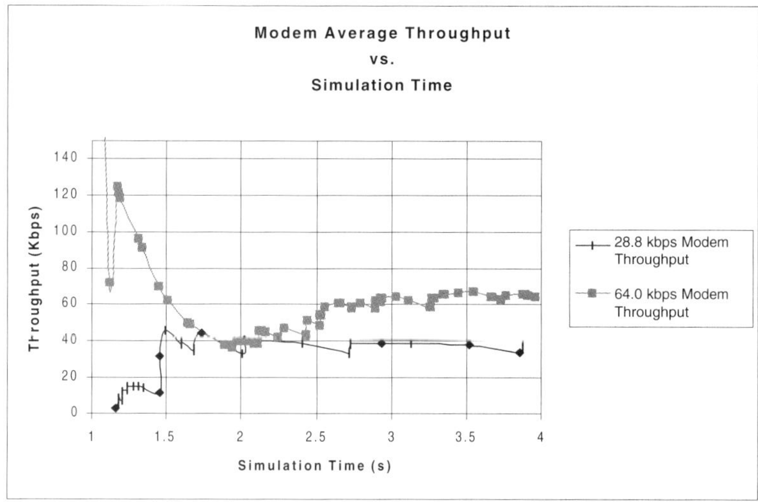

Figure 11: *Average Throughput from CIR Cable Modems vs. Simulation Time*

Figure 12 shows sizes of the maps transmitted during the simulation. The average size, once equilibrium is reached, is roughly 300 bytes. For a single upstream channel this represents 480 kbps of downstream bandwidth consumed by maps. Assuming that this will scale linearly for an eight upstream-channel system, the resulting bandwidth is 3.84 Mbps, less than 10% of the available downstream bandwidth.

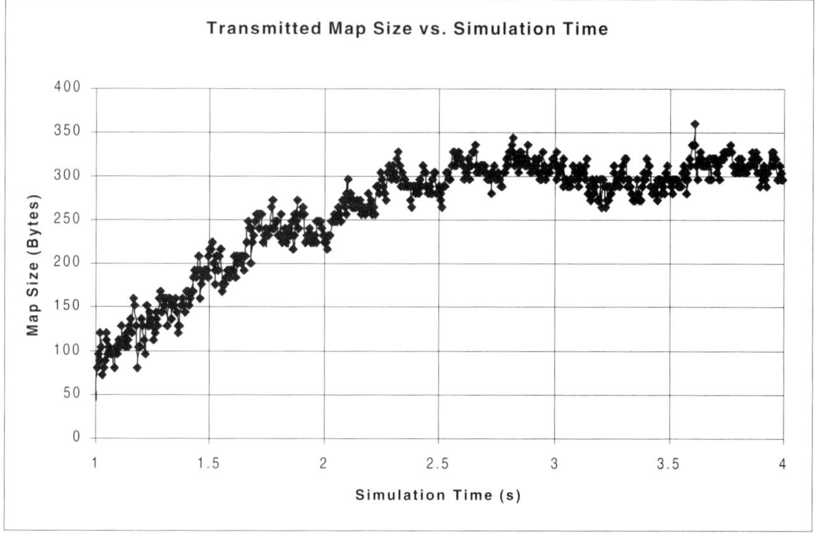

Figure 12: *Map Size for the CIR Simulation*

Conclusion

The simulation results show that a small amount of downstream map bandwidth and CMTS processor cycle time can buy tight control over frame duration, allocation of CIR bandwidth, and variations in latency for telephony traffic. Through the token bucket specification and dynamic allocation of upstream services, DOCSIS will have the tools necessary to support a variety of service types. This flexibility means that service providers will be able to furnish cost-based differentiated service offerings. The fact that the mechanism for establishing a modem's QoS requirements will be standardized allows designers of CMTS scheduling software to differentiate their products based on performance, while maintaining interoperability across all DOCSIS–compliant cable modems.

References

[DOCSIS1] "Data-Over-Cable Service Interface Specifications, Radio Frequency Interface Specification, SP-RFI-I02-971008."
[DOCSIS2] "Data-Over-Cable Service Interface Specifications, Extensions to Support DOCIS Quality of Service. TR-QOS-D02-980317."
[DOCSIS3] "Data-Over-Cable Service Interface Specifications, Extensions to Support DOCIS Fragmentation. TR-FRAG-D02-980414."
[OPNET1] Common Simulation Framework Opnet Model, and Application Note, www.cablemodem.com/vendor/docsiswgs/wgdocs.html, user ID and password required, contact CableLabs.
[RFC 1633] "Integrated Services in the Internet Architecture: An Overview." R. Braden, D. Clark, S. Shenker, June 1994.
[RFC 2205] "Resource Reservation Protocol (RSVP)," R. Braden, L. Zhang, S. Berson, S. Herzog, S. Jamin, September 1997.
[TCI1] "TCI.NET Interoperable Data Over Cable System, Request for Proposal, 1997, Section V–Technical Requirements."

Performance of Telco-Return Cable-Data Systems

Levent Gun
Vice President, Cable Access Business Unit
3Com

Chester A. Ruszczyk
Senior Software Engineer
3Com

Abstract

This paper reports simulation results of a telco return cable data system where users download files from a Web server using the transmission control protocol/Internet protocol (TCP/IP). The data from the server to the clients travels through a high-speed cable channel, while the traffic from the client to the server (TCP acknowledgments) travels through a 28.8-kbps dial-up connection. We simulate the appropriate link layer and TCP flow control protocols and report application level performance measures such as transaction throughput and response time. Our study indicates that the performance seen by a client is more sensitive to TCP window size and the distance between the clients and the server across the network than to the upstream bandwidth of the telco link. In fact, in most cases, the perceived performance of a telco return system is similar to that of a two-way system. Therefore, while many trade-offs exist in the decision to deploy a telco return or a two-way system for Internet access, performance viewed by a user is not one of them.

Introduction

Installed cable plants today are predominantly unidirectional, although some cable companies have aggressive plans for plant upgrades for two-way data communications over cable. With unidirectional facilities, it is still possible for cable operators to provide high-speed data to customers today using the telephony (telco) return architecture shown in *Figure 1*. This figure shows only one of the many potential configurations, wherein the cable company takes advantage of the dial-in ports already deployed by many Internet service providers (ISPs) today. Other configurations are possible, but are beyond the scope of this paper.

This paper addresses the performance issues related to the system shown in *Figure 1*. The study examined the impact of perceived system performance by an end user as a function of variables such as round-trip delay, TCP window size, and the upstream transmission speed over the telco line. The performance measures used in the study are the throughput and delay as seen by an end user downloading a 200-KB http page from a Web server over a TCP/IP network. In particular, the delay is the time to download the entire page, and throughput is the average rate at which the IP packets arrive at the cable modem over the cable network. The TCP acknowledgments are encapsulated over a point-to-point (PPP) protocol and are sent over the telco link. TCP header compression is not incorporated. The study assumed the framing proposed by Multimedia Cable Network System (MCNS) architecture over the cable networks. The paper begins with the detailed simulation model in section 2 followed by the results in section 3.

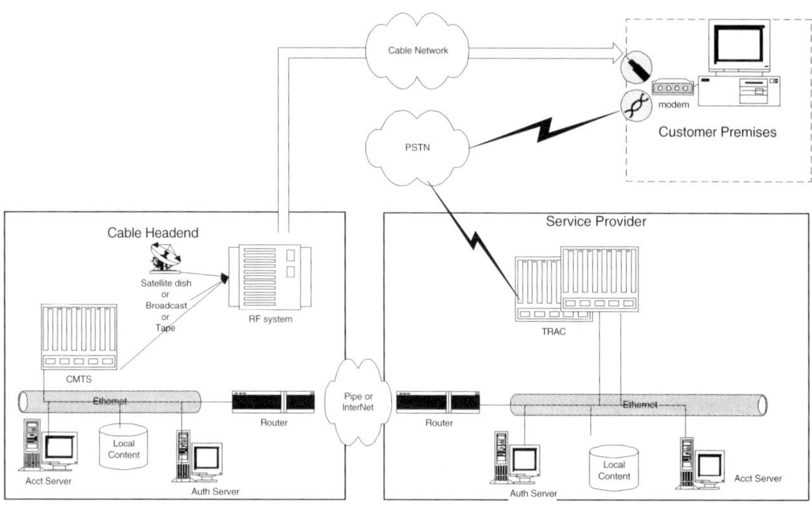

Figure 1: *Telco Return Cable Data System*

Simulation Model

The simulation model of the system under evaluation is shown in *Figure 2*. The actual simulation model contains 100 cable modems (CMs) that are all simultaneously active. Each CM is connected, via a PSTN upstream and cable coax

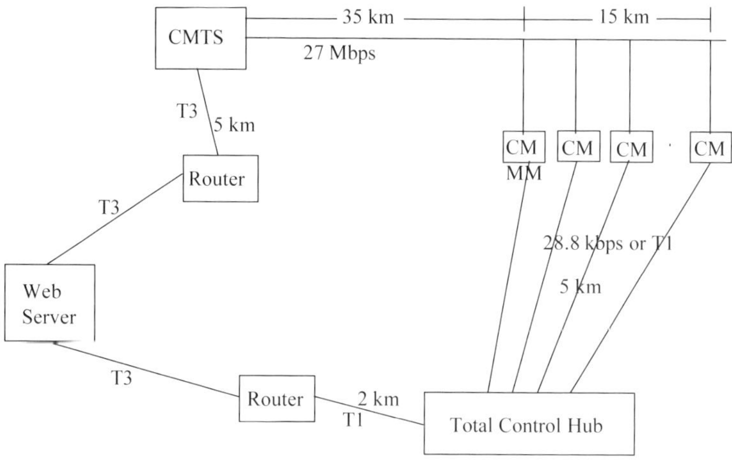

Figure 2: *Simulation Model*

downstream channel, to a Total Control Hub (US Robotics' dial-in network access server) and a cable modem termination system (CMTS), respectively. The upstream channel is used to send data to the Web server and the downstream channel to receive data from the Web server.

The distance between the CM and the Total Control Hub is assumed to be, on average, 5 km. The upstream channel data rate is fixed and for this model is either 28.8 kbps or a T1 rate. The distance between the CMTS and CM is between 35 km and 50 km, with the CMs uniformly distributed between 35 km and 50 km. The downstream channel data rate is 27 Mbps and excludes the physical layer overhead.

The Total Control Hub communicates with the Web server by traversing a single router. The Total Control Hub is connected to the router through a 2-km T1 line. The Web server transmits data to the CMTS through a single router. A 2 km line with a T3 data rate is assumed from the router to the CMTS. The distance between the upstream and downstream routers and the Web server is configurable with a T3 data rate. The study assumed the distance to be between 10 km and 4000 km.

Figure 3 shows the protocol layers simulated in this model. The CM client randomly generates a request to the Web server and waits for a response. Each request triggers the Web server to download a 200-KB Web page fragmented into a sequence of 1500-byte TCP/IP packets, except for the last packet, i.e., a total of 137 packets. The end-to-end TCP flow control mechanism determines when the packets are to be sent from the Web server toward the CM. There can be at most a window size (in bytes) worth of unacknowledged packets outstanding. After sending packets with the total byte count up to the window size, the server stops and waits for the CM to send acknowledgments over the telco link.

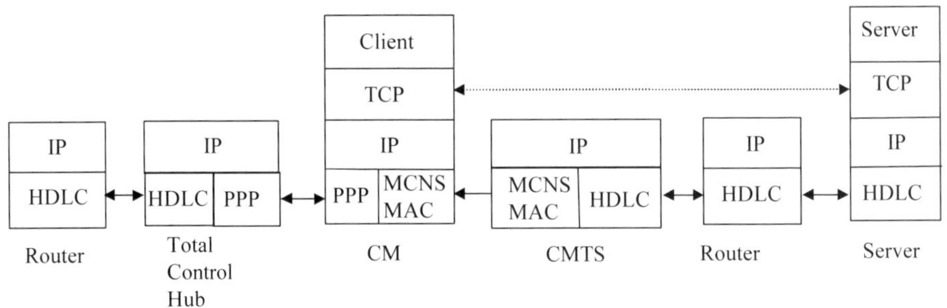

Figure 3: *Protocol Layers Simulated*

At the CM, a TCP acknowledgment (ack) packet is sent based on the TCP flow control mechanism. Ack packets are generated only when there are downstream packets received that need acknowledgments. Each ack packet acknowledges all packets received during a 10 msec window, i.e., the ack delay timer is set to 10 msec. Each ack packet is 40 bytes in length.

The PPP protocol is used to simulate the modem connection between the CM and the Total Control Hub. The PPP compression option is not incorporated. PPP overhead adds an additional six bytes to the TCP PDU.

The data link layer protocol, HDLC, handles the communication between the routers and Web server and between the router and CMTS.

The MCNS media-specific access control (MAC) stack, in *Figure 3*, simulates the downstream transmission from CMTS to CMs. Every packet transmitted on the downstream channel is encapsulated with a six-byte MCNS header and an 18-byte Ethernet/IEEE 802.3 overhead. We further assume that the simulation includes a 4 msec forward error correction (FEC) processing delay at the CMTS.

The two statistics evaluated in this simulation are the average transaction response time E[T] and the average peak transaction throughput E[R]. *Figure 4* aids in defining the statistics collected during the simulations. In this figure, a CM generates a request at time t_1. This request results in a sequence of packets transferred from the Web server starting at time t_2. The interdeparture times of these packets from the server, or the interarrival times at the CM, represent a random process whose statistics are determined by the TCP window mechanism. The last packet of this transaction (e.g., page download) occurs at time t_3. For each such transaction, t_3-t_1 represents the transaction response time T. The peak transaction throughput E[R] is the average rate at which packets arrive at the CM during the interval t_3-t_2. We use a transaction size (200 KB) / (t_3-t_2) to represent this random variable R.

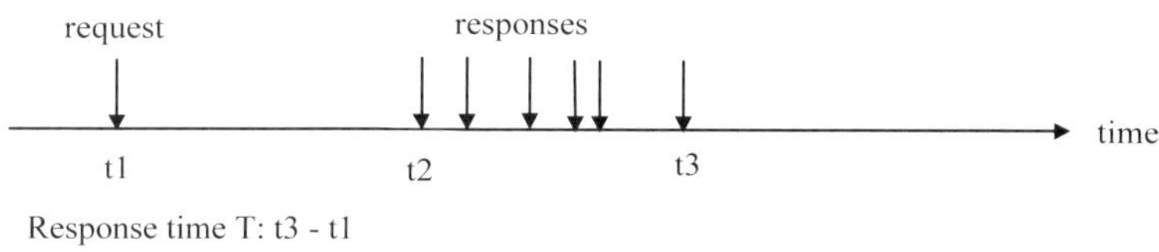

Response time T: t3 - t1

Peak throughput E[R]: E {# of bits received in [t2,t3] / (t3 -t2)}

Figure 4: *Response Time and Peak Throughput*

Each CM generates an initial request to initiate the TCP download, according to exponential random variable that is independent and identically distributed. The mean of the R.V. is ten seconds. With 100 CMs, and with the TCP/IP and MCNS overhead, the utilization of the 27 Mbps downstream cable channel is about 60%.

In all simulations, we have fixed the topology of *Figure 2*, except for the distances between the Web server and the two routers, which we varied but always kept equal to each other. Also, in addition to the standard 28.8-kbps dial-up telco link, we used an upstream rate of T1 (1.5 Mbps) to understand how sensitive the performance is to the upstream channel data rate. Note that the results on a two-way cable system, with a 1.5-Mbps upstream quaternery phase shift keying (QPSK) channel, should fall between the two systems simulated here. This is because here we assume there is a dedicated T1 to each CM and no delay due to collisions and contention resolution.

Simulation Results

Figures 5, 6, 7, and *8* show the average response time (E[T]) and peak throughput (E[R]) under changing round trip time, TCP window size, and the upstream data rate. In some of the newer TCP implementations the window size is negotiated at the start of the session and will dynamically change over time based on network congestion. Since we wanted to see how different window sizes affect the performance, we have used a fixed window size in each of the simulations.

Figure 5 and *Figure 7* show E[R] and E[T], respectively, for a 16 KB window size with two upstream channel rates for different round-trip delays between the routers and

delay	0.05 ms	20 ms	40 ms	60 ms	80 ms
T1 Rate	7.50 Mbps	2.27 Mbps	1.34 Mbps	0.94 Mbps	0.73 Mbps
28.8 kbps	4.38 Mbps	1.86 Mbps	1.18 Mbps	0.86 Mbps	0.68 Mbps

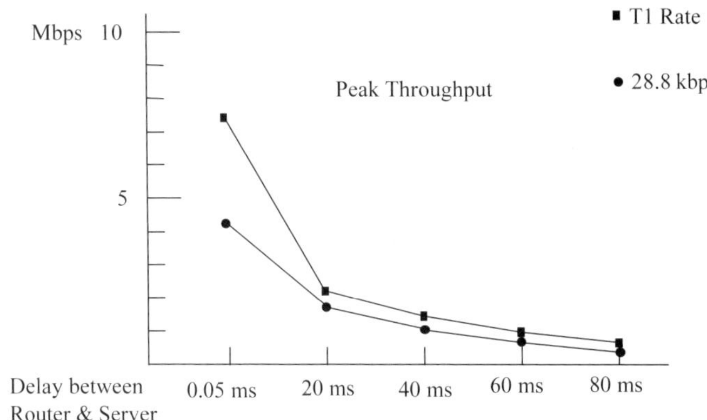

Figure 5: *Peak Throughput E[R] for 16-KB Window Size with Different Delay between Router and Server*

window size	64 KB	32 KB	16 KB	8 KB	4 KB
T1 Rate	8.90 Mbps	4.75 Mbps	2.27 Mbps	1.11 Mbps	0.42 Mbps
28.8 kbps	7.40 Mbps	3.91 Mbps	1.86 Mbps	0.91 Mbps	0.36 Mbps

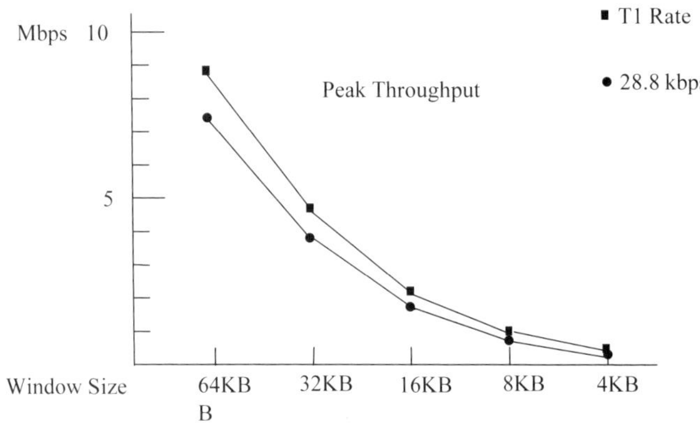

Figure 6: *Throughput E[R] for 20-ms Delay Between Router and Server with Different Window Sizes*

delay	0.05 ms	20 ms	40 ms	60 ms	80 ms
28.8 kbps	0.41 sec	0.96 sec	1.52 sec	2.08 sec	2.64 sec
T1 Rate	0.24 sec	0.78 sec	1.34 sec	1.90 sec	2.46 sec

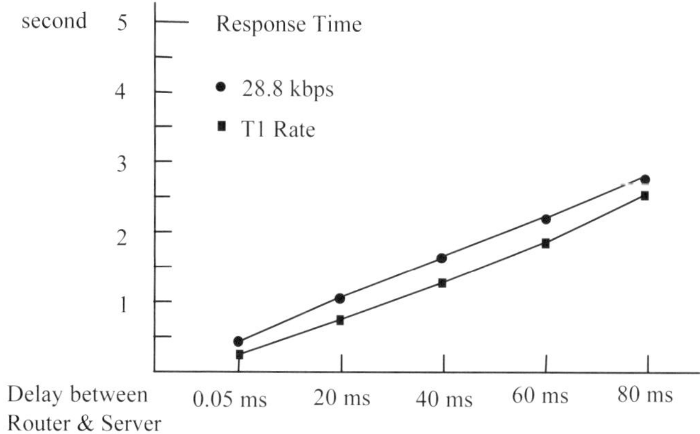

Figure 7: *Average Response Time E[T] for 16-KB Window Size*

window size	64KB	32KB	16KB	8KB	4KB
28.8 kbps	0.31 sec	0.50 sec	0.96 sec	1.91 sec	4.70 sec
T1 Rate	0.25 sec	0.41 sec	0.78 sec	1.55 sec	3.83 sec

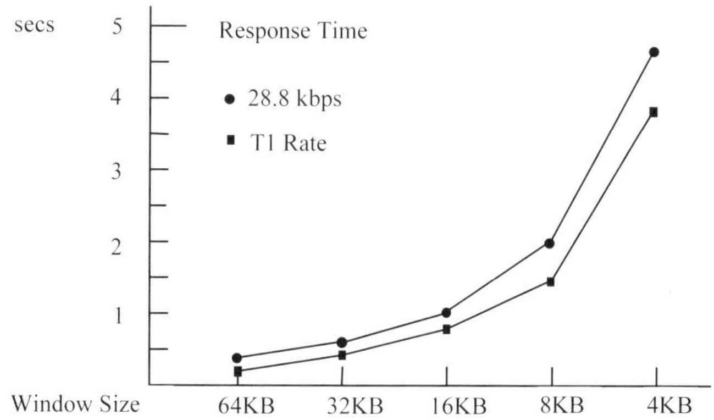

Figure 8: *Average Response Time E[T] for 20-ms Delay Between Router and Server*

the servers. The largest discrepancies in performance occurred with very low delays, 0.05 msec, with E[R] and E[T] providing the T1 rate a 41.6% and 41.5% performance advantage, respectively. At more realistic round trip delays, between 20 msecs and 80 msecs, the performance advantage for T1 over 28.8 kbps was between 18.1% and 7% for E[R] and 18.7% and 7.6% for E[T].

Figure 6 and *Figure 8* show E[R] and E[T] when the round-trip delay was assumed fixed at 20 msecs with a variable TCP window size from 4 KB to 64 KB. The major performance effects that the window size added were that E[R] was reduced by a factor of 20 when the window size was reduced from 64 to 4 KB. E[T] increased as the window size decreased. Between the 64- and 4-KB range, E[T] increased by a factor of 4. Note that in both cases the differences between the performance for the T1 upstream over the 28.8-kbps rate was between 18.1% and 14.3% for E[R] and 19.4% and 18.5% for E[T].

Conclusion

This paper evaluates a telco return cable system performance through simulation in the transfer of 200-KB frames using the TCP/IP protocol. Different system scenarios were considered to gain insight into the system's performance. From the simulation results, it is evident that a telco return system can provide the end user with several Mbps of throughput when downloading files and Web pages.

While several factors did impact the performance of the system, the upstream channel rate did not have a significant impact on the downstream performance, in terms of throughput and response time for reasonable round-trip distances/times between the CM and the Web server. However, factors such as the TCP window size and round-trip delays proved to have a greater impact on the performance. Note that the round-trip delay included all the processing delays as well as propagation delays.

Based on the results obtained using a dedicated T1 link for the upstream traffic, we do not anticipate any perceived performance differences between a telco return and two-way cable return system, assuming that the round trip time between the CM and the server is approximately the same.

In light of these observations, while there are many trade-offs for a cable company deciding whether to deploy a two-way or telco return cable data system for Internet access, performance should not be an issue in making that decision.

Part IV: Network Routing Layer

The Architectural Differences between DOCSIS–Compliant Bidirectional and Telco-Return Systems

Nurettin Burcak Beser
Software Architect, Cable Access Business Unit
3Com

Abstract

This paper discusses the differences between DOCSIS specifications for bidirectional and telco return cable systems from a network architecture viewpoint.

Introduction

The data over cable service interface specifications (DOCSIS) are managed by an organization called Multimedia Cable Network System (MCNS) Partners Ltd. MCNS consists of four firms: Comcast Cable Communications, Inc., Cox Communications, Tele-Communications, Inc., and Time Warner Cable. MCNS has partnered with Rogers Cablesystems Limited, Continental Cablevision, and Cable Television Laboratories, Inc. (CableLabs) to participate in and support the development of the necessary set of communications and operations support interface specifications for cable modems and associated equipment. This partnering of interests represents operators that provide cable service to a majority of North American cable subscribers (85% in the United States, 70% in Canada, and 20% in Mexico).[1]

DOCSIS project objectives are:[2]

1. Write interface specifications to facilitate a variety of data communications service offerings over modern hybrid fiber/coax (HFC) networks that are compatible with both typical customer-provided equipment as well as cable system hardware and business systems.

2. Ensure that these specifications are vendor-independent yet still accepted and supported by the vendors so that hardware from multiple sources will be available on a timely basis and yet allow for interchangeability.

DOCSIS defined two distinct architectures for the needs of its cable plants. The first, DOCSIS RFI specification (DRF), defines the cable system architecture for bidirectional cable plants.[3] The second, DOCSIS TRI specification (DTR), builds on the DRF specification and is intended to be used for cable plants that have only unidirectional capacity.[4]

Even though the DTR (DOCSIS TRI specification) builds upon the DRF (DOCSIS RFI specification), the architectural burden of having two unidirectional connections to possibly two different geographical locations makes the two specifications seem almost the same. Actually, they are considerably bigger.

In this paper the differences between DRF-defined bidirectional cable plant systems and DTR-defined unidirectional cable plant systems will be examined, and the reasons for these differences will be investigated.[5] It was not in the scope of this paper to investigate or examine the nonarchitectural issues regarding the physical layer or the data link layer.

Cable Network

The cable network is the cable television plant that would typically be used for data over cable services.[6] The DRF defines the cable network as depicted in *Figure 1*.

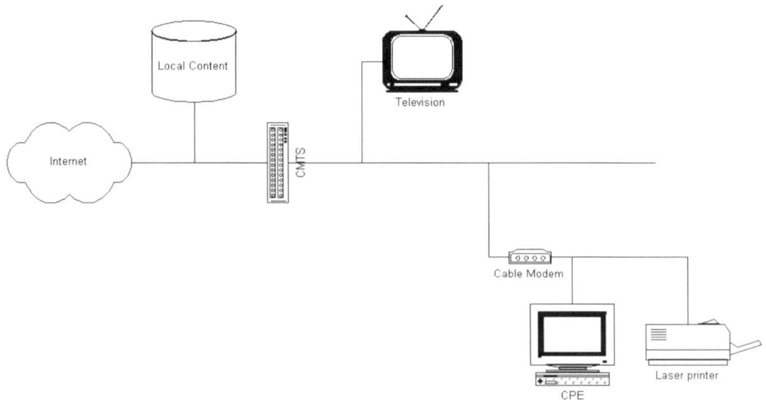

Figure 1: *Bidirectional Cable Network*

In *Figure 1*, the user (host computer) is connected to a cable modem via Ethernet. The cable modem is connected to a two-way cable RF path over a low-split HFC cable system. Thus, in the downstream direction (toward the subscriber), the cable modem receives signals in a predefined portion of the downstream passband, which will be between 50 MHz and a system-dependent upper limit that could be 750 MHz or more. In the upstream direction (away from the subscriber), the cable modem would transmit signals in a predefined portion of the upstream passband, which will be between 5 MHz and 42 MHz.

At the head end, corresponding cable modem termination system (CMTS) equipment will interface with the subscriber cable modems over the two-way RF path just described. The subscriber cable modems and the CMTS equipment will also interface and communicate with the required network, operations, and business support systems.

Since DTR assumes a unidirectional cable plant, the cable network considered is different, as shown in *Figure 2*.

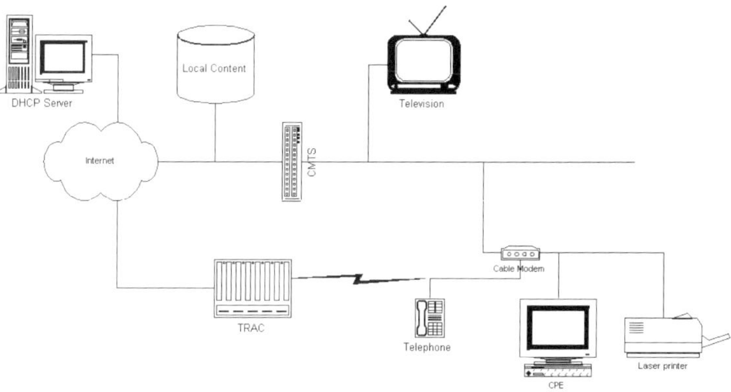

Figure 2: *Telco Return Cable Network*

The most important addition to the bidirectional cable network is the existence of a telephony remote access concentrator (TRAC). Using an analog modem, the cable modem sends IP datagrams upstream via the public switched telephone network (PSTN) in point-to-point protocol per RFC-1661.[7] The TRAC decodes the IP datagram and forwards it to its destination. When IP datagrams are directed for cable modems, the CMTS receives them and sends the datagrams via the cable link.

Cable Modem Termination System Requirements

The cable modem termination system (CMTS) is located at the cable television system head end or distribution hub. It provides complementary functionality to the cable modems to enable data connectivity to a wide-area network.[6] CMTS is typically connected to a backbone network, which in turn would be connected to local content servers and a gateway server to the Internet.

The DRF defines CMTS as an IP forwarder. For CMTS, the support of other network-layer protocols is optional. DRF states that address learning and address aging are vendor-dependent. The data forwarding through the CMTS may be a transparent bridge, a router, or an IP switch. For the purpose of showing the architectural differences from now on, the DRF CMTS will be considered to be a transparent bridge unless otherwise noted.

The DTR defines CMTS as a router per RFC-791.[8] Additionally, CMTS acts as a proxy ARP per RFC-1027 on the downstream cable network.[9] The construction of the ARP table is defined by gleaning through the dynamic host configuration protocol (DHCP) messages as described in detail later in this paper.

Cable Modem Requirements

The cable modem (CM) is a modulator-demodulator at a subscriber location intended to convey data communications on a cable television system.[10]

The CM in DRF is defined as a link-layer bridge following 802.1d general rules: [11]

- Frames addressed to unknown destinations must not be forwarded from the cable port to the host side.
- Broadcast frames must be forwarded to the host unless they are from source addresses at the provisioned or learned host CPE addresses.
- Multicast frames must be forwarded in accordance with filtering rules.
- Frames addressed to unknown destinations must be forwarded to cable.
- Broadcast frames must be forwarded to the cable.
- Frames originating from hosts other than the ones provisioned or learned must not be forwarded.

For DRF systems, address aging is defined as the following:[3]

- The CM must provision/learn the address of the CPE devices until the maximum number of CPE addresses are reached. Once the limit is reached, the new CPE addresses must not be learned.

- Host CPE addresses must not be aged out. If the CM is reset, then all the addresses must be discarded.

In sum, it can be said that DRF systems might be constructed with the CM and CMTS as transparent bridges; the CM can never forget the ARP entries once it learns; and the CMTS might forget them after a period of time. The effects of these rules will be discussed later in the paper.

The CM in telco return systems has two connections. The downstream connection is defined as the same as bidirectional systems, but the upstream connection is defined via analog telephone modem.

The DTR defines the CM to be a router per RFC-791,[8] with DHCP relay agent process per RFC-2131.[12] The specification also states that the CM cannot generate ARP messages to the upstream link. Instead, the CM must respond in a normal way as a default gateway to ARP requests from host CPEs.

Telephone Remote Access Concentrator Requirements

The telephone remote access concentrator (TRAC) is the equipment that terminates dial-up point-to-point protocol (PPP) sessions from the CMs. The TRAC may be located elsewhere than the CMTS and have routing associations to the CMTS. Together, the CMTS and TRAC are called the telephone return termination system (TRTS), whether or not they are colocated. TRACs may be in different geographic locations.

The TRAC in DTR is defined as a router per RFC-791,[8] and it must support the PPP per RFC-1661 and the Internet protocol control protocol (IPCP) per RFC-1332.[7,13]The TRAC must be capable of forwarding IP datagrams destined to an IP unicast address across the cable network or telephone link. The TRAC's IP broadcast capability is dependent upon the configuration of direct linkage, if any, between the TRAC and CMTS. If the TRAC and CMTS are in different geographical locations the TRAC might filter broadcast messages, or if desired might tunnel the broadcast messages to the CMTS.[14]

Before IP Connection

The CM and CMTS communicate using the media access control (MAC) protocol. The protocol enables data exchange between data link layers using the MAC frames. The IP packets are carried via the MAC frames. The DRF defines MAC management messages for communicating the CM and CMTS even before IP connection.

These MAC management messages can be divided into three groups: signaling messages, channel change messages, and registration messages. The signaling messages control the synchronization and upstream bandwidth allocation. The channel change message is used for forcing the CM to change the upstream channel on which it is transmitting. Both of the messages are not used in TR systems, since the phone line is not a shared medium. The registration message will be discussed later in the paper.

In addition to DRF, the DTR defines two new MAC management messages: the telephony channel descriptor and the termination system information.

The telephony channel descriptor (TCD) includes the information about service providers. The name "service provider" suggests the possibility of connecting different points-of-presence (POP) via telephone line, but it is also possible to define the same POP with different identities to provide different services.[15]

The TCD would include the name of the service provider (or the service for the reasons stated above), phone number(s) to connect, user name/password to connect, DHCP server IP address, and radius realm. The TCD enables users to choose the provider (service) they want to work with. This is different from the DRF systems in which the CM is only differentiated with its MAC address (the unique identifier necessary for network connections).

The good thing about the DTR is that it enables the system to have different personalities according to the person that logs in the network. For example, you can set different personalities for yourself (no restrictions), your child (game channel only), and your connection to work (a virtual tunnel to your company), and at the same time get different accounting levels for each service provided.

The second new MAC management message defined by DTR is the termination system information. The message defines the CMTS because without the information, the loop from telephone line to cable network cannot be drawn. The information in this message is used in closing the IP loop from telephone line to CMTS and registration.

Assigning IP Addresses for the Cable Network

The DRF dictates that the CM must use DHCP to get its IP address and filename and the TFTP server IP address for a configuration download. But the specification fails to define the assignment of IP addresses for host systems in cable network. Since DRF CM is a transparent bridge and CMTS can be a transparent bridge without any IP layer software, the DHCP DISCOVER broadcast would be sent out without any special treatment, making it impossible for DHCP and/or management software to define in a controlled way which IP address to assign.[16] The other difficulty arising from this is that since the CM and CMTS might be transparent bridges, controlling IP addresses would be a problem.[17]

The DTR defines the DHCP messaging in a rather detailed way, since it is the most important part for closing the loop from the telco side to the cable network. The process can be briefly described as the following:

1. The CM generates a DHCP DISCOVER message. In the message it fills the giaddr field with the CMTS IP address it gathered from the TSI message (see the *Before IP Connection* section). It sends it as a broadcast message or, if defined in the TCD message (see the *Before IP Connection* section), as a unicast to the DHCP server.[18]

   The most important event here is the filling of the giaddr field. The DHCP standard RFC-2131 states that giaddr field can only be filled by the first relay agent that sees the DHCP message.[12] By filling the giaddr field, the CM creates a "virtual" relay agent between the CM and TRAC.

2. The DHCP server sends the reply message to the first relay agent defined by giaddr field, which is the CMTS defined by the TCD message.

3. The CMTS relay agent sees the DHCPOFFER sent by DHCP server, notes that the downstream network interface IP address is addressed, and sends the packet through the downstream interface denoted.[19]

4. The CM at the downstream interface receives the DHCPOFFER. If there are multiple DHCPOFFERs, CM makes a selection and sends the DHCPREQUEST with giaddr field filled with the CMTS IP address.

5. The DHCP server sends the DHCPACK message to the "virtual" first relay agent CMTS.

6. This time, the CMTS creates an ARP table entry consisting of a MAC address of the CM and IP address assigned through the DHCP process. The process of adding an ARP entry using DHCP messaging is called "gleaning."[20]

7. From now on the CMTS will act as a proxy ARP for the IP address that is gleaned for the CM.[21]

The process described above puts some restrictions on the TRAC and routers between the TRAC and DHCP server. These restrictions will be explained in detail later in this paper.

Unlike the DRF, the DTR also defines in detail how a CPE can receive an IP address to connect the cable network: The procedure is a normal DHCP operation with the CM as a relay agent. There is no need for a virtual relay agent, since the CMTS advertises the relay agent CM (or the relay agent below CM) under the CMTS.

On the CMTS, all the crossing DHCP message traffic is listed. If the received DHCP message is destined for a cable network and is of type DHCPACK, then it associates the given IP address with the MAC address of the giaddr IP address.[22]This way, the host CPE packets would be sent to the CM.

The method described above has the advantage of using standard DHCP elements. In this way it is very easy to control the IP addresses to be given to certain CM's hosts, or even to set the DHCP resolved parameters such as the netmask difference per CM. The subnet behind the CM can be controlled using standard DHCP server software.

Registration Process

Registration is the last step of CM initialization. With the registration the CM presents itself to the CMTS showing the capabilities of the modem. At the same time, the CM rebounds the configuration file with the settings of privacy and class of service information for itself.

The registration process defined by DRF uses MAC management messages. The REG–REQ and REG–RSP messages are used for this purpose. The CM first loads the configuration file, which contains the privacy and class of service information set for itself. The configuration file name and the server to load it is given by the DHCP process. The CM first checks the CM message integrity check (MIC) field. The CM MIC is calculated using MD5 digest over the bytes of configuration setting.[23] This only ensures that the CM has received the configuration file intact. After it verifies that the configuration file is intact, the CM parses through it and obtains CM-specific information such as network access control (showing whether the CM is allowed to connect to the network or not) and adds its own information, like modem capabilities (e.g., does the modem support concatenation or not).

Upon reception of the REG+REQ MAC message, the CMTS checks the CMTS MIC. The CMTS MIC is calculated over a defined set of configuration settings in a predefined order. The MIC is calculated using MD5 and HMAC algorithms.[24] The use of HMAC enables the CMTS to check the authenticity of the configuration file.[25] After checking the authentication of the message, CMTS builds a profile as described in the configuration file and sends the REG-RSP message with the necessary parameters for data communication between the CM and CMTS.[26]

The DTR defines the REG–REQ and REG–RSP messages as UDP messages, since it is not possible to send MAC messages through the PPP link in a standard manner. The only other difference is that it is possible using the TR_HOSTS field to define the connected host CPEs in the REG–REQ message. This is necessary since it is possible that users would disconnect the phone and then connect at a later time without rebooting (or forcing them to make DHCP queries) all the network-addressable elements.

Using the TR_HOSTS option, once connected the CPE does not have to go through the DHCP process each time the telco connection is made.[27] The CMTS receives the host CPE IP addresses and builds the ARP table. After the successful build, it sends the REG–RSP through the UDP message (as it would in DRF case) but this time, if defined in the REG–REQ message, it includes the host's IP addresses that are added to the ARP table.

Resolving Addresses in the Cable Network

One of the main differences between the DRF and DTR systems is the resolution of addresses. The DTR specification makes sure that the ARPs are always proxied. The DRF defines that ARP messages (broadcasts) must be sent to the cable on the CM but leaves the point undefined for CMTS. If communication between two host CPEs is desired, the CMTS must rebroadcast the ARP messages. At this point the network security might be compromised. Consider the following scenario:

> The DRF CM is at a business site working as a transparent bridge. One user tries to print a document without realizing that the printer is off-line. The PC ARPs the IP address of the printer. The CM does not see the printer's IP address before sending the message upstream. The CMTS rebroadcasts the ARP message, which is sent through the CMs. A malicious user sends an ARP reply saying that the requested address is below his/her CM. The PC sends the file to be printed to the MAC address, which is defined under the malicious user's CM. All the malicious user has to do is to get the file printed on her/his printer.

The same thing can happen with DHCP and the other broadcast-oriented protocols. Since host CPEs would have no means of authenticating the replies, it is difficult to prevent spoofing.[28]

In telco return cable networks, the ARP is proxied at both the CM and the CMTS, with the CMTS building its ARP table through gleaning controlled by the DHCP server on which the network manager has control.[29] And the CM follows the default gateway response to ARP requests from host CPEs.

Data Privacy

Privacy provides user data secrecy by encrypting traffic that flows between the CM and CMTS. The downstream privacy is the same for both DRF and DTR systems. But unlike the DRF, the DTR systems do not have upstream privacy. The omission of the DTR privacy on upstream comes from the fact that the telco line (upstream) is not a shared medium as the cable network is.[30]

Additional Network Requirements for Telco Return Systems

The DTR poses additional requirements arising from the fact that the same network segment is not used both ways:

1. The TRAC must allow the DHCP allocated CM and CPE addresses to go out.

   Generally, TRACs are configured in such a way that the only addresses that might go out are the ones that it distributes by itself. This poses a problem for the host CPEs.[31] It should be ensured that the host CPE addresses can go out.[32]

2. The TRAC should allow the Martian source addresses address (at least for DHCP ports) to go out.[33]

   The DHCP messaging starts with the use of a Martian IP address (0.0.0.0) as a source address since the device has no legitimate IP address assigned to it. The TRAC should permit the use of Martian addresses as their source.

3. The routers between TRAC and CMTS should allow a DHCP unicast message with a Martian source address to go through.[34]

   Since the TRAC and the DHCP servers might be at different geographical locations, the DHCP unicast message with a Martian source address should arrive at the DHCP server.[35]

4. The CM should either have a legitimate IP address, or the routers between the CM and the CMTS should allow the CM address range to pass through (as source address).[35]

   The MCNS defines that the CMs must be SNMP manageable, meaning that the CMs must be able to send IP packets to the CMTS. For this to happen, the CM's source IP address should not be blocked by any of the routers between the TRAC and the CMTS.

Other Telco Return Architectural Differences

The DTR defines the service-recovery message (through the well-known MCNS UDP port) for effective fault tolerance. User station restart (USR) messaging is used to force any portion of the CM initialization to take place: restarting the PPP link, restarting the DHCP, re-performing transfer of operational parameters, and re-registering CMTS. At any time, network managers can force the CMs to reinitialize from any step they want.[36]

The TSI message (see *Before IP Connection*) includes an epoch showing the number of times the CMTS booted and reboot time with unspecified accuracy.[37] The boot-count epoch is also incremented each time the ARP table is flushed.[38] When the CM detects a change in both of the parameters, it should do DHCP messaging and should include in registration the host CPEs that are in its host's ARP table.[39,40] If the CM has just booted, it checks for the parameters to see if there is any change;[41] if the parameters are different, it goes through the initialization process but the CMTS response includes the host CPEs of the CM.

Conclusions

The DTR has advantages over the DRF in the areas of user service selection—management, network security, and network management. The weakness of the DTR comes from the fact that it has two unidirectional paths potentially leading to different locations.

Fortunately it is possible to have best of both worlds. The DTR also specifies a three-way modem that allows all the DTR functionality on a bidirectional communication between a CM and a CMTS.[42] In such a modem there will not be a use for the TCD-carried phone number, but the other TCD fields would make the user choose from available services.[43] Such a cable modem would also have network security via a proxy ARP, and easier management via service recovery. But it would not have the problems of a DHCP UDP port and source address filtering for CMs and host CPEs.

Cable networks have the potential to become the broadband communication medium of the near future. All that needs to be done is to make it happen.

Notes

1 DOCSIS FAQ, *http://www.cablemodem.com/mcnsfaq.htm*
2 DOCSIS, *http://www.cablemodem.com/*
3 DOCSIS, "Radio Frequency Interface Specification," *http://www.cablemodem.com/public/pubtechspec/rf/sp-rfi-i02-971008.pdf*
4 DOCSIS, "Cable Modem Telephony Return Interface Specification," *http://www.cablemodem.com/public/pubtechspec/tri/sp-cmtri-i01-970804.pdf*
5 The specifications examined in this document are the public releases of DOCSIS specifications as of November 11, 1997. The reader is encouraged to check the *http://www.cablemodem.com/public/pubtech.htm* for the latest releases. Also, the reader should keep in mind that the documents examined are Interim Specifications, not Released Specifications.
6 DOCSIS, "Cable Modem Termination System–Network Side Interface Specification," *http://www.cablemodem.com/public/pubtechspec/nsi/sp-nsi-i01.pdf*
7 Simpson, W., "RFC-1661 The Point-to-Point Protocol (PPP)," July 21, 1994, *http://ds1.internic.net/rfc/rfc1661.txt*
8 Postel, J., "RC-791 Internet Protocol," September 1, 1981, *http://ds1.internic.net/rfc/rfc791.txt*
9 Carl-Mitchell, S., Quarterman, J. S., "RFC-1027 Using ARP to implement transparent subnet gateways," October 1, 1987, *http://ds1.internic.net/rfc/rfc1027.txt*
10 DOCSIS, "Cable Modem to Customer Premises Equipment Interface," *http://www.cablemodem.com/public/pubtechspec/cmci/sp-cmci-i01.pdf*
11 ISO/IEC 10038 (ANSI/IEEE Std 802.1D), Information technology—-Telecommunications and information exchange between systems—Local-area networks—Media access control (MAC) bridges, 1993.
12 Droms, R., "RFC-2131 Dynamic Host Configuration Protocol," March 1997, *http://ds1.internic.net/rfc/rfc2131.txt*
13 McGregor, G., "RFC-1332 The PPP Internet Protocol Control Protocol (IPCP)," May 1992, *http://ds1.internic.net/rfc/rfc1332.txt*
14 The tunneling of broadcast messages is not defined in the DTR. The reason for this is to enable the specification to work with any TRAC currently installed. But the DTR does not prohibit the use of a broadcast tunnel between TRAC and CMTS. In

this respect it is possible for an implementation to tunnel all messages coming from a PPP link to the CMTS. The DTR allows the vendor-dependent TRAC– CMTS interaction.

15 The easiest way to achieve the same POP but to have different identities is to give different radius realms for each service and to give different DHCP server IP addresses for each service. This way all the other fields might remain the same, but the user would be authenticated using the service he/she chooses depending on the radius realm. The user would then be given an IP address and/or configuration file name depending on which DHCP server is connected. There are other ways to achieve this, but they are outside the scope of this paper.

16 The problem is impossible to solve using the current set of RFCs. But it is possible to solve it when one considers using a specially tuned DHCP server and a special aid from CMTS.

17 It is possible to exert some control over the IP filters that MCNS mandates. But in this case it would require a different configuration file for each CM.

18 According to the DTR, if the DHCP server IP address is not defined then the CM must broadcast the DHCPDISCOVER message. The behavior of the TRACs for this broadcast is not defined clearly. So it is always better to define the DHCP server IP address field.

19 Since a downstream IP address is used, it is possible to have more than one downstream interface per CMTS.

20 "Gleaning" is the single most important event in the process of closing the telco line to the cable loop. Without gleaning it is impossible (within the specification) for the CMTS to know for which IP addresses to proxy ARP.

21 How long would the ARP entry be kept in the ARP table? The DTR only defines that in the event of DHCPDECLINE and DHCPRELEASE the CM must copy the messages to the CMTS, and the CMTS would flush the IP address from the ARP table. The problem with this method is that it might not always be possible for the CMs to send DHCPDECLINE. For example, the phone line might be disconnected.

22 This part in the DTR is not correctly stated. Since CM is a router, the new IP address should have been paired with the routers (CM) MAC address. The author believes that it will be corrected in the next release of the document.

23 Rivest R., "RFC-1321 The MD5 Message-Digest Algorithm," April 1992, *http://ds1.internic.net/rfc/rfc1321.txt*

24 Krawczyk H, Bellare M., Canetti R., "RFC-2104 HMAC: Keyed-Hashing for Message Authentication," February 1997, *http://ds1.internic.net/rfc/rfc2104.txt*

25 The authenticity is done against the configuration file build procedure. But this does not prevent someone else from using the same configuration file for premium service, and then, instead of sending the one that was received (e.g., minimum service),

to send the premium service configuration file. The author believes that this will be fixed in the next release of the DRF document.

26 The information passed through in REG-RSP is the CMTS' response to modem capabilities and service ID-class of service tuples. The subject of service ID-class of service is very broad and a subject for another paper.

27 But DTR fails to note that the configuration file can also contain the TR_HOSTS IP addresses. If this were the case, then it would be possible to define passive hosts (e.g., printers) that have static addresses and no DHCP support. The author believes that this will be fixed in the next release of the DTR document.

28 One of the answers lies in the MCNS defined filters. The filters can be set in such a way that broadcast communications will not be sent downstream again, and the CMTS will proxy ARP the cable network. It is not possible to have proxy ARP capability at all CMTSs since it is not mandated in DRF. A side effect of this approach is that the cable network cannot be used in the manner that the Ethernet works. If you want to reach an IP device, you have to wait for it to send a packet upstream so that the CMTS would learn of its existence. A CM has to wait for the host it wishes to communicate with to reach out to the cable network in order to have an ARP entry that can be proxied (this has also other problems).

29 This prevents the other CMs and host CPEs from spoofing ARP replies, since the CMTS proxy ARPs and CMTS ARP table is constructed using DHCPACK messages.

30 It is possible to make virtual local-area networks using tunnels between the CM and the CMTS, with encryption on the data flowing through the tunnel. The problem with this solution is that the standards are not stable enough.

31 For CMs, one solution to this is to make the TRAC give out IP addresses, get the DHCP to approve the addresses, and set the TRAC not to proxy ARP for addresses given.

32 But in no case should the TRAC proxy ARP for CMs or host CPEs. Otherwise, the loop cannot be closed from the cable line.

33 One way to overcome this problem is to use a PPP–assigned IP address as a source IP address during DHCP messaging. But this requires the TRAC to give out legitimate IP addresses. Another way is to use a source address that is considered valid for all CMs connecting. Since the DHCP reply does not depend on the source address, there will not be any conflict if all the CMs use the same IP address for DHCP messaging.

34 The solution given in the note before this one might use a legitimate IP address for the whole network, overcoming the problem. Some routers have default filters against DHCP messages regardless of source addresses.

35 Another way to achieve this is to have a tunnel between the TRAC and the CMTS for the source addresses with a CM subnet.

36 Since this sensitive message is sent through the well-known MCNS service UDP port, the port must be filtered in the CMTS against hostile attempts.
37 If the CMTS does not have non-volatile storage, then the value should be set to zero.
38 From the CM viewpoint, the flushing ARP table would have the same effect as rebooting.
39 The DTR does not explicitly define the CM action.
40 The CM registration of host CPEs might be a threat to network security (see *Resolving IP Addresses in a Cable Network*).
41 The DTR CM must store the last epoch and time values it observed on the TSI message.
42 The three-way modems are the way to start with a telco return system, and later when the cable plant is upgraded, upgrade to a bidirectional cable line connection.
43 The user authentication mechanism for such a selection process is yet to be defined.

Future Directions for Telephony Return–DOCSIS Networks

John Fijolek
Chief Network Architect, Cable Access Business Unit
3Com

Abstract

"We believe that Viewdata is a major new medium according to the McLuhan definition; one comparable with print, radio, and television, and which could have as significant effects on society and our lives as those did and still do. Like them, it may well lead to major changes in social habits and styles of life, and have long-lasting as well as complex economic effects." (Fedida and Malik, 1979)

Fundamental telephony return data over cable television (CATV) concepts trace back to the Viewdata invention envisioned thirty years ago by Sam Fedida from the British Postal Research Laboratories. Products that have delivered data over CATV with return paths through a telephone network have been around in some form or another since that time. Naturally, the bandwidths are higher today, and the modulation methods are more sophisticated than they were thirty years ago. More important, for the first time, an interoperability standard exists for economical deployment of this technology. This standard enables mass-marketing of data delivery systems over cable. Now the burden of end user appliances can be retailed to the end user. PCs with interoperable retailed cable modems can replace specially adapted TV sets and provider-distributed set-top boxes.

The Multimedia Cable Network System (MCNS) Partners Ltd.'s data over cable service interface specifications (DOCSIS) cable modem telephony return interface (CMTRI) is an outgrowth of the original Viewdata vision. These CMTRI specifications have evolved from the original concept to include the increased data capabilities that have gained acceptance in the overall technical community. The DOCSIS CMTRI specification references many component data capabilities. These build upon applicable Internet Engineering Task Force (IETF), Society of Cable Telecommunications Engineers, Inc. (SCTE), and International Telecommunications Union (ITU) standards. In their own right, these component standards are also evolving within their respective standards-making bodies. This paper explores directions that DOCSIS CMTRI–compliant systems may take in the future as enabling data protocols and inherent technical capabilities mature.

Introduction

Books in the early 1980s described the commercial, social, and educational potential of viewdata and interactive videotex as the next estate in communications (Sigel, 1980; Woolfe, 1980; Chorafas, 1981; Winsbury, 1981). These systems introduced inventions like bulletin board systems (BBSs), advanced file transfer (AFT), and closed captioning. But the overall potential of on-line access never materialized until the advent of low-cost PCs and the availability of the World Wide Web. The high cost of videotex equipment and the lack of interworking between different vendors' products made these data services too expensive to attract consumers in any significant volume.

It was not until the 1990s with low PC prices and the popularity of the World Wide Web that cable operators projected the potential of cable data systems as effective mass-market cost centers. With that mindset, CATV providers created a means to realize open data over CATV specifications. Leading CATV providers and CableLabs formed MCNS Holdings L.P. to supply DOCSIS in a timely, market-oriented manner suitable to enable cross-vendor interoperability.[1,2]

While the eventual goal of MCNS member cable companies is to bring two-way broadband interconnections (radio frequency (RF) upstream and RF downstream) into homes, less than 15% of the cable plants in North America are capable of two-way traffic. Consequently, any massive, efficient deployment of cable data modems has to have alternative means to deliver upstream (cable subscriber to cable head end) data. In particular, telephony return cable data offers more than an interim technology. It has potential to be a long-term service that integrates the best features of both the public switched telephone network (PSTN) and cable plants. For telephony return, the MCNS project team solicited input from all interested parties and selected vendor authors for the specifications.[3] The CMTRI specification underwent rigorous MCNS and vendor review and was finalized by the DOCSIS working group. As distributed to the public in August 1997, this specification enables suppliers to confidently begin building interoperable, high-speed cable modems that can be marketed to most cable plants regardless of infrastructure.

The CMTRI specification extends cable data access to legacy systems. This is done by enabling existing cable plants, Internet service providers (ISPs), subscriber equipment, and software, instead of the more advanced RF infrastructures and more advanced CPE enabled by two-way systems. The CMTRI includes telephony modems and their associated dial-in remote access modem ports piggybacking on a huge existing ISP base. The goal of the CMTRI specifications is to extend cable-data opportunities to the majority of PC owners in North America.[4] By design, there are fewer physical limitations to preclude telephony return modem deployment than two-way RF cable modem deployment. In other words, if a cable subscriber's PC has dial-modem access capabilities, an opportunity to benefit from cable modem downstream capability is available through the CMTRI specifications. Thereby, CMTRI specifications create many market entry points to the overall Internet access market for providers. These entry points provide vital leverage for the providers to respond to aggressive competitive technologies in market windows that two-way cable modems may not be agile or timely enough to address.

The DOCSIS CMTRI specifications are based in part on existing IETF standards. As these standards and other standards-track IETF drafts evolve, aspects of the CMTRI will change to take advantage of expanded capabilities. This paper will look at standards track activity in the IETF with respect to new directions for DOCSIS CMTRI specifications.

Quality of service (QoS) and system maintenance are high prerogative considerations for any network. Telephony return cable systems, by virtue of the existence of separate physical plant connections, have the potential to enable special service support and diagnostic capabilities. This paper will discuss some telephony return system QoS and maintenance possibilities.

IP Address Administration

The IETF standards for the dynamic host configuration protocol (DHCP) form the base for all DOCSIS–compliant IP address administration. The DHCP standard RFC-2131 is referenced from the CMTRI for additional system functionality. In the CMTRI, a CMTS is required to create all cable modem termination system (CMTS) router address resolution protocol (ARP) table entries using the DHCP assignment process. The process of adding an ARP entry using DHCP messaging is called "gleaning."[5] Also, in the CMTRI, both the cable modem and the CMTS are required to be boot protocol (BOOTP) relay agents.

All methods described in the CMTRI use options that were standard DHCP elements when the CMTRI was published. There are drafts being considered in the

IETF DHC working group that could make the CMTRI DHCP interfaces more flexible.

PPP and DHCP

The point-to-point protocol (PPP) (IETF RFCs 1661, 1662, and 1663) is specified by the CMTRI for the upstream telephone connection. In PPP connections, automatic assignment of IP addresses is handled during PPP/IP control protocol (IPCP) negotiation. However, for various reasons, DOCSIS–compliant cable modems require DHCP IP administration. A telephony return cable modem with both a wide-area network (WAN) and cable connection can end up requesting IP addresses from both IPCP and DHCP.

There is no standard method specified in the IETF that integrates PPP, IPCP, and DHCP. Several manufacturers' remote access server (RAS) devices use proprietary means to deliver IP addresses through IPCPs that have been acquired from a DHCP server. This is sometimes referred to as a DHCP proxy function. Since PPP has strict timeout restrictions for messages, most of these implementations acquire a block of addresses from DHCP. The RAS then distributes addresses from that block to dial-in clients with IPCP. In these implementations the RAS, not the dial-in host, is the DHCP client.

For an MCNS cable modem initialization, there are two problems with the described proxy of DHCP by a RAS. First, contrary to the existing specification, the CM is not a DHCP client, and the DHCP server has no way of knowing or tracking the CM MAC address. The TRAC would be a proxy client for the CM and the DHCP server has no communication with, or knowledge of, the CM. Second, IPCP configuration options do not allow negotiation of desirable DOCSIS parameters. DHCP options that define the next server and the CM configuration file to be received from that server cannot be passed to the CM with IPCP. Without the DOCSIS parameters received from a configuration file, a CM is unable to register for service from a DOCSIS–compliant head end unit (CMTS).

One solution attempting to address both of these concerns is to have a cable modem send a DHCPinform message, in a proxy scenario, after receiving its IP address. This informs the DHCP server about actual client associations. DHCPinform is a DHCP message newly defined in RFC2131 but not yet widely used or supported. A DHCP server receiving DHCPinform would need to update its database to associate the CM MAC address with the IP address previously leased to the TRAC.[6] The DHCP server could then modify its persistent storage to indicate that the CM device's DHCP client can be made responsible for future IP address lease renewal. The subsequent DHCPoffer would contain the original IP address and options that point the cable modem to its TFTP server and configuration file. Hence, only one IP address is ever used to initialize a cable modem.[7]

Network Address Translation (NAT)

The major use of NATs today is to interconnect routing realms based on RFC1918 private addresses. NATs modify addresses in the IP header and adjust transport layer (TCP/UDP) checksum. A network address port translator (NAPT) translates a large pool of private addresses into a single address, but maps applications to different TCP/UDP port numbers. Using a NAT is contrary to DOCSIS requirements and to methods approved by most ISPs. However, the IETF has begun considering some kind of NAT standardization.

In a cable modem, a likely NAT implementation would have the cable modem spoof responses to DHCP queries from its CPE hosts. The NAT would appear to be a DHCP server to the CPE. This cable modem NAT (CM NAT) would need to make up for the loss of end-to-end address significance by maintaining a state for each session and managing the CPE host IP address lease state.

The CM NAT would obviate a lot of host management from the provider. This may not be an issue if the aggregate traffic billing model fits with the provider marketing policies. There are issues that need to be understood by NAT users and the providers that allow them. Since NATs are application-independent and do not do syntax and semantics of application data stream, they deal very poorly with application data that contain IP addresses, e.g., the file transfer protocol (FTP) PORT command. There can be policy issues with subscribers reselling services with NATs on either a CM or subscriber host. For the subscriber and provider there may be potential service, maintenance, and end-to-end security limitations created by NATs.

Tunneling

Tunneling supports the development of secure, multi-protocol virtual private networks (VPNs) over the Internet. One way to think of VPNs is as a virtual WAN. A virtual WAN is not physical, but forms on demand through a software call that sets up a point-to-point session between a tunnel client or front end processor (FEP) and a tunnel-enabled server. In a DOCSIS cable network, VPNs can give a corporate user end-to-end control from cable host CPE to corporate server. The tunnel connections are like private, controlled phone calls (rather than party lines) into the target servers because they can be set up, managed, and disconnected at will by either party. Tunneling is a solution when a company needs to make confidential all communications between two end points. Furthermore, it can facilitate a company's desire to let users roam between multiple Web servers, each of which houses confidential information.

From the cable plant perspective, tunneling between a host CPE and an outside corporate network should be transparent. However, potential tunneling between the cable modem and CMTS within the telephony return cable network itself is another possibility.

Layer 2 Tunneling Protocol (L2TP) is where most of the standards attention is being focused. Point-to-point tunneling protocol (PPTP) preceded L2TP in the IETF and formed a basis for L2TP signaling with some differences. PPTP separates the control and data channels into a control stream that runs over TCP and a data stream that runs over GRE. L2TP combines the control/data channels and uses UDP. L2TP, like PPTP, is protocol-independent, meaning it can run in milieus other than the Internet. L2TP as a layer two media carrier facilitates multi-protocol use (i.e., IPX, AppleTalk). Other tunneling protocols described in IETF include GRE, IPinIP, and IPsec.

For end-to-end client/server tunneling, PPTP currently has the advantage of a more ubiquitous distribution in the marketplace (to CPE clients) than L2TP. Any tunneling can generally be implemented in client/server in a manner completely transparent to the cable modem and CMTS. However, the TRAC and RADIUS may require tunnel support. Host software vendors that market tunneling solutions implement many multi-protocol tunneling techniques that combine proprietary and standards-based features.

If the goal of the tunnel is to use the cable modem client or CMTS as a tunnel endpoint, some explicit setup or signaling would be required. Software in the cable modem or PC could use DOCSIS TSI MAC management message parameters to pass down tunnel endpoint data. The software could set up the telephony return connection to appear as a local-area network (LAN) to the cable modem. The subscriber PC ideally would see only an Ethernet network interface and would not need software to support IP forwarding. The cable modem would handle all dial-out connections in a manner transparent to the PC. Alternatively, the PC may have to control the telephony return. In this case, an Ethernet driver would contain the telephony return software in a manner transparent to the PC. It would provide an Ethernet driver interface to the PC software and would have software interfaces below it to a PPP layer and to an Ethernet driver.

The terminal server in the ISP's point of presence (POP) acts as an L2TP access concentrator (LAC), and the router in the head end acts as an L2TP network server (LNS). The LAC and LNS are L2TP peers. L2TP runs over UDP using the same port number for both ends. Only one tunnel is built between an LAC and an LNS since it can carry multiple PPP connections. The terminal server builds an L2TP tunnel to the appropriate head end router if one does not already exist. Once a tunnel is built between the LAC and LNS, a new slot is allocated within the tunnel for the PPP connection. The setup notification allows the LNS to authenticate the subscriber and to accept or reject the connection. If the connection is accepted, a virtual interface is set up so that a PPP connection is made from the subscriber to the head end router through the terminal server. The terminal server strips off link layer framing and encapsulates the PPP packets inside L2TP packets and forwards them to the appropriate L2TP tunnel. The PPP connection appears to the router to be locally connected.

An architecture that presents all cable modems (two-way RF or telephony return) as Ethernet interfaces can be supported by using L2TP tunneling mechanisms or by a generic IPinIP mechanism. There may be performance implications with this architecture in spanning tree systems where the fluid topology of telephony return bridging limits the usefulness of transparent bridging. Using L2TP specifically as the tunneling mechanism may create performance and scaling concerns if the tunnel LNS connections are all backhauled and concentrated at a single RAS or CMTS. The CPU computational implications of a large network using L2TP would have to be carefully provided for.

The CMTRI specification currently requires simple IPinIP tunneling (RFC 2003) for multicast return paths. Tunnel set-up parameters are acquired automatically by the cable modem through the termination system information (TSI) MCNS MAC messages. This type of tunnel could be specified to encapsulate multiple protocols including Ethertypes with little effort. Like L2TP, this type of tunneling could have particular application in switched VPN configurations. Some VPN strategies could be configured the same whether they used two-way RFI or telephony return cable modems. An IPinIP tunnel does not incur the PPP computational overhead of an L2TP connection because PPP terminations are distributed to the TRACs. However, because IPinIP cannot set up a single connection with allocated slots as in L2TP, each tunnel requires a distinct initial set-up. Like L2TP, an IPinIP tunnel also calls forth similar questions regarding bridging performance.

Quality of Service

Quality of service will be needed for provider backbones transporting traffic of different types and of different importance to the enterprise. Some typical cases for quality of service are:

- Mixture of transaction-processing or real-time traffic with low-priority Web browsing and bulk file transfer

- Mixture of protocols that may require short transit times to avoid protocol timeouts and retransmissions with bulk traffic
- Integrated enterprise backbones, combining data traffic, teleconferencing, and telephony over a single cable channel
- Sharing of cable channel bandwidth by different organizations that want to apportion costs through service-level agreements

Service-level agreements allow the network administration to create various levels of service based on the network access speed and quality of service. Each user pays a fixed rate for a certain level of service. The use of service-level agreements also sets user expectations appropriately and allows management to control the use (and cost) of the enterprise's network resource. The network administration can design the network characteristics to each site based on the total number of user requests. They can then measure the delivered service levels to assure that the users are getting what they paid for. They can throttle bursty traffic to the limits paid for and set forth in the service level. Router priority queuing can be dynamically adjusted on the downstream system to reflect service level agreements.

CMTS Prioritization

CMTS routers can have a prioritization function: they can manipulate their internal queues to give preference to certain traffic because of that traffic's implied quality of service. For example, data traffic going to specific network addresses or to specific service-based addressing (associated with service IDs in DOCSIS) can be given priority or can have some internal buffers or bandwidth reserved for them. Although router prioritization is not as functional as latency and error rate quality of service guarantees, it is usually sufficient for legacy protocol traffic.

CMTS router prioritization can be usable for service level agreements when combined with restricted entry (e.g., attempted connections at high priority are rejected if they would increase the load through the routers to a point where traffic would slow down). This method can meet several of the service level agreement benchmarks mentioned in the earlier section. Policy-based control is exerted over the bandwidth available on the cable data network. Weighted round-robin queuing mechanisms with promotion allow fairness for on-demand bandwidth assignments to bursty traffic. Router prioritization, although not as functional as full quality of service guarantees, is sufficient to assure that most service-level agreements can function properly in a routed telephony return network.

Type of Service (IPv4 Precedence Bits)

IPv4, the current Internet protocol within the TCP/IP protocol suite, has an 8-bit type of service (TOS) tag that can be used to indicate simple classes and priority levels. Previously, the TOS tag had rarely been used, but some vendors are now introducing proprietary router software to handle it. IPv6 will standardize a flow label in addition to a priority tag for more detailed control.

The fact that TCP/IP can accept class-of-service requests has led to the creation of quality of service application programming interfaces (APIs). These permit users to request a specific quality of service for their application on a connection-by-connection basis and, possibly, to alter those quality of service requests during the lifetime of the connection. Some APIs are oriented towards complete quality of service, not just prioritization, as evidenced by some of the parameters in WinSock2 for the IP protocol family:

- traffic description, including peak and average bytes/second
- maximum acceptable transit delay
- level of quality of service guarantee
- cost

The primary disadvantage of using APIs for quality of service requests, other than the fact that applications must be rewritten to use them, is that the cable data network administration can lose control over the assignment of quality of service levels unless a policy administration method is put in place. If it is left up to the user or the user's applications to twiddle precedence bits for highest priorities on the cable data network, they will do exactly that. Both the IETF and DOCSIS respectively are developing policy-based enforcement specifications that can guarantee the integrity of ToS management. With such specifications, IP precedence becomes viable to prioritize traffic on cable data networks and can be used both in the cable loop and also mapped to network side interfaces.

Network Side Interfaces

Media Error Correction When the Internet (IP) is used to transmit digital information from some transmitting unit to one or more receivers, it can be assumed that these transmissions are usually subject to errors. The way TCP-IP is designed, these errors are, in general, inevitable. In a telephony return cable system where compressed real-time voice and video streaming may transit lower bit rates (at least in one direction), packet losses can have an extreme effect on the end-to-end quality of these services. A common way to deal with this is to utilize IETF real-time transport protocol work (RTP). RTP is the Internet-standard protocol for the transport of real-time data, including audio and video. It can be used for media-on-demand as well as interactive services such as Internet telephony. RTP consists of a data and a control part. The latter is called real-time control protocol (RTCP).

New RTP work is being studied in the IETF audio/video transport (AVT) working group, which is looking at ways to preprocess transmit data to deal with packet loss so that streams can be repaired or error-corrected on the receiver side. One method is to have an edge device inject redundant packet information into the packet flow to guarantee the integrity of a reconstructed packet for a given audio or video stream. These redundant packets give the receiver the capability to reconstruct a given sequence of packets even if only some of the packets of that given set are received. For example, with every eight packets, two additional packets are sent that algorithmically contain data to reconstruct any of the eight original packets. Thus, a large percentage of the time, if a packet is lost, that packet can be reconstructed without impairing the quality of the stream. Such techniques hold much promise for end-to-end QoS through the network, particularly for lower bit rates. This approach can have application in telephony return cable systems where the edges (both source and destination) of the transit path are known and controlled. As a closed, controlled environment, a cable loop architecture is a good candidate to implement some type of redundant media protection algorithms at appropriate network side interfaces. RTP does not address the issue of resource reservation or quality of service control; instead, it relies on resource reservation protocols such as resource reservation protocol (RSVP).

RSVP RSVP is an IETF signaling protocol under development that can be used by applications to request special treatment for particular IP traffic streams. Priorities can be set at the CMTS and buffers reserved at all intermediate routing or switching points. Using a connection separate from the data connection, the CMTS would need to map an RSVP request message to the network. The message identifies the IP data connection involved (using the IP destination address, port number, and protocol) and requests a level of service. RSVP can handle both point-to-point and multicast addresses, distributing the RSVP requests to all concerned routers in the path.

The RSVP IETF standards are still under development and unproven in large-scale networks. However, some vendors have begun to implement RSVP in routers and in some switches. This service is implemented through buffer reservation and prioritization using sophisticated queuing mechanisms. Use of these queuing mechanisms, combined with known flow rates across the cable network and TRAC, can provide predictable round-trip delays across router backbones.

System Maintenance Capabilities

Mission-critical networks needing minimal downtime and optimum maintainability require the capability of separate and distinct physical paths for maintenance data and revenue data, respectively. This is because the granularity of fault resolution is greater when a system is not forced to have a channel attempt to diagnose itself. A telephony return cable modem has two distinct physical connections. Potential uses of these connections include enabling extensive diagnostic capabilities in the modems and creating line backup reliability by utilizing dual homing techniques. From a system integrity and maintenance perspective, there are other potential value-added capabilities that can be developed for telephony return cable modems.

Dual Homing

The cable modem is connected to the cable plant head end by a coaxial cable, on which the data is flowing from the cable plant head end to the cable modem. It is possible to use the telco return line in a duplex mode so that additional data channels are utilized for revenue data. This would be analogous to a multi-link PPP (MP) (RFC-1990] connection.

MP–Based MP can be set up with a semi-static bundle head discovery mechanism. When a user dials into an RAS and negotiates multi-link PPP (MP) during the link control protocol (LCP) phase, the TRAC should create the secondary link on a bundle head. The CMTS as the primary link can be configured to discover the specific TRAC that is the secondary link of the MP connection. Both the mechanism by which a bundle head is discovered and the PPP fragment encapsulation should be specified. The L2TP could be used to transfer PPP fragments from a RAS containing a secondary link to the RAS containing the bundle head.

Failover Connection A cable modem could be configured to maintain two separate IP addresses for PPP and cable. A failure of the cable IP address could be sensed and mapped to at the socket level. This type of second connection only makes it possible to communicate with the cable modem when the coaxial line is broken. This provides a lifeline type of redundant service that can be branched when one traffic flow is sensed as faulty. When there is a problem in the data flow from cable plant head end to cable modem, it is possible to continue revenue data, and to receive diagnostic information, fault resolution, and service recovery. However, this type of mapping has the same application sensitivity as the NAT functions described earlier.

System Maintenance

Telephony modems and their connection capabilities are routinely testable by accessing diagnostic phone lines and modem ports. Maintenance methods exist to allow an ISP a high granularity of remote testability. Such tests can resolve faults to subscriber PC configuration, subscriber modem equipment, PSTN line quality, service provider equipment, and even application interfaces.

A telephony return cable modem has all the inherent diagnostic capabilities of a telephony modem test port connection. Additionally, telephony return has the security of being session-based, password protected, and providing a separate two-way physical path into the subscriber PC. With this point-to-point connection, a telephony return cable modem has the value-added potential for the network provider to use remote access to maintenance channels to perform line and functional quality testing of the modem's RF receiver.

Service Support Demarcation For both internal and external telephony modems, service demarcation tests are facilitated by software running on a host CPE. For telephony return, modem functions that are run on the host CPE enable additional resolution between modem faults and host faults. Determinate tests for host software interactions can be run on the host over the telephony maintenance channels. The telephony channel descriptor (TCD) MCNS MAC management message has the capability to provide all test connection facility information automatically to a subscriber. RF receiver tests over telephony channels have been demonstrated by vendor specific software tests running on the host CPE.

Performance tests can determine the actual data transmission speed since the quality of the line, the network traffic, and other factors can make data flow faster or slower than your modem's theoretical maximum speed. Software can automatically detect the presence of communications software like dialers and browsers to test support for them. A host software modem monitor can be implemented as a popup screen to record when the modem is sending and receiving data, to continually redial busy numbers and also to indicate the modem's actual transmission speed.

Safe Mode Diagnostics In addition, cable modems retailed for DOCSIS CMTRI can be required to supply safe mode diagnostics and monitoring software to run on the subscriber host. This type of tool can supply key decision tree information for cable customer support organizations. Examples are a modem connection problem, a forgotten password, getting the latest updates from a modem manufacturer, and customer service troubleshooting capabilities. Software programs on a telephony return cable modem could test the telephony or cable modem portions of the hardware, the phone line, and the communications software. When problems are detected, automatic step-by-step instructions could walk even novice modem users through a solution.

Safe mode diagnostics should perform the following:

- verify proper modem installation, cable RF reception, and phone line connections
- check internal modem wiring and external modem cable for faults and correct wiring
- determine source of connection problems and advise solutions
- track volume of data sent and received and the duration of each connection
- display connection activity by displaying the 'read' and 'send' data indicators
- test subscriber modem's actual speed
- dynamically determine the 'initialization string' for the installed modem.
- provide complete report of the computer's settings and maximum modem speed
- set up (and save for easy re-installation) multiple Windows 95 dial-up networking connections
- test both conventional and Internet connections by running on-line tests
- automatically re-dial and reconnect if connection is lost for any reason
- re-dial on a busy signal until subscriber gets the connection instead of reporting a failed attempt
- cut phone bills by allowing subscriber to choose to disconnect at the end of a file transfer
- log all your sessions, providing a cross-check against support logs and phone bills
- manage multiple connections from a single control
- give one-button access to subscriber modem manufacturer's tech support web

Conclusions

The data over cable industry has made an important step with the DOCSIS standards. This step, combined with low-cost retail subscriber equipment, can make the commercial, social, and educational Viewdata vision go mainstream over the PSTN/CATV mediums. With multivendor compatibility in downstream RF and the widespread availability of telephony upstream connections, providers have many options. They can implement CMTRI systems as a hedge against other broadband technologies, or they can implement it as a continuing service that obviates expensive cable plant upgrades and scrubbing of RF upstream channels.

This paper has given a snapshot of standards-track activity in the IETF with respect to new directions for DOCSIS CMTRI specifications. Some aspects of the CMTRI system changes that can take advantage of respective functionality are presented. Likewise, the potential to enable special service support and diagnostic possibilities is also presented. The reader should become familiarized with the existing specifications to understand how some of the capabilities presented here may diverge from current standard usage. In the future, providers will look to the marketplace and to equipment vendors to help determine when and how any new capabilities should be incorporated into the CMTRI specifications.

References

1 *TechnoBytes*, volume IV, issue XII, December 1997.
2 DOCSIS, "Radio Frequency Interface Specification," *http://www.cablemodem.com/public/pubtechspec/rf/sp-rfi-i02-971008.pdf*
3 DOCSIS, "Cable Modem Telephony Return Interface Specification," *http://www.cablemodem.com/public/pubtechspec/tri/sp-cmtri-i01-970804.pdf*
4 DOCSIS, "Cable Modem Termination System–Network Side Interface Specification," *http://www.cablemodem.com/public/pubtechspec/nsi/sp-nsi-i01.pdf*
5 Postel, J., "RC-791 Internet Protocol," September 1, 1981, *http://ds1.internic.net/rfc/rfc791.txt*
6 Droms, R., "RFC-2131 Dynamic Host Configuration Protocol," March 1997, *http://ds1.internic.net/rfc/rfc2131.txt*
7 Simpson, W., "RFC-1661 The Point-to-Point Protocol (PPP)," July 21, 1994, *http://ds1.internic.net/rfc/rfc1661.txt*
8 McGregor, G., "RFC-1332 The PPP Internet Protocol Control Protocol (IPCP)," May 1992, *http://ds1.internic.net/rfc/rfc1332.txt*
9 Minoli, D., "Video Dialtone Technology," 1995, *R.R. Donnelley and Sons Company*
10 Sklower, K. et al, "RFC 1990 The PPP Multilink Protocol (MP)," August 1996.
11 Perkins, C., "RFC-2003 IP Encapsulation Within IP," October 22, 1996.

Notes

1 Originally CableLabs, Rogers Cablesystems, Continental Cablevision, and the MCNS members Tele-Comunications Inc., Time Warner Cable, Comcast, and Cox Communications worked with selected vendor-authors to establish these open specifications throughout 1996 and 1997.

2 CableLabs is a research and development consortium of cable television system operators representing more than 85% of the cable subscribers in the United States, 70% of the subscribers in Canada, and between five and 10% of cable subscribers in Mexico.

3 U.S. Robotics (later 3Com) was selected by the MCNS working group as vendor-authors for the CMTRI specifications.

4 *TechnoBytes*, December 1997: Original sources: Software Publishers Association Statistics through July 1997; DataQuest Statistics through 1997. Note: These statistics are heavily weighted by a corporate users installed base. Average home users with advanced CPEs (WIN95, Pentium, PCI/USB) could be a significantly lower percentage than shown:

WIN95 users: 77 million; Win 3.x/DOS users: 164.6 million
Pentium: 91.3 million users; 486/x86: 114 million users
ISA capability: 205.3 million users; PCI: 121.3 million users; USB: 1.3 million users

5 On the CMTS, all the crossing DHCP message traffic is listened to, or wiretapped for the IP address/MAC address ARP tuple.

6 Usually in proprietary RAS DHCP implementations, the RAS supplies an arbitrary number to serve as the device network address when it obtains a block, or range of addresses from a DHCP server.

7 Although current CMTRI specification allows multiple IP addresses to be used to initialize a CM, only one IP address ultimately is used to source data traffic. Typically, the PPP IPCP address is only used for DHCP network connectivity and is not a globally unique IP address. Hence, the same number can be reused in other CM initializations.

Architectures for IP Services over CATV Networks

Enrique J. Hernandez-Valencia
Distinguished Member of Technical Staff
Bell Laboratories Lucent Technologies, Inc.

Abstract

Cable networks are preparing to deliver interactive Internet access, work-at-home access, small-office or home-office business access, packet telephony, local area network to local area network (LAN–LAN) connections, and LAN emulation services over community access-TV (CATV) networks. These services are anticipated from the ongoing convergence in the information, computer, and telecommunications industries and from the extension of the data networking capabilities in today's business sector to the residential consumer market. Although residential Internet protocol (IP) services are already being tested in the marketplace, substantive standardization efforts in this area have not materialized until quite recently. The reference service model for such data services is fashioned from an initial request for proposal (RFP) by Cable Labs. This effort has produced an initial set of recommendations by the Multimedia Cable Network Systems (MCNS) group, including the Data over Cable Interface Specification (DOCSIS) recommendation for a physical and media access control (MAC) layer interface over two-way hybrid fiber/coax (HFC) networks. In the meantime, other organizations have also been involved in the efforts of producing standards for two-way interactive services over CATV networks. More recently, the Internet Engineering Task Force (IETF) has created the IP over Cable Data Networks Working Group (IPCDN) to address data service architectural issues similar, if not identical, to those applying to Cable Lab's initial proposal. To ensure interoperability among vendors, a common data service framework that clearly specifies how IP–based services could be implemented over CATV and similar access networks will be required.

Introduction

Cable modems are one of the new broadband technologies preparing to deliver interactive Internet access, work-at-home access, small-office or home-office business access, packet telephony, LAN–LAN connections, and LAN emulation services over CATV networks. These new CATV services are the result of the ongoing convergence in the information, computer, and telecommunications industries and the desire to extend the data networking capabilities of the business sector to the residential consumer market.[1] Although residential IP services are already being tested in the marketplace, substantive standardization efforts in this area have not materialized until recently. The reference service model for such data services was fashioned after an initial RFP by Cable Labs.[2] This effort has produced an initial set of recommendations by the MCNS group, which include a Data over Cable Interface Specification (DOCSIS) recommendation for a physical and MAC layer interface over two-way HFC networks.[3] In the meantime, other organizations have also been actively involved in producing standards for two-way interactive services over CATV networks. The Institute of Electrical and Electronic Enginers (IEEE) Project 802.14 Working Group has been developing a set of specifications for physical and data link layer protocols that will be applicable to support data service over cable networks and other broadband access infrastructures.[4–6] The ATM Forum's Residential Broadband Working Group has been working on delivering asynchronous transfer mode (ATM) traffic to and within the home in a way compatible with the IEEE P 802.14 work.[7] Also, the Digital Auto Video Council (DAVIC), whose initial focus had been the international standardization of digital technologies applied to audio and video applications for set-top-boxes and video-on-demand servers, has also created recommendations for interactive data and Internet services over cable networks.[8] More recently, the IETF has created the IP over Cable Data Networks Working Group (IPCDN) to address data service architectural and management issues similar, if not identical, to those in the initial Cable Lab's proposal.[9,10] To ensure interoperability among vendors, a common data service framework that clearly specifies how IP–based services could be implemented over CATV and other similar access networks will be required.

In general, these CATV data service specifications attempt to develop the necessary set of communications, operations, and management interfaces for customer premises equipment (cable modems) and associated network equipment. Thus, they identify required hardware and software components, signaling interfaces, and required IP transport and routing services over the access network. Besides data transport issues, network provisioning and configuration and management tools for the specific needs of the CATV network environment must be addressed. Among the data networking issues, special attention must be paid to the following:

- physical and logical (IP) subnetwork configuration,
- unicast/multicast data forwarding/routing services,
- static/dynamic host address configuration options,
- subscription and accounting, and
- security.

In addition, besides the standard transmission control protocol (TCP), user datagram protocol (UDP), Internet control message protocol (ICMP), and IP Internet protocol suite,[11–14] any proposed access architecture for broadband residential data services should support other existing IP protocols and services. These may include the IP address resolution protocol (ARP) (or other suitable address resolution mechanisms over cable networks),[15] the dynamic host configuration protocol (DHCP),[16] the domain name system (DNS),[17] and the Internet group management protocol (IGMP).[18] For enhanced services other than the conventional "Best Effort" service, resource management capabilities being developed under the IETF working groups on resource reservation protocol (RSVP),[19] integrated services,[20] differentiated services,[21] and IP security,[22] without significant modification.

The paper will review general requirements to provide Internet residential services over the CATV infrastructure and discuss important network engineering, design, and operational issues. This material is a revised version of an earlier paper on the same subject.[23]

High-Level System Architecture

A high-level system architecture for residential data services over two-way CATV plants, also referred to as the cable data network (CDN), is depicted in *Figure 1a.* On the customer premise side, a cable data modem (CM) with either an Ethernet, ATM, or any other suitable PC interface provides the access interface to the CATV network. The CM connects the customer premise equipment to the standard CATV radio-frequency (RF) coax drop cable. On the carrier's head end (HE), the cable modem data termination system (CMTS) terminates the CATV RF link and implements the data link layer protocols supporting the residential service. Downstream, toward the CM, the CATV physical (PHY) layer would typically support a broadcast, one transmitter–multiple receiver mode with a raw bandwidth of about 10 Mbps to 30 Mbps, depending on the specific modulation scheme. Upstream, the PHY layer supports a one receiver–many transmitter mode with a raw bandwidth between 1.5 Mbps and 3 Mbps. Multiple upstream channels may be implemented to improve performance. Given the broadcast characteristics of the RF link, multiple residential customers and hence potentially many home-based LANs will be serviced from the same CMTS interface. The CATV PHY layer, however, is not expected to natively support the fully shared media model common in most LAN technologies. Such services are provided by the CATV MAC layer, which will also arbitrate access to the upstream receiver among CMs sharing the RF link.

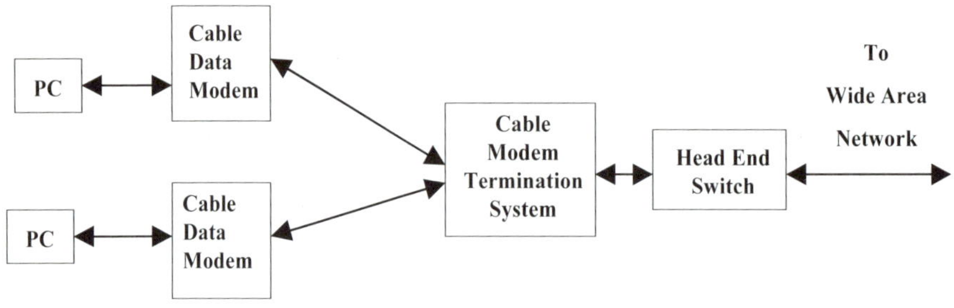

Figure 1a: *CDN over a Two-Way CATV Plant*

Nonetheless, over 65% of the CATV plants in North America are basically one-way distribution systems with a very noisy, low-bandwidth upstream channel used mostly for set-top box control purposes. To support interactive data services over this infrastructure, the return channel must be implemented over an alternative access network technology, such as the PSTN. Such a one-way system with PSTN return is depicted in *Figure 1b*. In this case the available upstream bandwidth would be limited to 56 Kbps (without data compression).

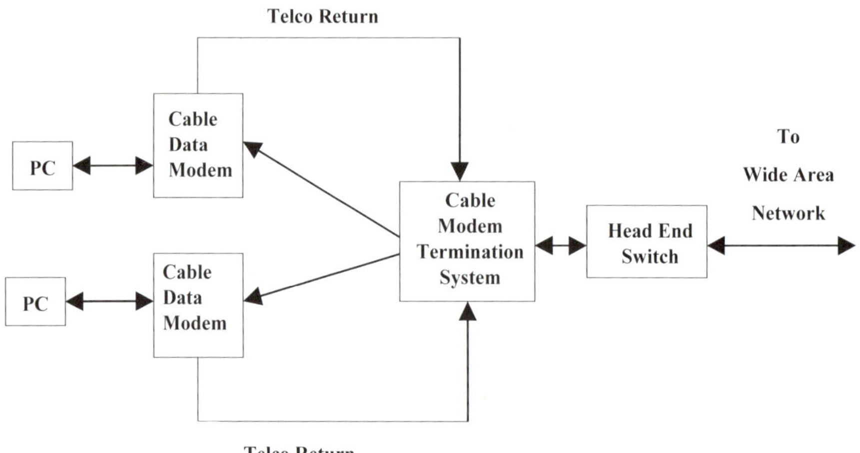

Figure 1b: *CDN over a One-Way CATV Plant*

The asymmetric nature of the CATV network, both in terms of access network topology and link rates, places special constraints on conventional point-to-point IP access models. In particular, for the so-called one-way CATV plant depicted in *Figure 1b*, the CATV access network can only deliver one-way (downstream) data transport services from the head-end equipment in the CDN to the subscriber's home LAN. In such a scenario, a PSTN link could transport any (upstream) traffic from the subscriber to the CDN service provider. From the CDN service provider viewpoint, it is imperative that most of the subscriber's downstream traffic is in fact forwarded to the home LAN over the CATV access network, rather than over the PSTN link (though some two-way traffic over the PSTN link may be required for connection initialization and control purposes). Additionally, and as an enhanced service for one-way CATV plants, it may also be envisioned that the two-way PSTN link could be a CDN's access backup in the event of a CATV plant failure. Moreover, the initial cable data service should not only be able to reuse exiting Internet technology, but also provide a simple migration path to a two-way CATV infrastructure.

A traditional point-to-point dial-up access service model can be readily adapted to the two-way CATV plant scenario. The main architectural difference between the CDN approach and the conventional PSTN dial-up service is in the broadcast nature of the downstream RF link and the associated communications privacy concerns. Although the potential lack of privacy has never been of much concern when surfing the Internet, it is troublesome when access to a company's private database is desired or when providing commercial transaction services. Communications privacy has not been an issue with plain old telephone service (POTS) or integrated services digital network (ISDN) access to private networks, since the link from the subscriber's home to the corporate intranet is over a dedicated circuit. For the existing CATV networks, enhanced data encryption capabilities will be required either at the MAC layer or at the higher layers (e.g., IP). For the one-way CATV plant, however, to provide such an extra level of privacy, a mechanism is required to identify which CMs are attached to a given RF link. This information would not be self-evident to the CMTS, as the upstream traffic from the CMs may arrive to the homing CMTS over an interface other than the RF link.

Data Networking Requirements for IP Services

For networking purposes, the specific format of the CATV network PHY and MAC layer protocol data units (PDUs) should be transparent to the higher protocol layers. Most CATV networks would either encapsulate Ethernet/802.2 frames (e.g., MCNS and IEEE 802.14) or emulate the Ethernet/802.2 services (e.g., ATMF/RBB). Hence, higher-layer PDUs would typically access the CATV network service via 802.2 (LLC), 802.1D (bridge), or AAL5. Operations and management of the head-end system requires some minimum standard IP data networking functionality. Below are typical requirements for Ethernet/802.2–based systems. In the case of ATM–based systems, additional facilities for IP over ATM, such "Classical IP over ATM" [RFC-2225] and NHRP [RFC-2332], will also be required.

Figure 2 shows basic Internet access facilities at the head-end site. At every minimum an Ethernet/802.2 bridge–based head-end system must include the following:

- Standard IP, TCP, UDP, ICMP, and ARP protocols to support basic unicast communications.
- File transfer protocol (FTP) [RFC-959] and trivial file transfer protocol (TFTP) [RFC-1350] for remote reconfiguration (under administrator control) , and
- Simple network management protocol (SNMP), typically the more secure SNMPv2 [RFC-1902] than SNMPv1 [RFC-1157].

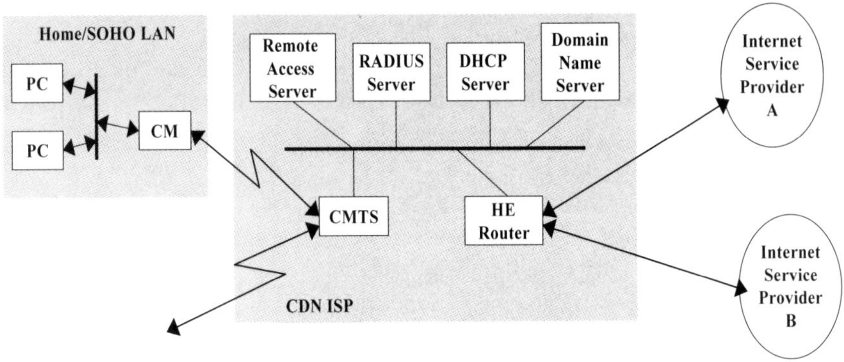

Figure 2: *CDN IP Services for a Two-Way CATV Plant*

Access to the CDN services may be granted on a per subscription basis (as in conventional broadcast CATV networks) or on an on-demand basis (particularly if the CMTS is shared among multiple CATV providers). A dynamic host configuration protocol (DHCP) server [RFC-1531], a remote access server (RAS), a terminal access concentrator (TAC), or a network access server (NAS) would be required for dynamic and static IPv4 address allocation to the CDN subscribers, as well as a RADIUS server [RFC-2139] for user authentication and billing and a domain name system (DNS) server [RFC-1034, RFC-2181] for domain name resolution outside the CDN. Additionally, routers in the CDN may also support various data link configuration and security functions such as:

- IP Path MTU Discovery [RFC-1191], particularly over the RF links
- IPsec AH and ESP [RFC-1825] for secure data transport and
- client-side support for network time protocol (NTP) [RFC-1305] to facilitate time distribution among components of the CDN

Route-based head-end systems will be required to support additional data networking functionality. Static IPv4 routing may not be adequate for large CDNs not only because of the scaling limitations imposed on routing tables and the scheduling policies, but also because of provisioning and management concerns. Efficient use of the IPv4 addressing space would also require support for classless routing. Hence, the minimum functionality can be expected as follows:

- variable-length subnet masks and classless inter-domain routing (CIDR) [RFC-1518] with IGMPv2 [RFC-2236]

- dynamic IP unicast forwarding based on either RIPv2 [RFC-1723] or OSPFv2 [RFC-2178]

- dynamic IP multicast routing such as DVMRP [RFC-1075] or PIM [RFC-2117]

CDN is expected to cater to the ISP market to support enhanced Internet, intranet, and virtual private networking capabilities. Hence, such systems will be expected to implement capabilities such as the following:

- IP tunneling features such as IP–in–IP, [RFC-2003], GRE [RFC–1701], and IPsec ESP tunnel mode[22]
- server-side layer 2 tunneling protocol (L2TP) decapsulation[24]
- proxy ARP for off-subnet address resolution
- BGPv4 [RFC-1771] for inter-domain forwarding

Requirements for customer premise equipment would be less stringent. For the small-office and home-office market, simple configuration features based on static IP routing would suffice. For the residential and entertainment sectors, operational features such as ease of configuration, privacy, and remote management from the CDN ISP will be a significant concern.

The next generation of IP (IPv6) is already being introduced. An even more ambitious goal would be applying the same broadband residential service architecture equally to networks based on the current version of IP (IPv4) and to IPv6 networks. Yet since the IP service models are somewhat different, it is natural to expect that their corresponding service framework over cable data networks may also differ. Most of the discussion later in this paper is directed toward IPv4 residential services. Any IP service model over one type of CATV plant, such as HFC access networks, will share significant commonality to other asymmetric access infrastructures. Therefore, it would also be expected that the same service framework will be equally applicable to residential access technologies such as the asymmetric digital subscriber line (ADSL), the passive optical networks (PON), wireless networks, and satellite networks. It could also be expected that the PSTN fallback mode would use IP encapsulation and routing techniques similar to those under Mobile IP.[25]

From a data service perspective, two important issues related to subscriber access need to be considered in any practical solution. First, some mechanism will be required to allocate IP addresses to the home appliances, particularly when the CATV provider is not also the Internet service provider. Second, mechanisms will be required to authenticate and secure data traffic to corporate intranets.

Access Models over CATV Networks

The generic access model for IP services over CATV networks is envisioned as a one-to-one high-speed replacement of conventional IP dial-up services. For the two-way cable plant, as depicted in *Figure 3a*, there is much resemblance between the dial-up and the cable-based access networks. Two exceptions are desirable traffic control features, such as traffic filtering capabilities at the CMTS to eliminate broadcasting messages over all RF interfaces (this is certainly not a requirement, but would be highly desirable to avoid significant throughput degradation), and data encryption at the MAC layer for improved data confidentiality.

For the one-way cable plant, however, the point of presence (POP) of the network access server (NAS, i.e., terminal server and related logical link control functions) controlling access to the networking facilities of an IP service provider (ISP) need not be collocated with the head-end equipment controlling access to the CATV network. If only one IP subnetwork address were given to a CM to support CDN service, an architectural choice must be made between placing the home LAN under the NAS's IP subnetwork or the HE/CMTS's IP subnetwork. Obviously, the choice would determine how data traffic is actually forwarded to the home LAN.

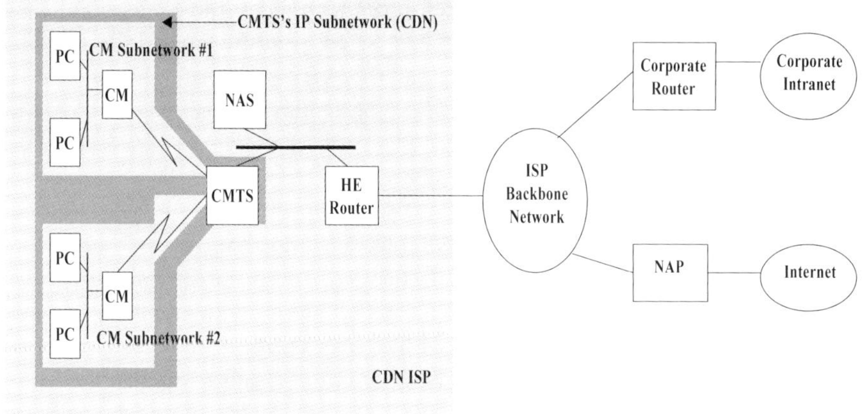

Figure 3a: *CDN Architecture over a Two-Way CATV Plant*

The CM in the NAS Subnetwork

In this model, traffic to the home LAN would be forwarded first to the homing NAS in the ISP network. Since the CATV network is the desired last hop to the home LAN, downstream traffic would need to be redirected to the CMTS on the carrier's head end. Thus, some additional forwarding functionality would be required to implement this access model. One obvious approach can be readily borrowed from the data tunneling techniques like those in Mobile IP and L2TP. Such approach requires the following:

- an IP tunnel from the homing NAS in the ISP network to the homing CMTS/HE in the CDN on which downstream traffic to the home LAN is forwarded to the CMTS,
- a (secure) mechanism to establish such tunnels,
- a (secure) mechanism to indicate to a NAS which users have access to particular CMTSs so their traffic can be forwarded over the appropriate NAS/CMTS IP tunnel.

Regardless of the additional software complexity associated with IP tunneling approaches, there are also performance drawbacks to be considered, including the potential for a large number of extra forwarding hops from the ISP to the cable data provider, double encapsulation, etc. The extra data forwarding hops will not only introduce more latency into the data forwarding path, but given the high throughput anticipated for the downstream path, may also significantly increase the total amount of network bandwidth and processing resources required at the NAS and intermediate routers to cope with this service. One potential advantage of the tunneling approach is that a fallback to dial-up access in the event of an RF link failure will be easier to implement, as most of the network configuration information required to forward the user data to the homing NAS is already in place.

The CM in the CMTS Subnetwork

In this model, traffic to the home LAN would be forwarded first to the homing CMTS on the HE LAN, which is already the preferred data forwarding path. As depicted in *Figure 3b*, this simple approach can support the minimum data forwarding functionality required to deliver IP services over the one-way CATV plant. Yet additional forwarding functionality will be required if a fallback path through the PSTN is desired in the event of a cable plant failure. In such a case the following observations apply:

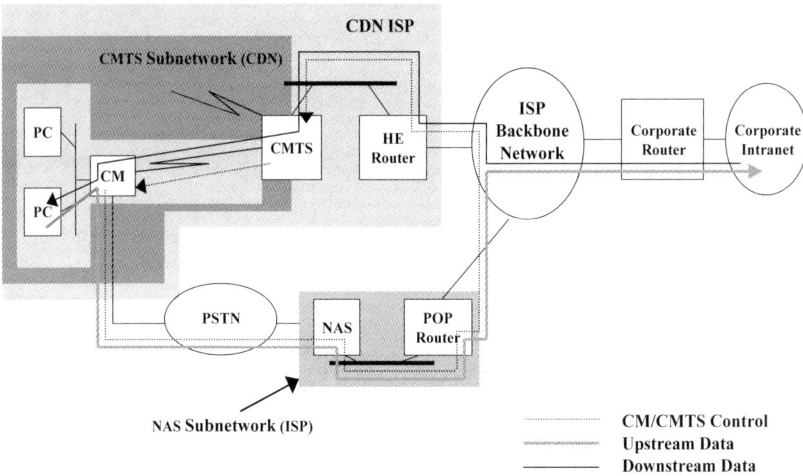

Figure 3b: *Single IP Subnetwork Model for a One-Way CATV Network with No PSTN Fallback*

- the CMTS needs to be able to identify homing NAS in the ISP network to open an IP tunnel to redirect downstream traffic; or
- the CMTS could source route IP packets to the CM (yet, source routing is not widely supported by router vendors); or
- the NAS in the ISP network needs a mechanism to map the incoming PSTN port to the CM's IP address for data forwarding purposes.

Hence, by the time fallback access over the PSTN is incorporated into the picture, the hardware and software complexity of both approaches may be about even. The later approach, though, gives the most direct data forwarding path to the CATV network. One potential concern is that upstream traffic may need to be redirected to the HE LAN under conventional (or in the absence of) security practices.

A Dual IP Subnetwork Model for One Way CATV Plant

A simpler CDN architecture that supports communications either over the access network or over a PSTN can be achieved by associating different IP subnetworks with each of the different CM interfaces outside the home LAN. Specifically, the CM's interface over the CATV network can be logically associated with the IP subnetwork of the homing CMTS in the carrier's head end. This assignment would establish the logical point of attachment of the CM to the CDN. Similarly, the CM's interface over the dial-up network can be logically associated with the IP subnetwork of the ISP's NAS. This assignment establishes the logical point of attachment of the CM to the PSTN. Given that the CATV access network is the desired data forwarding path to the home LAN, rather than the PSTN link, the IP addresses to be used by the home appliances to communicate with their target destinations should have the IP subnetwork address of the homing CMTS as a prefix.

Hence, under normal operation conditions, CDN–bound traffic from the home LAN would be relayed upstream by the CMs to the ISP network via the PSTN link. Traffic bound to the home LAN would always be forwarded through the homing CMTS, as the home LAN would appear to the external world as an IP subnetwork of the homing CMTS/HE router (assuming that classless interdomain routing [RFC-1518] is deployed in the CDN [see *Figure 4a*]). The CMTS would then forward

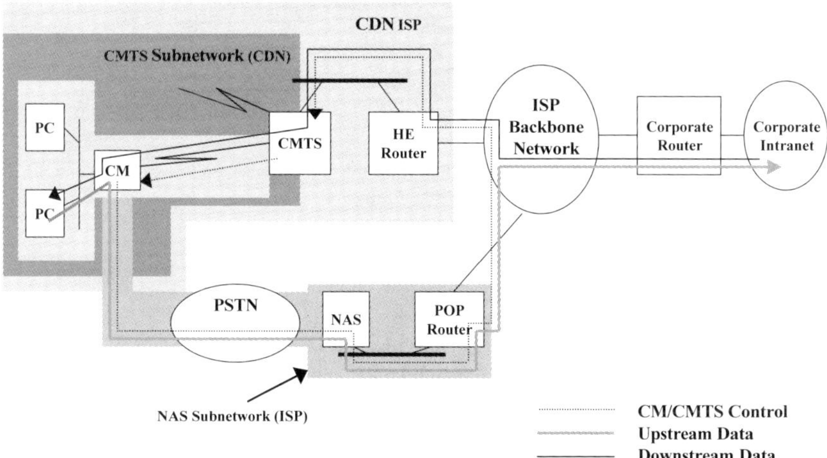

Figure 4a: *Dual IP Subnetworks for One-Way CDN Upstream/ Downstream Data Flows*

the traffic to the CM in the home LAN over the appropriate RF port, and the CM would then relay this traffic to the target home appliance. Implicit in this model is the need for mechanisms that 1) allow the CM to "discriminate" local traffic from CDN bound traffic, and 2) match an RF port on the CMTS to home appliances. Also, one must define the data relay function provided by the CM and the CMTS.

Under an RF link failure scenario, the CMTS would not be able to forward IP traffic directly to the CM. However, if the CMTS knows the IP address of the CM on the PSTN network, the CMTS can establish an IP tunnel to the CM to forward the home LAN–bound traffic. The CM would then extract the traffic from the tunnel and forward this traffic to the home LAN (see *Figure 4b*). The CMTS can learn the IP address that a CM is using over the PSTN network when the CM initiates the CDN connection request to the CMTS. At that point, it must already have an IP address assigned by the ISP. Notice that in a one-way CDN the connection to the ISP network must exist prior to any possible exchange of user data between a CM and its homing CMTS. Hence, besides the ability to establish a secure tunnel to the CM, there are not major functional requirements for this approach.

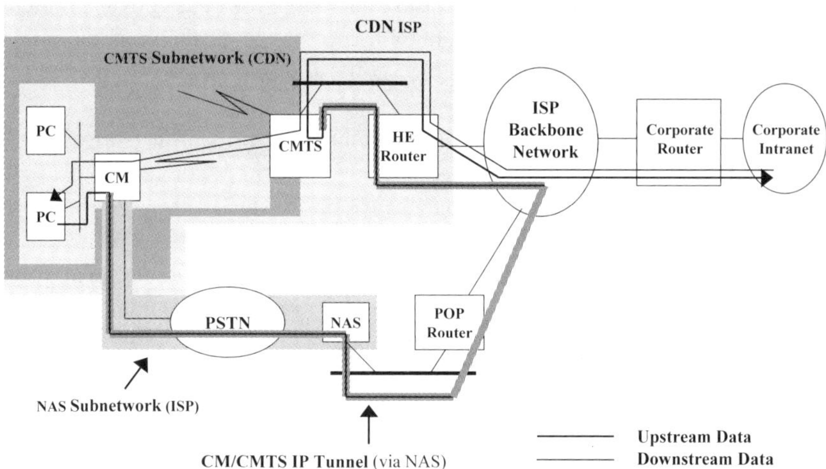

Figure 4b: *Dual IP Subnetworks for One-Way CDN Fallback Upstream/Downstream Data Flows*

Access to the Customer Premises

Both the single and the dual IP subnetwork models for the CM on the home LAN allow the CM and the CMTS to operate either as intelligent repeaters and bridges or routers. Clearly, the choice determines the routing and address resolution capabilities, as well as the scaling properties of the CDN service with respect to the number of subscribers.

The CM and CMTS as Routers

With the CM and the CMTS as routers, the CMTS would not need to be immediately aware of a home appliance's MAC address. IP address resolution must be supported, either by the CM or some other device in the home LAN. The CMTS can treat its connection to the CM as a virtual point-to-point link. Home LANs could be configured as stub IP subnetworks sharing the IP subnetwork prefix of the tending RF link. This approach significantly simplifies the operations, administration, maintenance, and provisioning (OAM&P) requirements for the CDN service provider, since supporting data services over the CATV network, including IP multicast, would not be different from supporting IP services over a conventional router network. Another benefit from this approach is that the IP routing table size for the CMTS can grow in proportion to the number of RF links, rather than in proportion to the number of CMs or home appliances, by sharing the homing CMTS's IP subnetwork prefix between all the home LANs. Similarly, the IP routing table entries at the HE routers can grow in proportion to the number of supported CMTSs. However, the total number of routing entries in a CMTS may still grow with the number of active CMs, if MAC–layer encryption is implemented to deliver some level of data confidentiality for broadcast tree and branch access networks such as HFC– and PON–based access networks. This may not be much different from the number of active home appliances.

Under this network model, the CM would need to support MAC address resolution on behalf of the home appliances. Thus, it should either 1) be able to ARP for the MAC address of the target home appliance, or 2) provide any other mechanism to associate MAC addresses to home appliances' IP addresses. Although ARP support is common on most routers and Unix-based systems, it is rather uncommon in most entry-level PCs for the residential market. This could become an important architectural consideration. For routing purposes, the CMTS would only be required to provide a mechanism to map the CM's IP subnetwork prefix to an RF port and exchange routing information with other routers in the same administrative domain, not in the CMs themselves.

Figure 5 shows the various communications stacks that could be involved in a communications scenario for CDN services over a one-way CATV plant. Clearly, the router-based approach places substantial data forwarding functionality, and hence, implementation complexity, on the CM. With the addition of support for IP tunnels to the CMTS, when a PSTN fallback in the event of RF link failures is desirable, almost any conventional data service can be easily delivered to the home LAN. The cost of the cable model and the scaling properties of the CDN with respect to IP unicast or multicast routing become the most relevant design and implementation concerns.

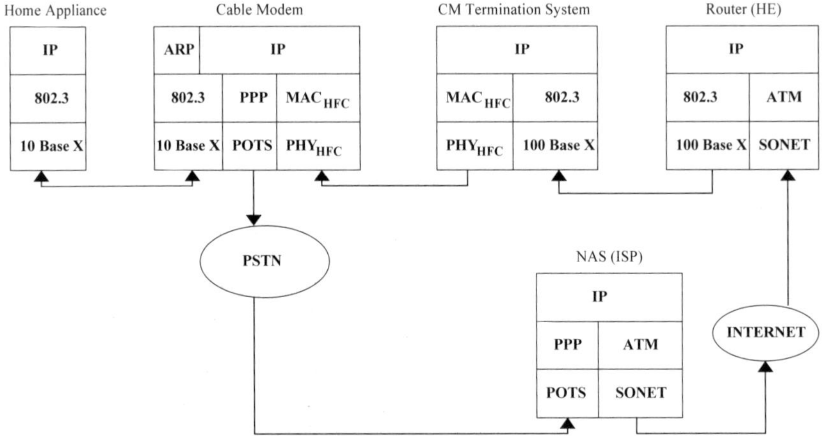

Figure 5: *A Plausible Communications Model for the CDN over a One-Way CATV Plant*

The CM and CMTS as Bridges and Repeaters

Bridging the home LAN to a carrier-based LAN in the head end is intrinsically appealing, since supporting data service to the home LAN then becomes similar to supporting those services over 802.3 LANs. Yet with the CM and the CMTS as a repeater or bridge, the routers on the carrier's head end must be able to resolve the home appliance's MAC address from its IP address. IP routing table requirements for head-end routers may grow in proportion to the number of home appliances needed to support, which could be a large number (anywhere between a few hundred subscribers to tens of thousands per head end). The CMTSs would at least be required to filter traffic on a RF link basis. Otherwise, they would needlessly broadcast traffic over all RF links, reducing the useful access network capacity. Forwarding cache tables on the CMTS may also grow proportionally to the number of active home appliances supported over its RF links. Additionally, the CATV link is not likely to be a true two-way shared link. More likely, according to most IEEE 802.14 proposals, is that it will be shared downstream but will be point-to-point upstream. Thus, the shared nature of the 802.3 MAC should be emulated by the CMTS.

On the other hand, the CM would only need to keep local home LAN traffic away from the access network. Thus, the CM's forwarding cache table only needs to support a few tens of entries per home. IP address resolution does not need to reside on the CM. Since the bridged network could be large, ARP traffic could also become a significant fraction of the CATV access network traffic. This issue can be partially managed by reducing the size of the logical IP subnetworks as seen by the routers at the head-end, at additional OAM&P complexity for the CDN service provider.

For one-way cable plants, the situation is more complex. First, the upstream and downstream paths are not the same, and thus an external mechanism is required to populate forwarding cache tables at the CMTS. Second, ARP responses to an HE router's ARP request, if supported, cannot be forwarded blindly over the PSTN link or they will probably be filtered out by the ISP network (ARP is not a routable protocol). Bridging increases the amount of overhead, which decreases end-to-end throughput. Additionally, the slow PSTN link and the additional forwarding hops from the ISP network to the homing CMTS/HE could make responsiveness of ARP rather sluggish. Supporting ARP efficiently may require that the CMTS either 1) acts as a proxy ARP server on behalf of its downstream home appliances, or 2) requires the CM to forward ARP responses (when supported by the home appliance) over the CM/CMTS IP tunnel, so the CMTS can relay them to routers in the carrier's head end. ARP proxies may be used in two-way CATV plans to keep ARP traffic of the cable plant. Notice that any information required to match a RF port on the CMTS to an IP subnetwork prefix for a CM could be either 1) dynamically acquired from the CM as part of the CDN connection setup phase or 2) statically provisioned.

The bridging approach allows for somewhat simpler implementation requirements for the CM at the expense of higher implementation and operation complexity at the CMTS and head-end routers. It also provides the least desirable scaling of routing tables at the CMTS and the HE routers with respect to the number of subscribers. Although the lower cost and complexity requirements on cable modems for two-way CATV plants are clearly appealing, it is unclear if those savings also translate to CMs for one-way cable plants.

Static vs. Dynamic Address Allocation

Given the explosive growth of the Internet over the last few years, the unassigned pool of public IPv4 addresses is quickly diminishing. Supporting millions of potential residential users would only exacerbate this problem. This design constraint applies to dial-up and cable data services alike.

Although the network address translators (NATs) have substantially helped to manage IPv4 address space scaling issues, extending the life of network access solutions that rely on static address allocation policies, the preferred long-term solution is the use of dynamic address.

Statically Allocated Addresses

Draft proposals of CDN services still envision allocating IP addresses to home appliances and even CMs, statically. That may be adequate for service demonstration purposes or for initial service introduction, but it would not be feasible for large user populations to use IPv4 as these addresses are already scarce. IP service providers are already required to justify any additional IP address allocation requests to the Internet Assigned Numbers Authority (IANA) and any static address allocation is strongly discouraged by the Internet address registers. Because of this practice, very few ISPs could afford to statically allocate public IPv4 addresses to their customers for free, particularly when considering the number of subscribed customers to most dial-up services is about 100 to 1000 times higher than the number of simultaneously active customers during peak business hours. Thus, any static allocation of IPv4 addresses from the public address space would most likely be deemed a liability.

A separate but relevant observation about the allocation of IP addresses is that neither user traffic constrained within the boundaries of a given ISP, nor intermediate networking devices such as routers, CMTSs, and CMs within a given CDN, require public IP addresses. Private IP address allocation is an appealing approach to alleviate the reliance of data service providers on public IP addresses, even for relatively large subscriber populations, and extend the lifetime of static address allocation. Such implementations necessitate the deployment of NATs to translate private IP addresses into public IP addresses for traffic flowing into the Internet, as well as private address filters in the edge routers of the CDN to avoid advertising such private IP addresses onto the Internet. Clearly, this approach would be highly intrusive on any ISP network not already supporting private IP addressing, as it would require a significant overhaul of the transport network to provide the required access gateways for address translation and route filtering. This approach may be deemed as shortsighted. NATs do not enjoy much support within the IETF community, as a long-term solution for the depletion of IP address space. Given the lack of other mature IP addressing options, though, its popularity has been growing steadily. Notice, however, that implicit in the usage of private IP addressing is the dynamic allocation of public IP addresses at the edges of the CDN. Hence, the combination of statically–allocated IP addresses to end users and NATs could be used as a convenient mechanism to transition to more dynamic address allocation from the network access gateways to directly to the end user.

Address space concerns are moot for any data services relying on IPv6. This approach will require even more significant infrastructure upgrades to data service providers, given the obvious need to interwork with IPv4–based networks. Although this functionality will likely be available in most ISP in a not too distant future, there are still multiple interoperability concerns that must be ironed out before such an approach may be considered commercially mature.

Dynamically Allocated Addresses

Most ISPs are already architected for dynamic allocation of either public or private IP addresses. With an IP address from the public block, no further addressing or routing functionality is required from a CDN service provider to support access to the Internet. Under IPv4, though, a CDN must still face the problem of address space depletion. For private intranet access services, such as work-at-home, the IP address allocated to home users may come from the private address block, since most intranet firewalls will perform address translation to satisfy some of their internal security concerns.[26] Thus, CDN service providers can leave to corporate network policy the decision on whether such users require a public or private IP address, or whether the address should come directly from the corporate IP address block. For instance, CDN service providers could allocate IP addresses from the public block just for Internet access services. Such addresses would more likely be taken from the CDN's allocated public block. For intranet access services such as work-at-home, the allocated IP addresses could come from the intranet's private block (assuming that intranet-specific services are only allowed). If the corporate employee was also allowed to demand concurrent Internet and work-at-home access services, an IP address from the intranet's public block could be allocated so the employee can access the Internet directly through the company's intranet. The use of IP addresses from the private address space does not necessitate the deployment of network address translators, but the use of route filters is still required.

For the two-way cable plant, it would be expected in most situations that the CDN service provider would also be the subscriber's ISP. For a one-way cable plant, the ISP may want to establish agreements with local CATV providers to deliver data services to the home. Either way, when the CATV provider and the ISP provider are not the same, there will be a more pressing need to allocate IP address dynamically to simplify CDN operations. Otherwise, a significant amount of provisioning may be required to configure CMs, CMTSs, routers, and network access servers to add new subscribers into service.

From the previous discussion, the following IP addressing/routing functionality in the provider network requires the following to support dynamic allocation of public/private IP addresses:

- dynamic host address configuration server for CDN attached hosts,
- network address translators at the edge of the CDN and/or the edge of a private Intranet,

- private IP address filters at the access points to the Internet (to be implemented by the service provider), and

- suitable security procedures, and associated hardware or software, such that the CDN user can interact with both the NAS, CMTS, and the intranet's security server, if any.

From the end user's perspective, dynamically allocating public or private IP addresses to CDN users (home appliances) requires the following:

- the dynamic host address configuration client must IP addressees from the CDN. The CMTS or a configuration server/adjunct could interact with a DHCP server in the ISP network to acquire the IP addresses to be allocated to the CDN subscribers during the duration of a connection to the CATV access network.

- A yet-to-be-defined mechanism must either pass IP addresses from the CM to the home appliances, if the addresses are initially allocated to the CM, or administer IP addresses directly between the CDN and the home appliances via the CM.

Other enhancements may also be needed to the POP software to provide differential treatment to Internet and work-at-home users.

Secure Corporate Intranet Access

Traditionally, home users have gained access to terminal servers on corporate intranets by dialing up their remote access server over point-to-point links (POTS or ISDN). This approach allows most corporations to reduce security concerns to an appropriate user authentication procedure if all data exchange occurs over a dedicated circuit-switched connection. Data confidentiality was assumed from the dedicated circuit-switched nature of the phone service. When corporate intranet access takes place over the Internet, the data exchanges take place over a distributed (shared) and open data networking infrastructure. By default, there is no intrinsic security afforded by the Internet data transport service. Such security must be provided as an add-on service by the CDN or ISP.

It is useful to draw a distinction between public and private IP networks. When the corporate intranet has public IP visibility, that is, when any IP host on the corporate network can be readily reached over the Internet, it will always be up to the corporate network to support strong authentication, encryption, and other related security mechanisms at each host. Security in this case would be provided above the data transport layer, and hence, can be deemed as corporate or application-specific. Protocols such as Keberos [RFC-1510], a user-to-host authentication protocol, would fall within this category. A more sophisticated implementation of the same principle would direct incoming corporate network traffic to a security gateway at the edge of its private intranet, where network address translation may be performed. Both types of services would be functionally indistinguishable to the end user. Established secure Internet access approaches such as secure socket layer (SSL) and secure hypertext transfer protocol would also fall within the general category of host-based security services.[27,28] Yet the SSL protocol only addresses the issue of authenticating end-systems (host-to-host authentication) rather than users. The secured hypertext transfer protocol, however, is only applicable to Web-based applications.

Network Layer Security: A more comprehensive approach to providing flexible data services to residential customers applicable to public and private IP networks alike is to supply the desired security services at the data transport layer. This solution can coexist with any of the host- or user-based authentication procedures mentioned previously. Although the Internet community has been working in this direction for some time, no mature standard delivering transport layer authentication and confidentiality services over the Internet currently exists. IPSEC, an IETF draft proposal currently under development, is the logical candidate to fill this role. Unfortunately, IPSEC is missing key components, such as a public key management system. IPSEC will also require software upgrades to the communication stacks of any home-based appliance or corporate host, as well as the deployment of a public key management infrastructure. Of course, proprietary solutions are also viable.

An outcome of the current lack of a complete security framework for IP has been that, without significant hardware/software upgrades, most private networks today can only afford marginal support to external corporate traffic from unrecognizable IP addresses. This is prompted by the large variety of attacks to which they would be exposed if they did not filter such traffic. Also, conventional networking practices strongly encourage large private companies to allocate their own IP addresses and to keep the internal details of their IP addressing plan away from third parties, including any external service provider. Hence, the ability to allocate IP addresses to end users from the corporate's address space, as well as the ability to manage the address allocation directly rather than through the IP service provider, will be critical to the deployment of secure private corporate network access over the Internet.

Link Layer Security: Most authentication services (as well as IP address allocation services) for residential users are implemented at the data link layer via point-to-point protocols such as PPP. Any equivalent approach that preserves the current dial-up user authentication interface over other access technologies, such as CATV access networks, would be inherently appealing, for it would require little if any modification to the existing base of customer premise hardware/software. Two very similar IETF proposals (the point-to-point tunneling protocol [PPTP] from Ascend, Microsoft, and US Robotics, among others, and the Layer 2 Forwarding [L2F] from Cisco Systems) attempt to do so by defining a secure mechanism to tunnel PPP traffic over IP networks. Merged into a single proposal, L2TP,[26] they extend the current dial-up access framework to allow virtual dial-up services to be supported over the Internet. The proposals provide mechanisms to establish a secure link between an NAS in an ISP network and a foreign access gateway, as well as the framework under which all standard services available under dedicated access configuration (authentication, authorization, address allocation, and accounting) could be provided by the foreign network separately from the equivalent services provided by the ISP.

Approaches that provide virtual dial-up access to remote users, such as PPTP/L2F/L2TP could be extended to the two-way CATV plant. An example is depicted in *Figure 6a*. In *Figure 6a*, the conventional NAS functions are split into two separate components: one provided by the NAS on the ISP network and performing PHY and MAC control, configuration, and management functions; the other, provided by a remote access gateway on the foreign network, performing logical link control, network layer configuration, and management functions, including PPP termination. The CMTS could either provide only PHY and MAC layer functionality to support data services over a CATV plant, for which a colocated NAS will be required, or be subsumed as one of the line interfaces of an NAS. The latter approach may be more convenient as an ISP will prefer that managing the CATV RF link not be significantly different from managing a PSTN or an ISDN link. An IP tunnel would be established between CMTS and the corporate access gateway, over which PPP traffic from the home LAN is forwarded. Minimal software upgrades would be required for current NASes, remote access gateways, and to the home appliances since, for instance, they do not rely on a public key management system like IPSEC. Hence, in principle, remote user access over a two-way CATV plant could easily reuse virtual dial-up access technology currently being developed by IETF.

It is reasonable to anticipate that any remote user access solution for a one-way CATV plant will also provide a simple migration path to a two-way CATV plant. This can be done by allowing the CMTS to support the L2TP as illustrated in *Figure 6b*. Current IETF draft proposals for L2TP expect all home traffic to flow over the CMTS-to-access gateway IP tunnel. Hence, a mechanism is required to forward corporate-bound traffic back to the CMTS, instead of sending it directly to the corporate access gateway from the NAS. Since the CM may have already established an

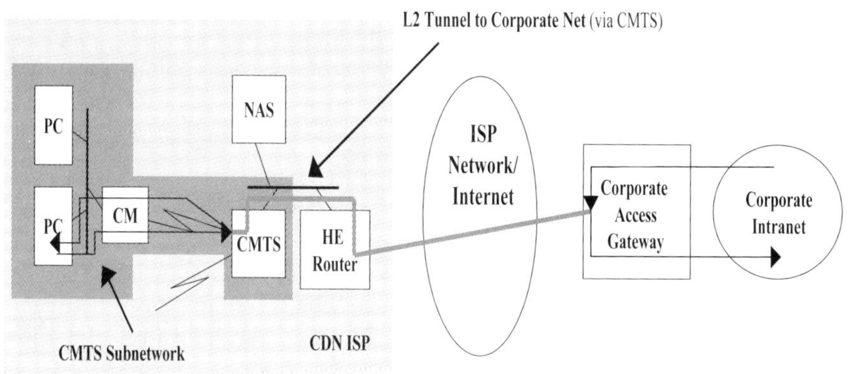

Figure 6a: *L2 Tunnel for Remote User Access over a Two-Way CATV Plant*

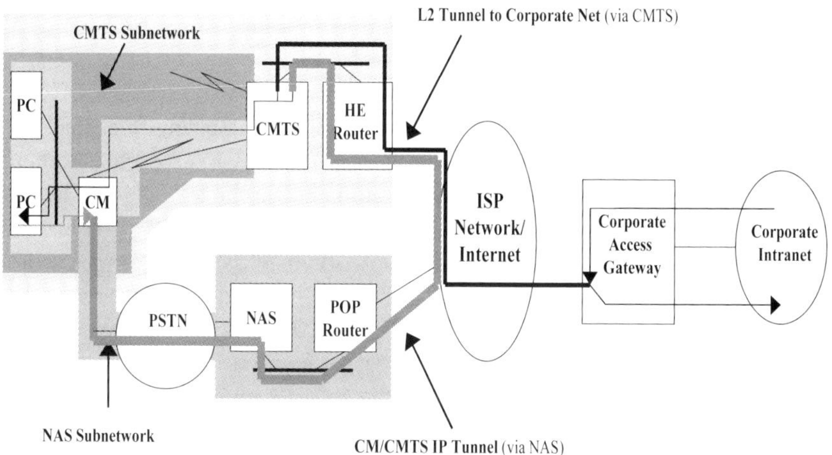

Figure 6b: *One-Way CATV Plant with an L2 Tunnel for Remote User Access (Dual IP Subnetwork Model)*

IP tunnel with the CMTS for backup purposes, the CMTS-to-CM communications protocol for a one-way CATV plant could be easily enhanced to support this additional functionality. Namely, the CMTS or CM must be able to multiplex and demultiplex traffic to and from the home appliance and traffic to and from the CMTS or CM itself. If necessary, two separate communication channels could be established for data relay and connection control functions. Although this seems convenient as it reuses most of the functionality already anticipated for the CMTS, it does have some obvious performance drawbacks. For instance, the most direct path between the home LAN and the foreign network would not be employed. Using a direct path would require either a mechanism to authenticate the CMs with the foreign access gateways, which would deviate significantly from the current draft proposals for L2TP, or a mechanism for the CM to act as an L2TP proxy for the CMTS, which would add additional complexity to existing remote access server proposals.

Practical Performance Considerations

In any reliable data transmission, end-to-end throughput is a function of the bandwidth-delay product, the error characteristics of the communications path, and the sender and receiver buffer sizes. The TCP transmitters regularly probe the data path and adjust their transmission window sizes to match the available network bandwidth. Most TCP receivers are expected to acknowledge every IP segment or every other segment if implementing delayed acknowledgment messages (acks).[29] Although the raw link rate for cable data services is anticipated to be about 30 Mbps for downstream traffic and about 1.5 Mbps to 3.0 Mbps for upstream traffic, actual data rates can be constrained by configuration options for the various TCP/IP stacks.

The maximum window (WIN) and segment (MSS) sizes applying to a specific connection are negotiated during connection setup. TCP/IP hosts exchange the MSS and WIN values they are willing to support and choose the lowest MSS as the one applying to the connection. Since the MSS is typically derived from the maximum transmission unit (MTU) on the outgoing network interface, this process could lead to very inefficient use of the bandwidth when the upstream interface has a much smaller MSS value than the downstream interface. Also, since the data forwarding path changes after a reroute, there is a further potential for IP segment fragmentation. At a minimum, path MTU discovery [RFC-1191] must be supported to avoid unnecessary IP segment fragmentation. Ideally, MSS renegotiation should also be supported (but most TCP/IP implementations do not support this).

Because of the asymmetric nature of the data forwarding paths over a one-way CATV plant, the reduced network bandwidth over the upstream PSTN link will throttle the rate of TCP acknowledgments to the TCP transmitter and further constrain the end-to-end goodput over the downstream link. *Figure 7* depicts the maximum end-to-end goodput as a function of the packet size for common PSTN access speeds. End-to-end throughput, or goodput, can be improved by using standard techniques such as TCP/IP header compression [RFC-1144]. Further goodput gain can also be obtained by also taking advantage of the PPP's various packet compression mechanisms.

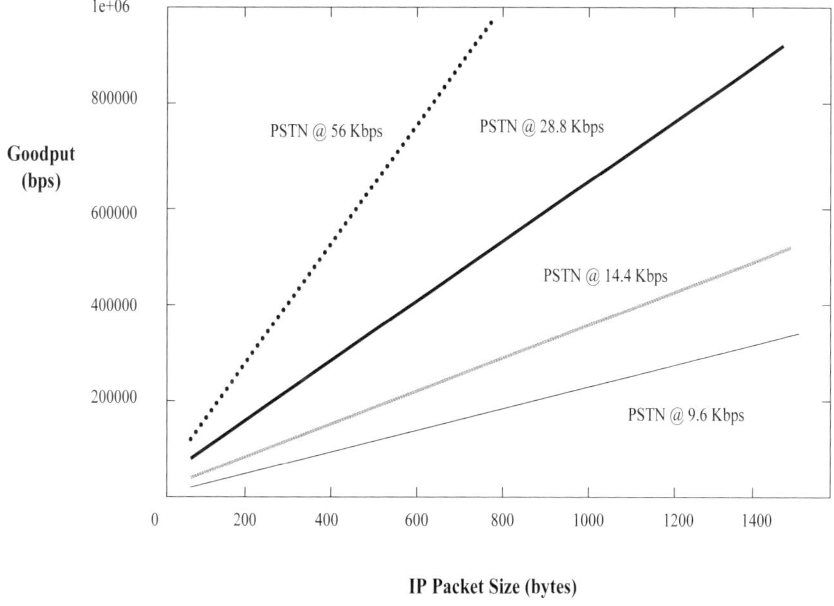

Figure 7: *Downstream End-to-End Goodput as a Function of Packet Size*

RF impartments and limited buffering capacity may require fine-tuning of the TCP/IP protocol stack parameters to optimize end-system performance over asymmetric HFC networks.[30]

Conclusion

This paper reviews high-level architecture design options available to deliver cable data services over one-way and two-way CATV plants. In the process, some of the performance trade-offs among design options are also identified. At this time, it seems feasible to deliver residential data service for Internet access, work-at-home, small-business access, LAN–LAN interconnect and LAN emulation services over two-way CATV networks with little modification to existing data networking protocols for conventional dial-up services. Although slightly more work is required to extend such services over one-way plants, which constitute a significant fraction of the deployed CATV distribution systems, it is anticipated that cable data networks will become more prevalent over the next few years as the by-products from the recent standardization efforts reach the market.

Acknowledgments

The author would like to thank Donald Blahut, James Bobeck, Robert Dalton, Barat Doshi, Roger Levy, Harrison Miles, Robert Rance, Behrokh Samadi, and Ronald Sharp for their contributions to many of the views expressed in this paper.

References

1 Grzelak, S. A., Miles, H., Szurkowski, E. S., and Weber, W. P., "Residential Data Services via Broadband Local Access Network," *Bell Labs Technical Journal*, Vol. 1, No. 1. August 1996.
2 Cable Television Laboratories, "High Speed Cable Data Services Request for Proposal." April 1995.
3 MCNS Data Over Cable Service Interface Specification (DOCSIS) RFP. December 1995.
4 IEEE P802.14 Working Group, "Cable-TV Functional Requirements and Evaluation Criteria." IEEE 802 Committee. February 1995.
5 IEEE P802.14 Working Group, "MAC Convergence Proposal." IEEE 802 Committee. Work in progress. September 1996.
6 Eng, J. W., "Standards for HFC-based Residential Broadband: IEEE Project 802.14, Its Mission, Charter and Status," *SPIE Proceedings*, Vol. 2609 (October 1995): 2–9.
7 ATM Forum's Residential Broadband Group. Work in progress. See www.atmforum.org.
8 Digital Audio Video Council (DAVIC). Work in progress. See www.davic.org.
9 White, G., "Logical IP Subnetworks over MCNS Data Link Services," IETF IPCDN WG. Work in progress.
10 Roeck, G., "Cable Device MIB for MCNS Compliant Cable Modems and Cable Modem Termination Systems. IPCDN WG. Work in progress.
11 Postel, J. "Internet Protocol," RFC-791. September 1991.
12 Postel, J. "Internet Control Message Protocol," RFC-792. September 1991.
13 Postel, J. "Transmission Control Protocol," RFC-793. September 1991.
14 Postel, J. "User Datagram Protocol," RFC-768. 28 August 1990.
15 Plummer, D., "An Ethernet Address Resolution Protocol," RFC-826. November 1982.
16 Droms, R., "Dynamic Host Configuration Protocol," RFC-1531. October 1993.
17 Mockapetris, P. V., "Domain Name: Concepts and Facilities." RFC-1034. November 1987.
18 Deering, S. E., "Host Extensions for IP Multicasting." RFC-1112. August 1989.

19 Zhang, L., Deering, S. E., Estrin, D., Shanker, S., and Zappala, D., "RSVP: A New Resource Reservation Protocol," *IEEE Network Magazine*, 1994.
20 Braden, R., Clark, D., and Shenker, S., "Integrated Services in the Internet: An Overview." RFC-1633. June 1994.
21 Black, D., Blake, S., Carlson, M., Davies, E., Wang, Z., and Waiss, W., "An Architecture for Differentiated Services." IETF Differentiated Services WG. Work in progress.
22 Atkinson, R., "Security Architecture for the Internet Protocol." RFC-1825. August 1995.
23 Hernandez-Valencia, E., "Architectures for Broadband Residential IP Services over CATV Networks." *IEEE Network Magazine*. Vol. 11, No. 1 (June 1997): 36–43.
24 Valencia, A., Hamzeh, K., Rubens, A., Kolar, T., Littlewood, M., Townsley, W., Taarud, J., Pall, G., Palter, B., and Verthein, W., "Layer Two Tunneling Protocol," ETF PPP Extensions WG. Work in progress. 1996.
25 Perkins, C., (editor), "IP Mobility Support." IETF Mobile IP WG. Work in progress. 1996.
26 Cheswick, W. R. and Bellovin, S. M., "Firewalls and Internet Security." Addison-Wesley. 1994.
27 Netscape, "Secure Socket Layer Protocol." March 1996.
28 Rescorla, E. and Schiffman, A., "The Secure Hypertext Transfer Protocol." IETF Web Transaction Security WG. Work in progress. 1996.
29 Stevens, W. R., "TCP/IP Illustrated, Volume1." Addison-Wesley. November 1994.
30 Cohen, R. and Ramanahan, S., "TCP for High Performance in Hybrid Fiber Coaxial Broadband Access Networks," IEEE Trans. on Networking. Vol. 6, No. 1. February 1998.

Referenced RFCs:

[RFC-791] Postel, J., "Internet Protocol," RFC-791. September 1991.
[RFC-792] Postel, J., "Internet Control Message Protocol," RFC-792. September 1991.
[RFC-793] Postel, J., "Transmission Control Protocol," RFC-793. September 1991.
[RFC-768] Postel, J., "User Datagram Protocol," RFC-768. 28 August 1990.
[RFC-826] Plummer, D., An Ethernet Address Resolution Protocol," RFC-826. November 1982.
[RFC-959] Postel, J., "File Transfer Protocol," RFC-959. October 1985.
[RFC-1034] Mockapetris, P. V., "Domain Name: Concepts and Facilities," RFC-1034. November 1987.
[RFC-1075] Waitzman, D., Partridge, C., and Deering, S., "Distance Vector Multicast Routing Protocol." RFC-1075. November 1988.

[RFC-1112] Deering, S. E., "Host Extensions for IP Multicasting," RFC-1112. August 1989.

[RFC-1144] Jacobson, V., "Compressing TCP/IP Headers," RFC-1144. January 1990.

[RFC-1157] Schoffstall, M., Fedor, M., Davin, J., and Case, J., "Simple Network Management Protocol," RFC-1157. May 1990.

[RFC-1191] Mogul, J. C. and Deering, S. E., "Path MTU Discover," RFC-1191. April 1990.

[RFC-1305] Mills, D., "Network Time Protocol (V3)," RFC-1305. April 1992.

[RFC-1350] Sollins, K., "The TFTP Protocol (Revision 2)," RFC-1350. July 1992.

[RFC-1510] Kohl, J. and Newman, C., "The Kerberos Network Authentication Server (V5)." RFC-1510. September 1993.

[RFC-1518] Fuller, V., Li, T., Yu, J., and Vahradhan, K., "Classless Interdomain Routing (CIDR): An Address Assignment and Aggregation Strategy, RFC-1518. September 1993.

[RFC-1531] Droms, R., "Dynamic Host Configuration Protocol," RFC-1531. October 1993.

[RFC-1633] Braden, R., Clark, D., and Shenker, S., "Integrated Services in the Internet: an Overview," RFC-1633. June 1994.

[RFC-1701] Hanks, S., Li, T., Farinacci, D., and Triana, P., "Generic Routing Encapsulation," RFC-1701. October 1994.

[RFC-1723] Malkin, G., "RIPv2," RFC-1723. November 1994.

[RFC-1771] Rekhter, Y. and Li, T., "Border Gateway Prorocol," RFC-1771. March 1995.

[RFC-1825] Atkinson, R., "Security Architecture for the Internet Protocol." RFC 1825. August 1995.

[RFC-1902] Case, J., McCloghire, K., Rose, M., and Waldbusser, S., "Structure of Management Information for Version 2 of the Simple Network Management Protocol," RFC-1902. January 1996.

[RFC-2003] Perkins, C., "IP Encapsulation within IP," RFC-2003. October 1996.

[RFC-2117] Estrin, D., Faranacci, D., Helmy, A., Thaler, D., Deering, S., Handley, M., Jacobson, V., Liu, C., Sharma, P., and Wai, L., "Protocol Independent Multicast—Sparce Mode (PIM-SM): Protocol Specification," RFC-2117. June 1997.

[RFC-2139] Rigney, C., Rubens, A., Simpson, W., and Willens, S., "RADIUS," RFC-2139. July 1997.

[RFC-2178] Moy, J., "Open Shortest Path First (OSPF) Routing Protocol, Version 2," RFC-2178. July 1997.

[RFC-2181] Elz, R., "Clarification of the DNS Specification," RFC-2182. April 1997.

[RFC-2225] Lauchback, M. and Harpern, J., "Classical IP and ARP over ATM," RFC-2225. April 1998.
[RFC-2236] Fenner, W., "Internet Group Management Protocol, Version 2," RFC-2236. November 1997.
[RFC-2332] Luciani, J., Katz, D., Piscitello, D., Cole, B., and Doraswamy, N., "NBNA Next Hop Resolution Protocol (NHRP)," RFC-2332. April 1998.

Wireless Broadband Networks

Robert Furniss
Director
Hybrid Networks, Inc.

Keith Ross
Product Manager, Microwave Systems Division
ADC Telecommunications

Introduction

Wireless broadband Internet access uses many frequency bands. Wireless broadband modems offer similar performance to cable modems operating over wired cable television systems (CATV), but the business is easier to enter as the transmitters cover an entire city or region.

A broadband wireless system can deliver up to 30 Mbps data capacity in a 6-MHz channel. The wireless system's strength is that it can quickly provide high burst-speed Internet access to a 10-mile, 20-mile, or 35-mile radius depending on the frequency band used. This allows the service provider to work with or compete with CATV to serve small-sized and medium-sized business and high-end users. The CATV coaxial cable plant typically services residential neighborhoods and may not serve these premium customers. Moreover, wireless broadband Internet access imposes special requirements on the relationship between technology and business.

Internet Transmission Basics

Though it is not necessary to understand how the Internet works, the basic differences between the Internet and continuous data transmission should be understood. There are two key differences:

- Continuous data-transfer concepts such as the bandwidth assigned to each user become meaningless, at least for Internet traffic. The Internet provides "bandwidth on demand" except it really transfers packets in bursts. There will be delays in receiving a file or burst of data if there are many users. These delays, which increase with traffic load, are a measure of the quality of service.

- Most forms of Internet file transfer require Internet protocol acknowledgements (ACKs). Delays or latency in returning ACKs back to the information source slow the downstream file transfer.

The downstream communication uses the transmission control protocol/Internet protocol (TCP/IP). The customer's computer acknowledges the receipt of packets by sending an ACK signal upstream. TCP uses a form of data flow control called a sliding window protocol. This allows very fast downstream bursts suited to the high bit rate of the downstream channel (but lower average rates scaled to shared channels and the computer's ability to receive data at 1 Mbps to 2 Mbps). It optimizes burst communication where the latency, or the delay time in receiving the ACK, impacts the downstream speed.

Listed below are the effects of TCP/IP on cable modem systems:

- TCP/IP is much more tolerant of occasional brief losses of downstream communication caused by interference than a television signal because it can request retransmissions.

- A return path is needed to acknowledge the transmitted packets as well as the requested files.

- Traffic is asymmetric—that is, a 10 Mbps wireless downstream path still yields fast downstream file transfers even with a low bandwidth return path such as a 14.4-kbps phone connection. A wireless downstream path with a telephone return path is a viable way to provide service. (Wireless return options are in their infancy due to the cost of the equipment and limited available bandwidth. They offer lower latency and, in most cases, a higher upstream speed for file transfers in business applications.)

- Both telephone and wireless return paths can introduce additional factors that cause high latency resulting in slower downstream speeds. For example, v.34 28.8 kb telephone modems have greater latency than v.32 14.4 kb telephone modems. Similarly, some cellular wireless and satellite data services have extreme latency and, as a result, give slow file transfers.

The broadband modem system can contribute round-trip latencies that vary from 25 msec for wireless return and upwards of 125 msec for telephone return. ACK is about 40 bytes long for telephone return systems and about 80 bytes long for wireless return due to their different protocols.

Basics of the Cable Modem System and the Cable Modem

Cable Modem System

Figure 1 is used to bring out points applicable to wireless. The client server modem architecture provides the operator control of the modems (customers) from the head end.

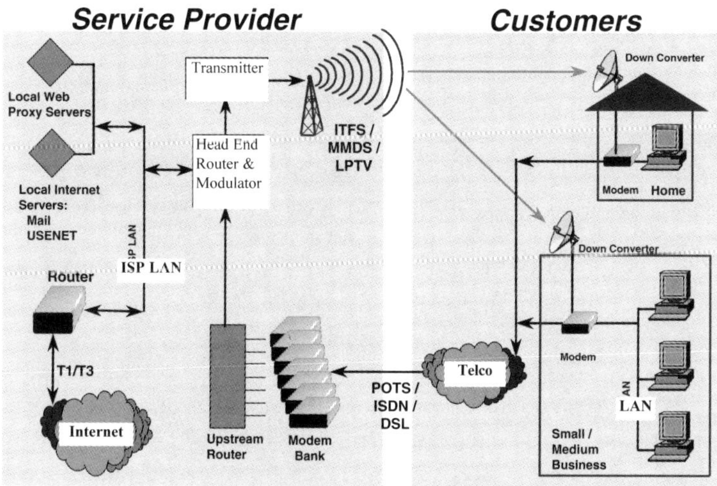

Figure 1: *Simplified System Diagram*

Starting with the service provider, a connection to the Internet is required in addition to a number of local servers, such as e-mail and a caching server, to save frequently accessed pages instead of requesting them repeatedly over the network. Connected to this local-area network (LAN) is some form of downstream router. The digital output is converted by a 64–quadrature amplitude modulation (QAM) modulator, which provides a 44-MHz IF signal to the transmitter. This signal occupies a 6-MHz bandwidth. (The modulation type and channel size varies according to the vendor.) The customer receives the line-of-sight signal, and the cable modem is connected by way of Ethernet to the computer or multiple computers on an Ethernet LAN. The cable modem return path in this diagram is over the dial-up telephone network to a modem bank and an upstream router. This return path carries both upstream requests and acknowledgements of the downstream packets. Connecting the return path directly to the provider's LAN in this way gives the fastest file transfer from the local servers. The modem banks could also be at a remote location connected by traditional digital facilities. (The modem banks could also be connected to another Internet access point. This makes no difference to the traffic from the Internet but would slow the transfer from the local servers due to Internet delay of ACKs.)

Each 6-MHz channel can support about 9000 subscribers. The number varies according to the type of traffic and the type of user. It can be increased by using directional instead of omnidirectional transmitter antennas so that different sectors are served by different transmitters.

Cable Modem

Wireless broadband modem service builds upon the simpler system architecture of wired cable service, so it is helpful to first understand operation in a cable system, as it has fewer variables. A modem cable television system has downstream channels from 50 MHz to 806 MHz. The cable modem tunes to all or some of these channels depending on its design. The signal level is nominally 0dBmV (1 mVolt in 75 ohms) at the interface to the customer and is stable, varying slowly with temperature and time. USA CATV channels are 6-MHz wide and with 6-MHz spacing. A 64–QAM modulator can fit 30 Mbps of data in a 6-MHz channel.

A cable modem consists of a standard digitally controlled "cable ready" television tuner, a 64–QAM demodulator, and an Ethernet connection to the customer's computer (see *Figure 2*). The return path from the computer carries ACK messages or upstream file transfers such as e-mail. The upstream connection may use an RS232 connection to a telephone modem or a modulated upstream signal such as QPSK for return over the cable system. An internal processor with memory controls the tuner and the return path as well as more complex functions such as downstream packet filtering.

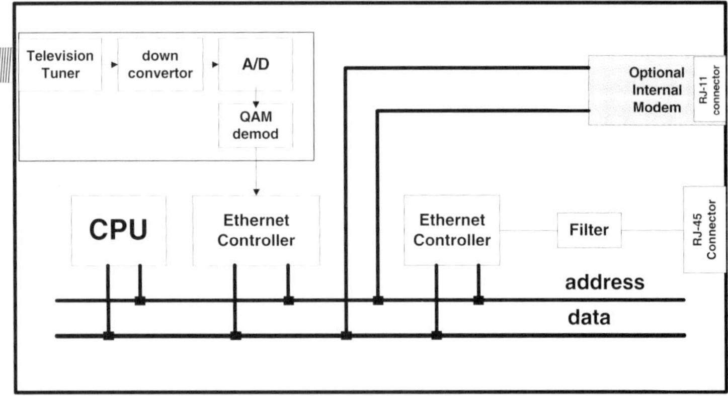

Figure 2: *Simplified Block Diagram of a Cable Modem*

There are some points to note with regards to modem tuning:

- The ideal modem operated both for wired CATV and wireless, although wireless has some different requirements due to the propagation of the signal and the manner in which it is received.
- A standard digitally controlled TV tuner has fixed tuning steps of 62.5 kHz.
- The cable modem vendor makes a modem that tunes only to fixed frequencies. Cable modems use discrete digital, not analog, tuning. Many modems tune in 6-MHz steps, which is not sufficient for all wireless bands.
- The 64–QAM demodulator also drives the tuner. The demodulator has to be able to correct frequency errors that occur in the television tuner/downconverter and the received signal. The input frequency stability is normally in the range of +/– 50 kHz.

Architecture of a Wireless Downstream System

Engineering of a wireless system is complicated by several factors:

- The physical wireless transmitter location may be different from the Internet head-end location.
- Line-of-sight transmission is required.
- The signal power falls with distance.
- Different fixed-transmission frequencies are used according to license availability.
- Multipath distortion exists.
- A receiving antenna is required on the roof, the side of the building, or sometimes in a window.
- Two-way operation is limited by available licenses, bandwidth, and FCC regulation.

Some of these requirements interact in the following ways:

- If one had complete freedom, the transmitter would be at a convenient site for an Internet service connection, yet have line-of-sight transmission to customers. These two requirements conflict about half the time, so the transmitter and the Internet head end cannot be colocated.

- The site for the wireless antenna and the transmitter is usually chosen to provide line-of-sight television to a metropolitan area or several rural communities. The frequency bands were licensed for this wide-area coverage—they are not personal communication service (PCS) cells. As a result, the transmitter site may be physically inaccessible—e.g., on a mountain.

- The nearest point of connection to the Internet is often downtown. This is often the best place for the Internet head-end maintenance.

- As can be seen from *Figure 3*, the microwave multipoint distribution system (MMDS) transmitter may not cover the whole area. A low-power booster may be needed to fill in areas where line-of-sight does not exist.

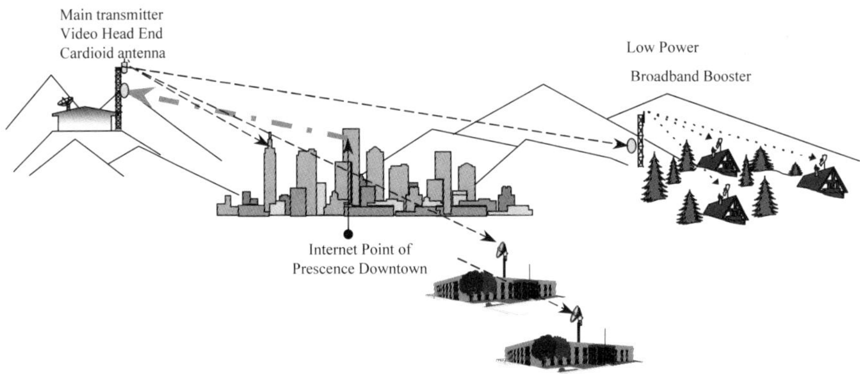

Figure 3: *MMDS Transmission Configuration: Distant Transmitter with the Internet Head End Downtown Colocated with Point of Presence*

Frequency Bands and Limitations

The frequency band determines the type of receiving antenna and the coverage, or range, of the transmitter. MMDS, instructional television fixed service (ITFS), and multipoint distribution service (MDS) analog-television transmitters were the first bands used for Internet access. The bands were underutilized for television and one, two, or even three digital 2-MHz subchannels could be made to work with adjustments to the analog transmitter diplexers and filters. Digital transmitters have recently been made available from several vendors.

Wireless communications service (WCS) is a newly available band with wireless transmitters and downconverters under development. Ultra-high frequency (UHF) and very high frequency (VHF) low-power transmitters are in service for Internet downstream access using experimental licenses. Most bandwidths are 6 MHz downstream, whereas WCS is 5 MHz (see *Table 1* and *Table 2*).

Name	Full name and basic usage
MMDS	Multichannel Multipoint Distribution Service. Most of these TV transmitters are analog and require upgrade to digital unless sub-channelization is used.
MDS	Multipoint Distribution Service. Analog TV are not used.
WCS	Wireless Communications Services. New
ITFS	Instructional Television Fixed Service. Educational service includes Internet Access.
LMDS	Local Multipoint Distribution Service. New
ISM	Instructional Scientific and Medical. Unlicensed bands used for LANs and for return path of two-way modem systems.

Table 1: *Frequency Band Designations*

Sub-channelization is a way to increase the power of the adaptive equalizer in the cable modem by applying it to a narrower bandwidth. A standard equalizer can equalize three times the amplitude tilt, or triple the delay, if it operates over a 2-MHz channel instead of a 6-MHz channel. The narrower bandwidth may allow operation without upgrading the diplexer or removing sound subcarrier stop filters of an existing analog transmitter. In the latter case, only one or two subchannels may operate instead of three.

Multipath is the reception of two or more signals over different paths. The direct signal may combine with the reflection from a roof, wall, or other surface, or refraction off trees or an atmospheric layer. The received signal is the vector sum of the two signals creating both an amplitude and phase change. This distortion may move rapidly across the frequency band.

Name	Frequency	Notes
MMDS	2500–2686 MHz	Thirty-one individual 6 MHz television channels (transmitters) including ITFS. Some operators have only four channels, a range of 35 miles; requires line-of-sight and is affected by multipath.
MDS1	2150–2156	Single channel 6 MHz. See MMDS
MDS2	2156–2162	Single channel 6 MHz. See MMDS
MDS2A	2156–2160	MDS2 truncated on one side to 4 MHz
WCS	2305–2320	5-MHz or 10-MHz blocks (New in May 1997)
WCS	2345–2360	5-MHz or 10-MHz blocks (New in May 1997)
ITFS	2500–2690	6-MHz channels shared with MMDS

Table 2: *Frequency Bands*

LPTV	54–72 78–88 174-216 470–806	Low-Power Broadcast 6 MHz channels, experimental licenses Low power can be 50 Kw effective radiated power. (This includes the antenna gain). Line-of-sight operation is advised.
LMDS	27500–28350	Short range, 3 miles, 20-MHz channels (New)
	31000–31300	Propagation is affected by rain
ISM	902–928	Short range 0.5 miles spread spectrum omni-directional
	2,400–2,483.5	Short range, similar to 900 MHz but can also be engineered beyond 15 miles point-to point as the return path for a cable modem system

Table 2: *Frequency Bands (continued)*

Receiving the Signal at the Subscriber

The frequency band determines the type of receiving antenna and downconverter (if needed). Low-power television (LPTV) signals are received by the cable modem using a normal outside TV antenna, possibly with an amplifier, by paying more attention to the signal level than one would with an analog television set. The level must be close to the high end of the range acceptable to the modem, as the signal is more likely to fade than increase in level.

MMDS, MDS, and WCS require a small antenna integrated with a downconverter mounted on the roof or the side of the building. These low cost units are one-third the price of the cable modems. The antennas may be flat arrays, one foot square with 17 dB gain, or dome-shaped designs with similar characteristics. Other products use partical parabolas, two feet in diameter, or cast Yagi antennas, two feet long. Most downconverters allow a choice of internal gain as well as antenna gain so that output signal level can be as close as possible to 0 dBmV for the modem.

The following are points to note in a wireless environment:

- Locate the antenna to minimize multipath.
- Make sure the antenna is located to receive the direct signal rather than a stronger reflected signal off a building, which may vary in level.
- Set the signal level close to 0 dBmV.
- Work with the downconverter vendor to achieve the simplest selection of amplifier and antenna gain for the coverage area to minimize the number of different models needed.
- Use the appropriate path-loss calculations for microwave links. The calculation for multiple modulators through a common transmitter (sub-channelization) is a little different from a single signal on a transmitter.
- A downconverter installation engineered for a cable modem can also support video service. The reverse may not be true as the cable modem has specific level requirements.
- The nominal signal in a wired cable environment is 0 dBmV in 75 ohms. The cable modem tuner/demodulator will accept a limited variation in input level. Depending on manufacturer, model, frequency of operation, and MMDS environment, the path loss is stable except for multipath distortion (ideally removed by the cable modem adaptive equalizer), and the antenna or downconverter gain can be selected to provide the right level to the cable modem. In the case of LPTV, path fading is more common, but an outdoor antenna with line-of-sight operation will minimize the level variations from multipath.

Getting the Signal to the Broadcast Wireless Transmitter

Most wireless transmitters require high towers to get the coverage area. These are often out of town, so the problem is to get the downstream signal from the Internet point-of-presence to the transmitter. This path is as important as the rest of the system and must be properly designed. There are two basic transmission media, wireless and fiber-optics, and two transmission technologies, amplitude modulation and frequency modulation. Amplitude modulation offers more channels. The choice between the two depends on whether there are existing lines for television signals and how far it is possible to plan ahead as the business grows (see *Table 3*).

	Amplitude Modulation	Frequency Modulation
Wireless	AML (Amplitude Modulated Link) to 31 channels	STL (Studio Transmitter Link) 1 data channel
Precautions	Noise filter may be needed to baseband	Downconvert from 44 MHz Noise filter used
Fiber Optics	Standard CATV Fiber Optics link 36 to >69 channels	FM fiber link 1 to 16 channels obsolescent
Precautions	Standard	Downconvert from 44 MHz to baseband. Noise filter used

Table 3: *A Choice of Uplink Types*

It is important to note that in a wireless environment, frequency drift in the uplink degrades the stability of the final transmitted downstream signal. In the studio transmitter link (STL) case, the downconverters and upconverters convert the 44-MHz intermediate frequency (IF) signal to near-baseband stability. In the case of amplitude-modified links, the conversion to the CATV band and back to IF degrades stability unless the devices can be locked to a stable reference.

Wireless Return (Two-Way Wireless) Options

Two-way wireless operation is in its infancy due to the availability of spectrum, equipment, and changing FCC regulations. The technology, regulations, and business issues are changing very quickly. Thus, a service provider must use a mix of telephone, ISDN, and wireless return as necessary to meet customer needs.

Use of Industrial Scientific Medium (ISM) Bands

It is possible to use commercially available spread spectrum modulators in any of the ISM bands. These can be driven from the RS232 output of a cable modem the same way as an external phone modem for telephone return. The advantage is that the spectrum is free and available now. It allows an operator to understand the deployment issues. It can be a viable long-term solution, given the cost of spectrum in the traditional frequency bands, and the highly directional 2.4-GHz antennas minimize the probability of interference with other users (see *Figure 4*).

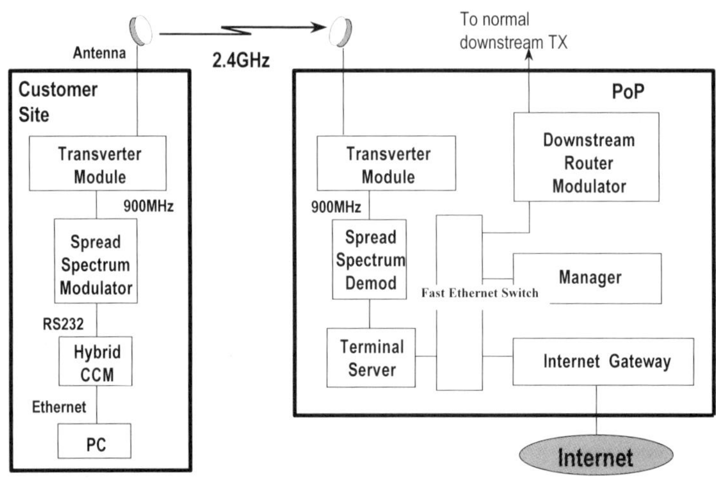

Figure 4: *One Way to Use the ISM Band for Wireless Return 900 MHz/2.4 GHz*

Use of the Quaternary Phase Shift Keying (QPSK) Signal from a Cable Modem

Transverters and upconverters are under development from various vendors to take the 5-MHz QPSK signal output of a cable return modem and translate it to the WCS, MDS1, MDS2, and MDSA bands. A transverter is simply a combined downconverter and upconverter. These units are similar in appearance to downconverters and generate 50 mW or 100 mW of transmitted power. This is enough to reach back to the downstream transmitter site or to the Internet head end if it is closer and a line-of-sight exists. The QPSK signal bit rate can be set to different values, from 256 kbps to 5.12 Mbps. Lower values allow more users (see *Figure 5*).

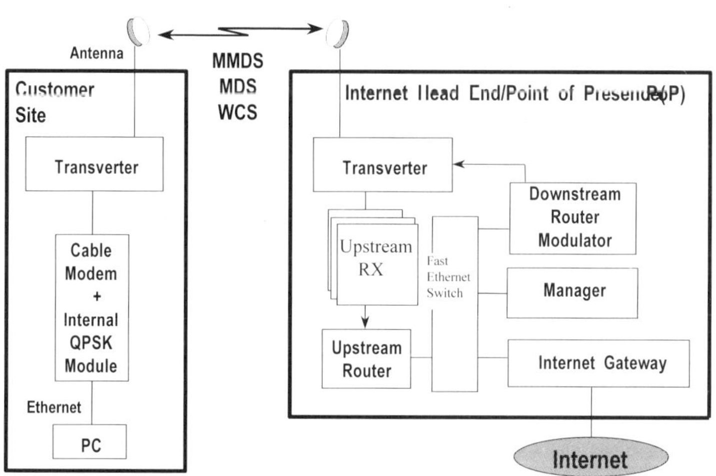

Figure 5: *Wireless Return Using the QPSK Signal from a Cable Modem*

The following are issues facing two-way operators:

- If the transmitter is separated from the Internet head end, a return path must be licensed and purchased similar to the downstream.

- The system cost, including the opportunity cost of spectrum, could be used for downstream signals rather than upstream.

- The bandwidth/bit rate used for the upstream. For example, the 12 MHz of bandwidth in the combined MDS1 and MDS2 bands could be sliced into 200-kHz channels with 256-kbps or 320-kbps rates. This gives channels supporting around 3000 subscribers with omnidirectional non-sectored antennas. This means that the return path receive antennas must be sectored as the business grows to achieve the same upstream subscriber capacity as the downstream.

- The QPSK modulator and the transverter upconverter portion determine the frequency stability and spurious signal output of the return path transmission.

- Transmitting back to the downstream transmitter site allows integrated transmit and receive antennas at the customer site. These cost less than separate units and, since they are pointing in the same direction, the transmit antenna is automatically aligned when the receive antenna is installed and aligned.

- Transmitting back to the Internet head end (PoP) wherever possible requires separate antennas with individual alignment, but sidesteps the requirement to get the return signal from the downstream transmitter back to the PoP.

Business Background for Wireless Internet Access

Typically the wireless operator does not offer Internet security directly but partners with an Internet service provider (ISP) with an existing customer base and the necessary data-networking skills. Some larger ISPs even provide service using telephone access, integrated services digital network (ISDN), cable modems over wired CATV plants, and service over wireless. One result of this is that ISPs need cable modem systems that work for both cable and wireless; another is that telephone return systems usually use the ISPs' existing phone modem banks.

Technologies such as asymmetric digital subscriber line (ADSL), T1, or wired cable may require a larger capital investment and more time for construction. As a result, some CATV- or telephone-based ISPs exercise a deliberate test-marketing strategy using wireless. Once the customer response is known, it is possible to make a business case for longer-term investment to build out the coaxial cable for higher-service penetration.

The larger wireless cable operators are finding high-speed Internet service an aid to competition with the local CATV companies as well as businesses in their own right. Internet service on one or two wireless channels is compatible with both analog and digital video on adjacent channels, as there is no loss of capacity or interference.

There are several reasons why businesses, libraries, hotels, and schools are the first marketing targets:

- They usually pay a premium price, reflecting a greater need for speed.

- It is easier to locate an antenna on the roof of a large building than on most homes.
- They require less support, as they have their own management information sys-

tem (MIS) people.

- The service organization has time to be trained before handling large numbers of customers.

In addition, larger businesses may prefer wireless for Internet access. They may not wish to pay for an actual T1 connection to the Internet even though they may have T1 (1.544 Mbps) local digital facilities installed for telephone service. This initial business focus drives the following two requirements:

- There is a need for a modem that serves multiple users.

- The ISP must use some form of technology to preserve performance for businesses at the same time as adding residential service. One technique is to assign two subchannels to a few businesses and one subchannel to a larger number of residential users.

Business Applications of the Internet

- The Internet is a worldwide technical library at a time when most companies cannot afford an in-house library. Search engines beat microfilm readers for most applications. (IBM has its patent library and the FCC its wireless orders and rulings on-line for everyone to use; NTIS, the federal bookstore, provides on-line access to Department of Commerce regulations for a fee.)

- Most technical suppliers provide a Web site with sufficient information to produce at least the starting point of a product catalog or data search.

- Some cities have the building permit process on-line and provide access to regulations.

- Realtors have their listings. It is possible for salespeople to attract prospects by showing pictures of listings. This allows comparison of what the customer says he wants with what he wants and can cut down the time to search. It avoids excessive time driving around and exposure to other realtors.

- So you want to buy a car? Salespeople know that they have one and know where it is located so that they can close the sale right there. (This is especially important in the new era of mega-dealers such as Republic Industries.)

- On-line banking, accounting, and outsourcing accounting services are available.

- Farms are small businesses, requiring data ranging from weather updates to price trends. The physical isolation of farm families can make the school, educational, catalog shopping, and communications applications of the Internet more important. Many wireless operators cover rural areas.

- Electronic data transfer supporting purchasing, billing, and accounting functions is especially important for the large number of smaller businesses providing just-in-time service for larger businesses. The concept started with manufacturing but extends to the medical and professional fields.

MMDS/WCS Transmitter Systems for Internet Access

MMDS is presently used in the United States to deliver video program content for entertainment and, in cooperation with instructional television fixed service (TFS) operators, to delivery video for distance learning activities. Most systems now use analog transmission under the National Television Systems Committee (NTSC) standard to deliver one video program per 6-MHz radio frequency (RF) channel. The available downstream spectrum includes the following:

- 2 MDS channels 2150–2162 MHz
- 16 ITFS channels A through D Group 2500–2596 MHz
- 8 MMDS channels E through F Group 2596–2644 MHz
- 4 ITFS G Group interleaved with 3 MMDS H1/2/3 channels 2644–2686 MHz

In many other countries, a similar amount of downstream spectrum has been assigned for MMDS use within the range of 2 GHz to 3 GHz.

A major change is occurring in the MMDS industry with the transition to digital video compression and transmission. Digital technology enables compression of at least five video streams of similar resolution to NTSC analog video into one 6-MHz RF channel. In the digital environment, an operator who has access to most of the downstream channels listed above can offer a selection of program streams that can aggressively compete with either direct broadcast satellite (DBS) or CATV entertainment video delivery systems with a few channels to spare.
With the spectral efficiencies of digital video compression, a few RF channels per

MMDS system can be dedicated to provide broadband (>10 Mbps) high-speed data service to Internet users and provide an additional revenue stream for MMDS operators. This is already authorized by the FCC for the case of RF downstream with telco line upstream. To meet FCC guidelines, the downstream transmission must be substantially similar to one of the presently authorized digital video modulation formats for co-channel and adjacent-channel interference reasons. A single 6-MHz RF channel using 64 QAM can deliver a raw data rate of 30 Mbps or 27 Mbps after forward error correction (FEC).

Several operators have applied for developmental licenses for various cities to test data access schemes with RF return path from subscriber to hub. A consortium of several MMDS industry participants has been formed and is funding theoretical studies and conducting lab and field tests. These activities are intended to support a formal request to the FCC for rulemaking to authorize standard rules to permit RF two-way data access via MMDS.

Assuming that most of the MMDS RF channels are dedicated to an entertainment video delivery service, only a limited number of channels are available for downstream data. A conversion to digital video may be an attractive way to increase the capacity of systems and the channels available for data. It is an alternative to antenna sectorization.

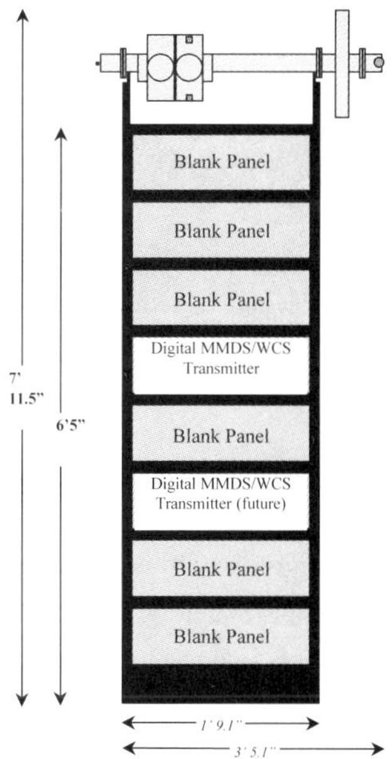

Figure 6: *A Typical MMDS/WCS Transmitter Block Diagram*

MMDS/WCS transmitter systems for Internet access include channelized transmitters, channel combiner, automatic backup, and network management equipment. Digital transmitters accept a 44-MHz–centered QAM IF; the transmitters then upconvert and amplify the signal (see *Figure 6*). Transmitters are available with 5 W to 100 W of average output power. Crystal oscillators are used to provide excellent phase noise. Feed-forward amplifiers and equalization techniques provide good gain and phase linearity while minimizing power consumption. To improve system availability, an automatic backup system with primary and backup transmitters can be used (see *Figure 7).* Spectral shaping filters are required for WCS systems to meet stringent FCC specifications for out-of-band power. Channel combiners utilize waveguide directional filters to combine either non-adjacent or adjacent microwave channels for transmission. Transmitter systems are supplied with signaling network management protocol (SNMP) network management capability for remote configuration and fault management.

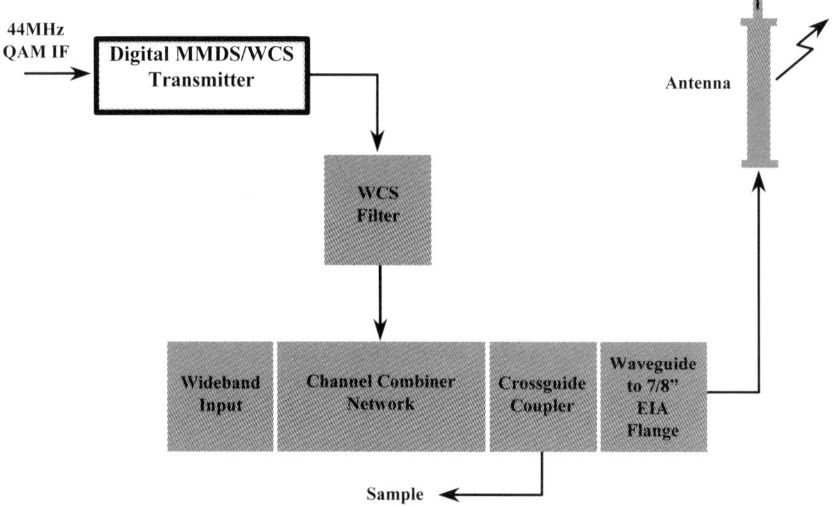

Figure 7: *A Typical Transmitter Rack with Two Transmitters*

MMDS Path Engineering: An Example

As this is only an example, actual figures will vary according to the specific downconverter vendor's product. Sample coverage assumes the following design criteria (see *Table 4*):

- 2.5 dB typical channel combiner loss
- 2 dB typical waveguide/jumper loss

- 13 dB transmit antenna gain
- 32 dB Block downconverter gain
- 2.5 dB Block downconverter noise figure
- 6 dB typical receive line loss
- 15 dB typical fade margin

Tx Output Power	Rx Antenna Gain	Maximum Coverage
5 watts avg.	18 dBi	12.5 mi.
	24 dBi	25.5 mi.
10 watts avg.	18 dBi	12.5 mi.
	24 dBi	25.5 mi.
15 watts avg.	18 dBi	15.5 mi.
	24 dBi	31.0 mi.

Table 4: *Sample Coverage*

MMDS Path Engineering: The Math

The critical design factor for a digital system is the required signal to noise ratio of the receiver. For this case, 25 dB S/N. Fade margin is an estimated value that will vary at each site. Factors that can contribute to fade include heavy rain, multipath, terrain, and foliage blockage.

Signal Strength Link Budget

Txp – CL – TLL + TAG – PL – FM + RAG + BDCG – RLL = signal strength @ rx site in dBm.

S/N Link Budget

Txp – CL – TLL + TAG – PL – FM + RAG – nf = S/N @ rx site in dB. Note that this calculation applies for a modem with 2-MHz channels and 4.5 dB transmitter additional headroom to allow for the transmission of the peak signal arising from three modulators on one channel. The numbers will vary for a single modulator and for more subchannels.

Additional Sources of Information

For additional information on wireless broadband modems, please consult the following sources:

1 TCP/IP Illustrated, Vol. 1: The Protocols, written by W. Richard Stevens and published by Addison-Wesley
2 Cable Datacom News: http://cabledatacomnews.com
3 FCC Web page: http://www.fcc.gov
4 National Science Foundation Wireless Field Tests (2.4 GHz): http://wireless.oldcolo.com

Part V: Security

Cable Modem Security

Robert Rance
Distinguished Member of Technical Staff
Secure Technologies Department
Bell Laboratories Lucent Technologies, Inc.

Introduction

This paper introduces the multifaceted nature of cable modem (CM) security, and will examine in detail some of the topics that particularly influence cable modem design.

Historically, introduction of new kinds of media and systems have engendered new types of attacks based on their inherent vulnerabilities. Hybrid fiber/coax (HFC) systems are no exception to this rule. They provide current and new types of data and voice services and digitally-encoded premium entertainment that is delivered to CMs. The architecture of HFC systems provide new opportunities for electronic piracy and malicious "hacking." This paper employs the shorthand word "attacks" to refer to both piracy and hacking.

When one million or more customers use a common HFC system design, the system becomes particularly vulnerable. Conversely, smaller HFC systems with unique designs are significantly more secure because the return accrued to the attacker would be outweighed by the cost of attack. The most vulnerable scenario is a single system serving several million customers. For example, if the system provides telephony for a city, the impact of an attack here would be huge. This essay will cover attacks relevant to both large and small HFC systems.

Overview of Security

Initially we introduce an HFC system in terms of its components and their respective security issues. Next we narrow our focus to the link between CM and head end, and then to other system pieces where relevant. This focus is not intended to minimize other important areas with security concerns, however, such as server security and physical plant security.

Many of the attacks on mutually-authenticated links arise from "cloning" either a legitimately registered cable modem with or without its owner's consent, or from cloning a head end. A clone is hardware and/or PC–based software that emulates the functionality of a cable modem or head end and additionally clones its identity by copying its secret keying information. By emulating functionality and identity, the clone becomes indistinguishable from a legitimate unit as seen by the head end or CM unit being attacked.

Threats to data come in various forms. The pirate can masquerade as a registered CM and steal data service. This is called a "theft of service" attack and is typically only detected when legitimate customers receive spurious charges on their end-of-the-month bill. A second threat is one against privacy. Here, a pirate merely listens to the HFC channel and copies information of a sensitive personal or company proprietary nature, or copies banking or credit card information. Finally, if electronic commerce becomes widespread, a more dangerous attack becomes remunerative for the pirate. Here the pirate masquerades as a head end to the CM and gains access to the user's PC, which can contain a significant amount of electronic cash as well as banking and credit card information.

In addition to these attacks, voice traffic is vulnerable to additional significant threats. In addition to the privacy and theft-of-service threats as with data traffic, a large deployment of CMs that offer telephony becomes subject to malicious hacking or extortion via large-scale denial of service. Basically, the attacker permanently disables all cable modems within a state-wide area, thus denying phone service to those customers.

In the case of digitally-encoded premium entertainment paid for on an individual basis, a new type of threat occurs. An individual may now, for an appropriate amount of money, release their keying information to a pirate. The pirate programs many clones with this and with keying information from other CMs so that these clones are well provisioned to receive free any premium entertainment of the clone owner's choosing.

Finally, CMs and, to a lesser extent, head ends are susceptible to attack via a bogus software download and control signal spoofing. These can enable many of the attacks already described, or can merely be of a malicious nature.

There are several ways to mount an attack. The pirate can make a clone by merely copying keying information from a legitimate CM. Perhaps more easily, the pirate can break into databases where the keys are stored, either at the head end or at a point of presence (POP). Such a break-in can be achieved by exploiting a weakness in the service provider's firewall, by an insider attack, or by "recycling" a discarded hard disk. A particularly damaging form of attack occurs when the pirate modifies a new software version to be downloaded to a large installed CM base, or otherwise initiates a new download.

The replay attack is subtler. Here the pirate or hacker need not have a copy of the keying information, but merely copies and replays later a given payload or control message. If the pirate can attach significance to the timing of a message, significant damage or monetary loss can occur. For example, the pirate can replay an encrypted broadcast message downstream granting every subtending CM the permission to watch entertainment to which they have not subscribed. Another example would be to replay a set of messages that effect a transfer of funds into the pirate's account.

Exhaustive key search is yet another possible attack. The attacker tries all possible keys in the encryption algorithm being used until he or she gets a match of input and output. Currently, exportable 40-bit keys are widely considered insecure. Various sources have cited that major governments are financially capable of developing dedicated key-cracking machines that routinely and efficiently break single-DES (Data Encryption Standard) 56-bit keys. A business case can be made for an off-shore piracy concern to have the same capability. Twenty years in the future, due to Moore's Law,[1] all PCs equipped with some dedicated hardware will also be able to crack 56-bit DES keys. Thus if a single-DES is employed, it must be used in a manner commensurate with the value and lifetime of the data that it protects.

We should not discount future advances in algorithms that crack public keys and thus break public key-based key management systems. At present, cryptologists feel that a 2048-bit modulus (the length of a public key) is probably sufficient for a twenty-year information lifetime. Most encryption systems use 1024 bits or less and appear not to be anticipating algorithmic improvements.

An important distinction here is that even though a fielded system may not have an operational life of twenty years, the information that it secures may need to be protected for that long. Decrypting data that is copied today from a "secured" HFC link using technology twenty years hence may be advantageous for the pirate. For example, such information might comprise sensitive medical information or financial data pertaining to public figures.

Encryption algorithms are potentially vulnerable even without resorting to exhaustive key search techniques. The only provably secure encryption algorithms are those that are totally inpracticable to build. Thus, the successful encryption algorithms in use today are those that no one knows how to break. A very large number of encryption algorithm topologies are possible. Thus we have only scratched the surface in defining encryption algorithms and exploring their attack suite.

Once an HFC security system is designed to withstand all of the above attacks, an additional implementation-dependent attack analysis must be done to ensure the implementation itself does not present weak links in the security chain.[2] As an example, keying may be accomplished by either a secret key-based or public key-based key distribution system. The impact of the former is that the CM manufacturer and/or the service provider may need to retain large databases of secret keys with both high security and high redundancy. The clash of these two opposing objectives yields a rather user-unfriendly key-distribution system.

Solutions

One method for reducing this vulnerability of the POP is to restrict encryption keys to those areas where they are specifically needed. In particular, there is no significant advantage in exporting session keys (short term keys) from the head ends to the POP.

If the developer has elected to use public keying algorithms and, in particular, public key certificates for key management in large-scale HFC systems, which type should he or she choose? Various algorithms are possible: RSA, Rabin, Diffie-Hellman, and Elliptic Curve, to name a few. However, independent of the choice of an actual algorithm, there is the differentiation in capability between "global certificates," such as RSA and Rabin, and "point-to-point certificates," such as Diffie-Hellman. Point-to-point certificates cannot service a distribution model where an individual buys, owns, and crypto-initializes (keys from an external source) a CM from a distributor. In contrast, global certificates can accommodate both this model and another model when the cable service provider-owner security linkage is done at the modem factory.

Some of the attacks previously mentioned are enabled by a bogus software download. CMs or NIUs may be always downloadable for two reasons: their firmware is not sufficiently mature, and the service providers might want an easy path to install new features. There are two basic methods to protect against bogus downloads. One is to partition the cable modem or NIU into three areas: ROM-based download software, NVM-based operational code, and a separate security area. The second method does not bifurcate the functionality into a separate and possibly expensive security area, but authenticates all downloads. A question then arises as to who will be the authenticator: the manufacturer, the service provider, or both?

This question brings up broader issues: Which parties have security interests in the link? Who will install and own the CM? Four parties could have simultaneous security interests in a given user's CM: the service provider, the user, the content provider, and the corporation in work-at-home applications. One method of serving this mix would be to establish a somewhat tamper-proof security partition within the modem in which the partition is enforced and secured by a publicly accessible code developed by a manufacturer.

Link security is complementary to end-to-end (typically application-based) security. Most HFC systems will need both. The two securities complement each other in two ways: different services require a varying mix of privacy afforded by the two securities, and the two securities protect against different types of active attacks. As an example of the former, there are services, such as telephony, that are physically secure until the link, and thus require only link security. Proprietary data secured at the application layer, however, does not require link encryption of its payload. As an example of the latter, the theft of E-cash and other sensitive information from an individual's PC requires link security. In fact, link security here will protect against revealing of passwords and other secrets used for end-to-end security.

The emergence of multifaceted HFC systems provides a new fertile ground for piracy. An HFC security system must address an HFC-specific suite of potential attacks as well as more generic attacks. One must choose security components with some thought based upon a study of possible attacks.

Often an HFC system can be attacked even though the provider has a mutually authenticated link encryption system in place. Mutual authentication means that both sides, the head end and CM, actively confirm that every encrypted message arriving from the other side has in fact been transmitted by that entity and not by some third attacking party. The term "encryption" means that the transmitted information is totally obscured to anyone not holding the appropriate keying information. Finally, there is a difference between payload and control. Payload comprises the three services mentioned earlier, while control messages are those meant to either configure the CM, or otherwise communicate operational status in both directions between CM and head end.

The Domain to Be Secured

Data and voice services and digitally encoded premium entertainment sent over HFC media to CMs cover the last leg of their journey over optical fiber and coax (see *Figure 1*). This leg is often called the "link" and is particularly vulnerable to attack. Its most vulnerable portion is the coax that runs between the owner's home and the fiber node where the radio frequency (RF) signals are converted to light (see *Figure 1*).

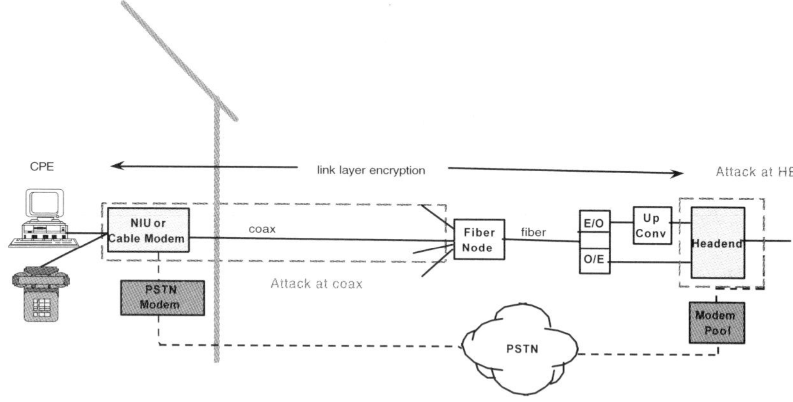

Figure 1: *Link and Attack Points*

The coax is particularly vulnerable because it enters both legitimate and illegitimate user's homes, and carries services of a large number of neighbors: up to 4000 for cable modem service. The fiber portion of the link is much less accessible because it is both physically hard to couple to the optical signal and also physically inaccessible, as it typically is run via telephone poles.

In a one-way system, the PSTN (telephone twisted-pair) return offers sufficient physical protection against most attacks on an upstream link. However, in a two-way system, the upstream is nearly as vulnerable as the downstream. We say "nearly" because the upstream bandwidth is typically much smaller, and because an attacker midway on the coax link can only access those modems downstream from the attack point.

There are some limitations to what can be protected. For example, it is relatively easy to bombard the cable with noise from a home. Also, one cannot protect against physical theft of a modem from a home.

Another dimension to the domain of protection regards the types of payload needing protection. In particular, some examples of these are data (IP, ATM), voice, video, other limited distribution material, and media access control (MAC) control messages.

Threats

The design of a security architecture should start with a threat model: a listing of possible attacks, the players associated with each attack, and the monetary or political loss resulting from a given attack. The following table lists attacks and their remedies for each of three contents: access, voice, and control information (see *Table 1*).

Threat	Examples	Prevented by
Access (Link Data)		
Data Privacy	Proprietary, Credit Card #s, etc.	Encryption
Theft of Service	Masquerading as another user	Modem authentication
CPE Intrusion	Pinhole in cable, 2000 PCs raided of e-cash	Headend authentication
Voice		
Privacy	Listening to neighbors	Encryption
Theft of Service	Long distance calls	Modem authentication
Large Scale Denial of Service	Eliminating most telephony in state-wide area for 3 months	Headend and download authentication
Control		
Intrusion at Headend	Collecting shared secret keys for all modems	Modem authentication; careful security design
Intrusion at Modem	Malicious reconfiguring of modem	Headend and download authentication

***Table 1:** Attacks and Remedies for Access, Voice, and Control Information*

Threats to data come in many forms. The pirate can masquerade as a registered CM and steal data service. This is called a "Theft of Service" attack and is typically only detected when the legitimate customer receives spurious charges on their end-of-the-month bill. A second threat is one against privacy. Here, a pirate merely listens to the HFC channel and copies information of a sensitive personal or company proprietary nature, or copies banking or credit card information. Finally, if electronic commerce becomes widespread, then a more dangerous attack becomes remunerative for the pirate. The pirate masquerades as a head end to the CM and gains access to the user's PC, which can contain significant amounts of electronic cash as well as the aforementioned banking and credit card information.

Lest the preceding cause worry, the aforementioned vulnerabilities are easily fixed by either encryption, authentication, or both. By definition, encryption obscures data to unintended recipients. Authentication ensures communicating parties that the message source is who they think it is.

Messages can be encrypted but not authenticated (e.g., if you had sent your secret key to an unknown party and this party used it to encrypt messages to you). Messages can also be authenticated without encryption (e.g., an unencrypted message with an appended message authentication code).

The following outlines two particular types of attacks that could be collectively described as clone attacks:

1. Theft of Service (Data, Voice, or Entertainment) in the Absence of Encryption and Keying Material

- The pirate either sells clone box or modifies owner's cable modem (CM) software.
- The CM clone listens to a downstream channel and collects addresses of 2000 CMs.
- When the owner wants free long distance or data service, his CM masquerades as another CM by using one of the collected addresses.
- Subsequent accesses to free service use different addresses to minimize detection.

2. Theft of Service via Bogus Download on an Otherwise Secure System

- Pirates replace CM code update with their own design.
- The CMs send their secrets over the Internet encrypted with pirates' keys.
- Pirates accrue database of all CM secrets allowing mass manufacture of clones.
- Free long distance or data service is then obtained as described earlier.

Many of the attacks on mutually authenticated links arise from "cloning" a legitimately registered cable modem with or without its owner's consent.

Without security, theft of service becomes trivial: a pirate merely needs a clone box that assumes the identity of another entity on the same cable run. By assuming multiple identities, the pirate can avoid detection.

The second clone attack is from a particularly powerful class of attacks enabled by a bogus software download. The bogus software could be substituted near the source, or sent individually to all of the head ends over the network that connects them.

Bogus software downloading is a particularly potent threat. Some CMs and NIUs (telephony Network Interface Units) often have their program memory contained in Flash (a particular form of nonvolatile memory). These units are designed to be downloaded over the air to add new features and/or to correct software "bugs." A bogus download inserted at the right place could affect up to twenty million CMs or NIUs (see *Figure 2*). This would allow a significant return on investment to the pirate.

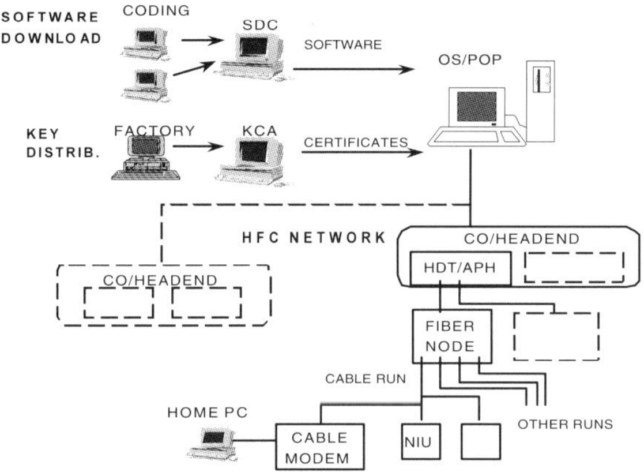

Figure 2: *The HFC Network*

There are two more dramatic types of attacks that may occur if a given CM system becomes very large:

1. Telephony Nullified: Scenario: twenty million homes are supplied with HFC telephony from one provider.

- The hacker or terrorist infects NIU code update with a software "time bomb."
- After all NIUs have been updated, the time bombs begin to explode.
- All twenty million NIUs are disabled and their updated software prevents further downloads.

Result: Phone service is nullified in a wide area, possibly including a major city, for about three months until all NIUs have been reprogrammed by technicians or at the factory.

2. Attacker Gets Rich: Scenario: electronic commerce and cash are prevalent in the coming years.

- The attacker places a pinhole in the cable, upstream of 2000 CMs.
- The attacker blocks the head end signal and masquerades as the head end.
- 1000 home PCs are entered and raided of a net ten million dollars in the form of credit card and bank account numbers, and anonymous e-cash.
- The attacker then repeats with the next cable run.

These dramatic examples are more interesting than the more probable attacks in *Table 1*. However, they are nonetheless feasible for sufficiently large systems because of the one-to-many topology inherent in HFC networks. As shown, the payoff to the attacker can be huge.

Encryption and Authentication

Two basic enablers of security for many systems in general and CMs in particular are encryption and authentication. Encryption can be defined as the process of obscuring data to make it unintelligible to an unintended listener. Authentication is the process that enables one to know that received information originated from the expected source. Systems can have encryption without explicit authentication. As an example, many telephony privacy systems do not explicitly tell you who the other party is; here, voice recognition provides implicit authentication if one knows the other party.

Other systems can have authentication without encryption. One such scenario would be control messages in HFC systems, where, in many cases, it is not important to obscure the nature of the messages but is important that the receiving system knows that the message is legitimate.

Encryption

A generic cryptographic algorithm is represented in *Figure 3*.

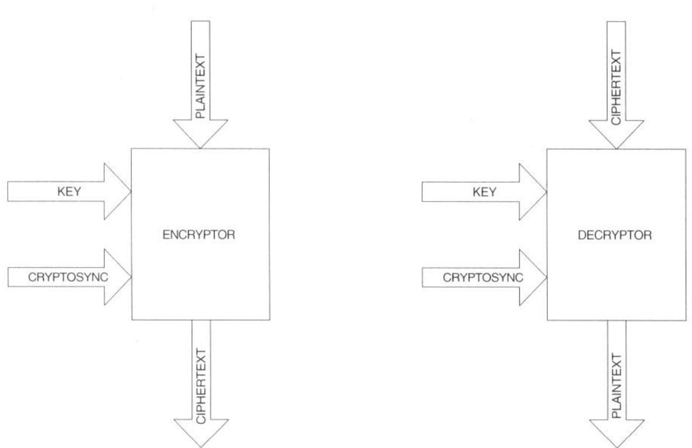

Figure 3: *Generic Encryptor and Decryptor*

An encryptor renders a plaintext message into ciphertext, so that no plaintext information is discernible to the attacker. The decryptor reverses the process and outputs plaintext, but (ideally) only to the legitimate receiver. *Figure 3* shows the plaintext and ciphertext as a "block" of data, which typically ranges in length from 1 bit to 256 bytes.

The secret key is fixed, secret data shared only by the intended transmitting and receiving entities, and are commonly termed "shared secret keys." Typical key lengths range from 40 bits to 128 bits, although lengths in the 56-bit to 64-bit range have, until recently, been considered sufficient to secure most commercial applications. A given key secretly selects a set of mappings from a very large space of possible mappings. Shorter keys can be deemed secure if they are changed frequently.

Although the ciphertext length may be greater than the plaintext length, most popular encryptors avoid such bandwidth expansion for practicality. One exception is public keying. However, public keying is seldom used for encryption of significant amounts of data, and instead is used to create shared secret keys.

There are many different types of encryptors. The most popular and widely-used is NIST's data encryption standard (DES), which encrypts a 64-bit plaintext block with a 56-bit key.

A secure encryptor protects its plaintext and its key from revelation. *Figure 4* is a simple example of encryption.

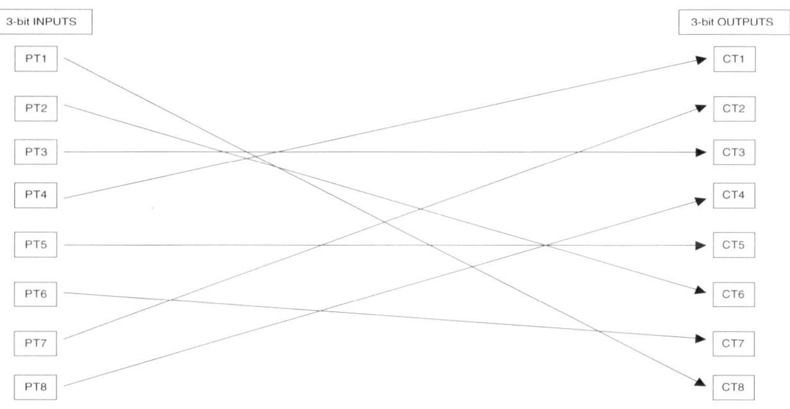

Number of Possible DES–like Encryptors

- DES encrypts a 64–bit block via a 56–bit key.
- A general 64–bit encryptor would have a maximum key size of $2^{64}!$ or about 2^{70} bits (not that one could ever be built).
- Thus, the general encryptor comprises about $2^{2^{70}}$ possible mappings, falling some where between a googol and a googolplex—a large number!
- DES (or any other 64–bit encryptor keyed with 56 bits) is a collection of a mere 2^{56} of these mappings.
- Overall, there are about $(2^{2^{70}})^{2^{56}} \simeq 2^{2^{62}}$ possible encryptors of this type. This is much greater than a googolplex—a very large number!

Conclusion: We have just scratched the surface in inventing cryptoalgorithms.

Figure 4: *A 3-bit Encryptor*

Encryption Algorithm Example: A 3 bit Encryptor In *Figure 4*, there is an implicit key that maps the 3-bit plaintext to 3-bit ciphertext. Different keys will create different mappings. By reversing the mapping, we decrypt to derive the original plaintext. Even in this simple example, there are 8! = 40320 possible mappings.

Key Stream Generators

Encryption algorithms like DES are termed block algorithms because they encrypt a block of plaintext to form a block of ciphertext. A second very important class of cryptoalgorithms is key stream generators (KSGs), also known as stream ciphers. Instead of encrypting a block of plaintext directly, they output a stream of secret bits called a keystream. These bits are then typically XORed with the plaintext to form ciphertext. Unlike block encryptors, KSGs have no error extension due to the XOR process. They have a disadvantage in that they mandate cryptosync, which may not be available in a given system.

Cryptosync

An important aspect of encryption is cryptosync. Earlier, this paper described a secret input to the encryption algorithm, called a key, that secretly selected a mapping from a very large set. Cryptosync also selects one mapping from a very large set, but it is publicly known and changes with each new plaintext block.

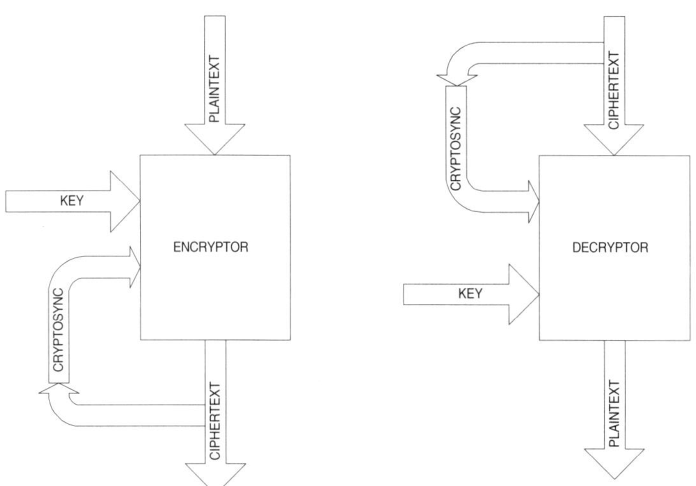

Figure 5: *Data-Dependent Cryptosync*

Cryptosync can be either data-independent or data-dependent. Data-independent cryptosync takes the form of an external, public synchronizing signal, as shown directly in *Figure 5*. Data-dependent cryptosync uses prior known data (e.g., ciphertext) to act as a synchronizing signal for the current encryption with the disadvantage of bringing back error extension.

There are several needs for cryptosync:

- preventing codebook and replay attacks
- enhancing cryptoalgorithm-inherent security
- allowing strong encryption without error extension

To describe these needs, one first must define two of the more well-known methods of attacking encryption: known plaintext and chosen plaintext. In the known plaintext attack, the attacker has either total or partial knowledge of what is being encrypted. This knowledge enables the attacker to partially or totally break the encryption scheme so that future ciphertext may be easily deciphered. The chosen plaintext attack is typically more powerful in that the attacker has control over what is sent, allowing him or her to choose those plaintexts most advantageous for the attack.

The codebook attack is enabled by either the known or chosen plaintext attack. Here the attacker forms a table of plaintext-ciphertext pairs. If the number of possible plaintexts is small, then such a table can be easily completed. Thereafter, the attacker merely needs to address the table with a newly received ciphertext to obtain the corresponding plaintext. By adding cryptosync to the plaintext, we defeat this attack because formerly identical plaintexts now become unique and, thereby, encrypt differently.

Some cryptoalgorithms are designed to work best with cryptosync applied to their internals. One example is a hybrid design in the form of a KSG that continuously reconfigures a block algorithm.

Other Encryption Facts and Definitions

DES and all other known reasonably implemented cryptoalgorithms have not yet proven secure. Nonetheless, DES has not been significantly weakened below the exhaustive key search limit when reasonable amounts of data are encrypted under a single key. An exhaustive key search is defined as attempting all keys in order to break a cryptoalgorithm. Another useful term is a "replay attack," in which an adversary resends an encrypted message and has it accepted as valid. It is a fact that most cryptosystems fail due to protocol design errors and not due to cracking the algorithm itself.

Table 2 is based on "Minimal Key Lengths for Symmetric Ciphers to Provide Adequate Commercial Security" by Matthew Blaze, et al.

Breaking (Single) DES in 20 Years
$100 ASIC-equipped plug-in board for home PC: 1hr/key

Breaking (Single) DES in 10 Years
$1000 ASIC-equipped plug-in board for home PC: 10hr/key

Breaking (Single) DES **Now**
Offshore pirate business invests $10M in DES cracking hardware
US-based customer sends single PT-CT pair via PGP over the Internet anonymously along with $100
At $38 cost to them, pirate business returns DES key to customer in 6 minutes

Table 2: *More Facts on DES*

Currently, exportable 40-bit keys are widely considered insecure. In fact, it appears likely that major governments can routinely and efficiently break the single-DES algorithm that uses 56-bit keys. A business case can be made for an off-shore piracy concern to have the same capability. If single DES is employed, it must be used in a manner commensurate with the value and lifetime of the data that it protects. In fact, just recently, the Electronic Frontier Foundation released details of a DES–cracking machine that they had built to win RSA Laboratory's "DES Challenge II" contest. Their machine, built for less than $250,000, found a DES key in three days.

Encryption Modes

There are three encryption mode examples for use in MAC systems: electronic codebook (ECB), cipher block chaining (CBC), and keystream generator (KSG) (see *Figures 6, 7,* and *8*).

Implications for Cable Modem Systems

If encryption occurs at the MAC layer or below, then a certain type of encryption methodology is particularly advantageous for CM systems. Broadcasting the cryptosync, for example, greatly conserves cryptosync bandwidth. By keeping encryption at a lower layer, cryptosync is easily conveyed up from the physical layer.

Such cryptosync allows encryption via a KSG. KSGs are typically easier to build (more security for a given amount of processing power) than block encryptors. It is particularly important to note that they allow encryption without error extension, a feature particularly valuable for voice encryption. Other advantages of a KSG–based approach are no error extension, no delay, and minimal waste of processing for typical KSGs that output their keystream in byte or bit format.

Modes

Three encryption mode examples for use in MAC systems: ECB (electronic codebook), CBC (cipher block chaining), and KSG keystream generator

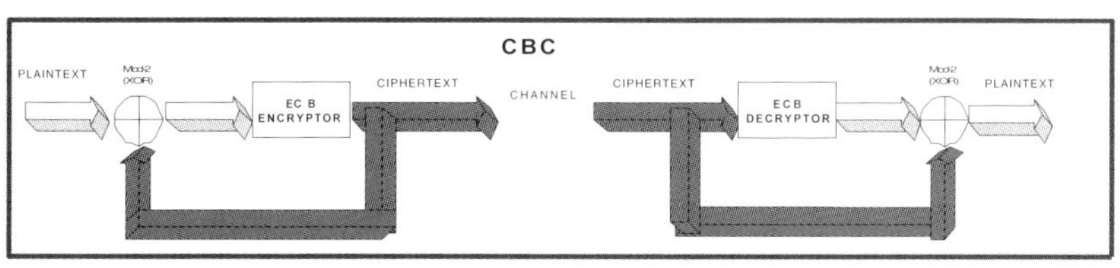

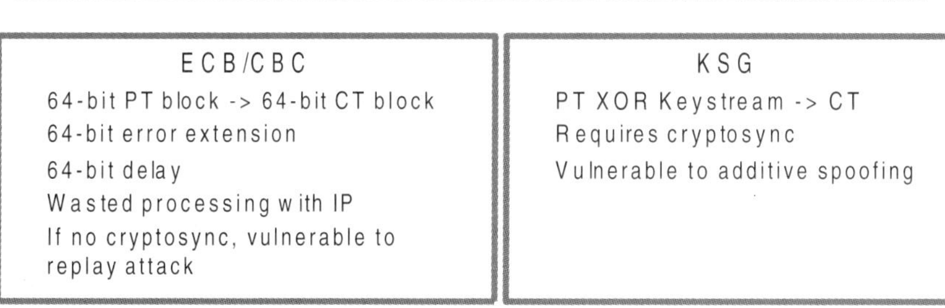

Figure 6: *Three Encyption Mode Examples for the MAC System*

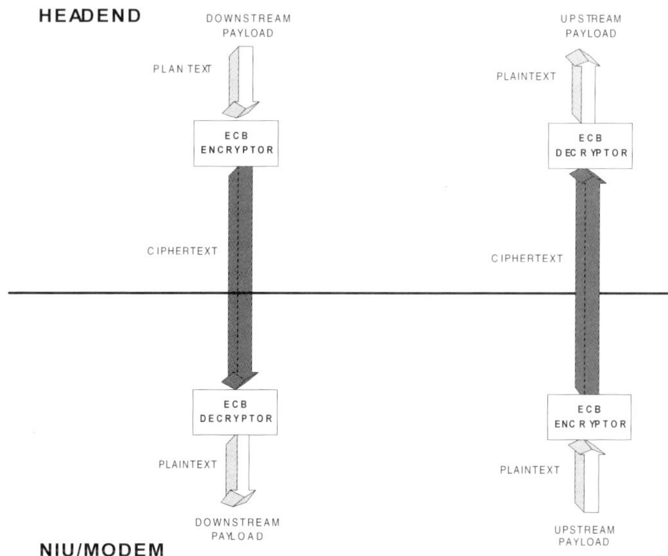

Figure 7: *ECBin Action*

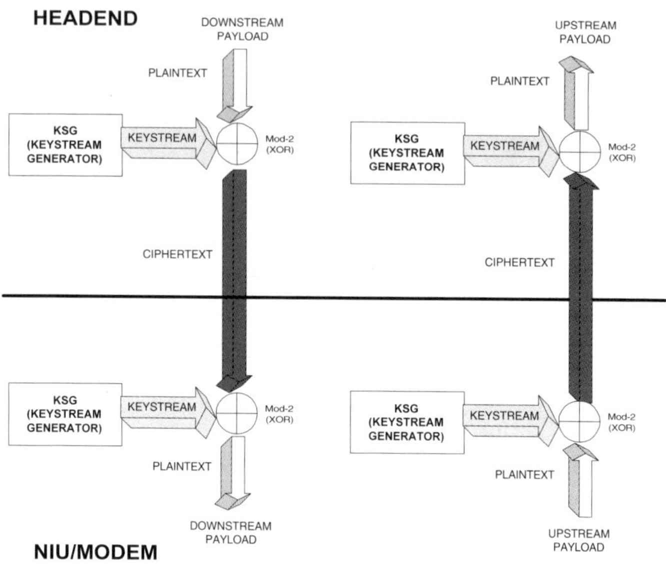

Figure 8: *KSG in Action*

Choosing an RNG

A random number generator (RNG) or randomizer generates bits or numbers that cannot be predicted by any other entity, adversarial or friendly. RNGs are needed to create keys and give cryptoalgorithms unique and unpredictable starting states. Without an RNG, a cryptosystem would always start at the same state at all levels of its hierarchy and thus would be vulnerable. Some types of RNGs include a sampled noice source; a sampled oscillator; chaotic, cryptoalgorithmic, and radioactive disk drive variations; and keypad timings (see *Figure 9*). RNGs are perhaps the least understood components that comprise a good cryptosystem. The irony here is that in contrast to encryption algorithms, many RNGs can be proven secure in that the quality and secrecy of their output can be directly tied to physical principles. Systems have been broken due to insufficient RNG design.

Authentication and Public Keying

Authentication assures communicating parties that the identity of the message source is correct. Messages can be encrypted but not authenticated, and also can be authenticated without encryption. A message authentication code is a message appendage formed by applying a keyed hash function to a message. It is important that both the head end and the NIU/modem engage in mutual authentication to

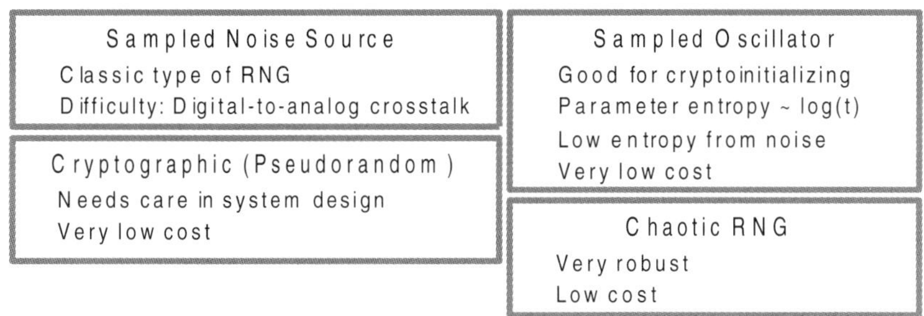

Figure 9: *Examples of RNG*

prevent attacks. If processing cycles are at a premium, the message content may dictate whether a given message is authenticated or encrypted.

Certificates allow Party A to convince Party B of the authenticity of a third item in a public manner akin to a notary public's function. Because they are fully public, certificates can be sent via the Internet. If the "third item" is the public key of a third party, the certification is hierarchical. Typically, but not always, certificates arise from public key algorithms. When they arise from secret key algorithms, they are called message authentication codes. Using public certificates avoids a messy secret key–based key distribution system in which large databases of secret keys require redundant storage at both factory and service provider.

We may separate certificates into two classes: point-to-point and global. An example of a point-to-point certificate would be a factory sending a certificate to the home owner's NIU telling it which telephone company's OS to trust. A global certificate would be a certificate sent by the factory to all CMs, linking the name of Provider X to X's public key. In the example in *Figure 11*, only a global certificate can be sent, because the factory does not know *a priori* the provider-to-modem linkage. For example, RSA and Rabin's method can form either type of certificate, but Diffie-Hellman can only implement point-to-point certificates.

Two Examples of Key Management

In this part, we show two examples of how one might secure an HFC system via public keying. Many other approaches are possible.

In *Figure 10*, the factory programs the NIUs with secrets. Once an NIU is shipped to a given telephone company, the factory's KCA forms a certificate that links the NIU, via its serial number, to the telephone company via that company's public key. When that certificate is conveyed to the NIU at installation, the NIU is directed to trust that telephone company and no other. The trust is then conveyed hierarchically, and the NIU is instructed by the telephone company to trust a particular head end.

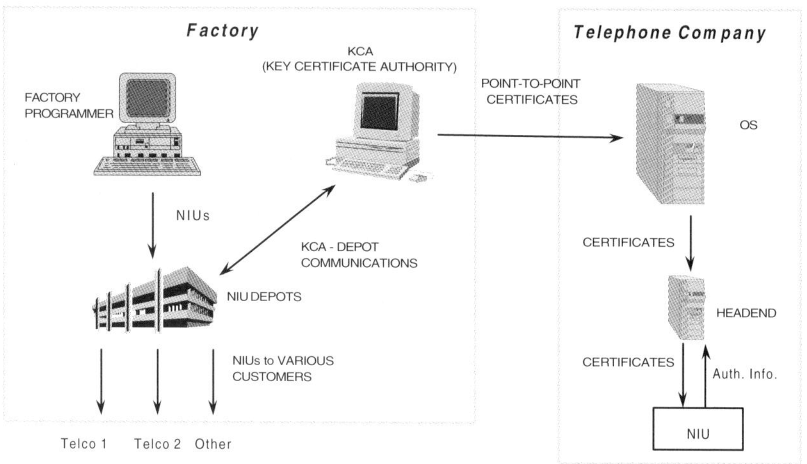

Figure 10: *Public Key-based Management via Factory Crypto-initializing: HFC Telephony*

Similarly, the telephone company and finally the head end are given public information about the NIU secret information, which allows the head end to authenticate the NIU. In this manner, the NIU and the head end arrive at mutual authentication.

In *Figure 11*, the scenario is slightly different in that the factory has no knowledge of the service provider at the time of shipment. It thus creates global certificates for the providers, which include the provider's name and their public key certified by the factory's public key. Such certificates can either be part of the installation software or, as is particularly the case of the newer providers, can be passed via the POP and the head end.

When the owner installs his or her CM, he or she selects onscreen the provider's name. The software then provides the appropriate public key to the modem. Mutual authentication then proceeds as in *Figure 10.*

Remaining Issue of Trust

One remaining issue is trust: can Corporation Y trust Cable Provider X with its proprietary information? Even if information is encrypted end-to-end between the CM and Corporation Y, the provider still may have access to the keys (see *Figure 12*).

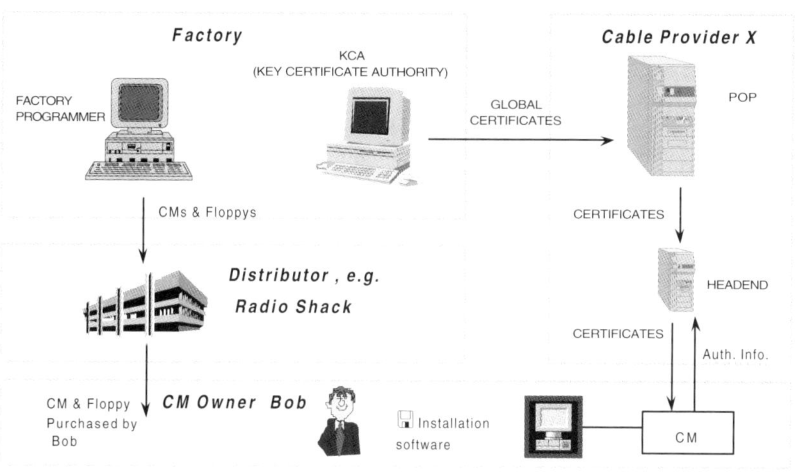

Figure 11: *Public Key Based Key Management via User Crypto-Initializing: Cable Modems*

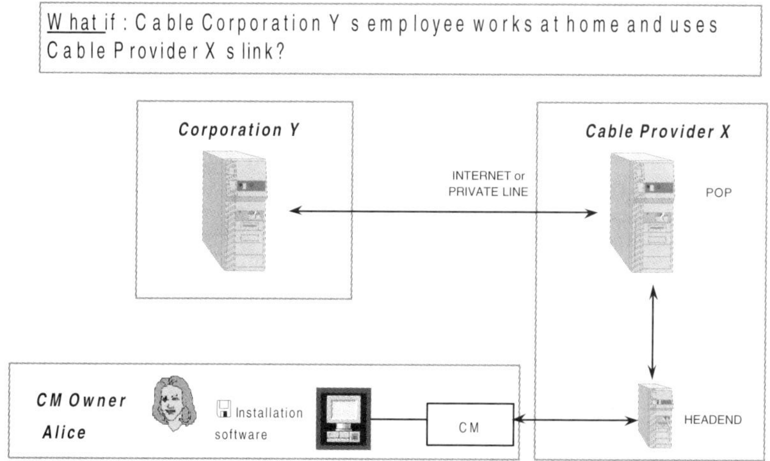

Figure 12: *Possible Confliction of Security Interests*

In fact, there are several parties with possibly conflicting security interests. The CM owner wants to protect items like medical information and credit card numbers. The service provider wants to protect against theft of service, intrusion into the head end, and liability for other compromises. The content provider wants to ensure that its entertainment data is sent only to paying customers. Corporations need to protect their proprietary data (see *Table 3*). The most attractive solution appears to be

Security Terminus in:

PC

Encryption speed will not be a factor in near future

However, work-at-home keys are vulnerable to service provi

CM Hardware

Security partition(s) separating work-at-home, service provid

Partition enforced by manufacturer

Design of CM security is public

Security updates or other downloads via manufacturers sign certificate

Files optionally encrypted on PC

Thus service providers liabilities are minimized

Table 3: *Review of Security Options*

manufacturer-enforced security partition(s) in the CM where the security design is public knowledge. This relieves the service provider from guaranteeing security.

Further Reading

Two good references on security are Schneier[3] and Menezes[4]. The former is a good introductory and general text, while the latter explores many cryptographic isues in significant mathematical detail.

Two developing cable modem standards are as follows: Data Over Cable Service Interface Specifications (DOCSIS) managed by Cable Labs at http://www.cablemodem.com/, and the IEEE Project 802.14 at http://www.walk-ingdog.com/. A significant portion of each of these standards addresses security issues akin to those raised in this essay. Also, they are important to read since they show alternative implementations of CM security.

References

1 Moore's Law: Computing power approximately doubles every 1.5 years. In ten years, PCs will have 100 times the processing speed. In twenty years, this factor will be 10,000. It appears that computer memory, hard disk and RAM, also follows this curve.

2 A security system has often been compared with a chain; it will break at its weakest link. Pirates are often adept at finding that link and concentrating their energies there.

3 Menezes, A., van Oorschot, P., and Vanstone, S., *Handbook of Applied Cryptography*, CRC Press, Boca Raton, Florida, 1997.

4 Schneier, B., *Applied Cryptography Protocols, Algorithms, and Source Code in C*, John Wiley & Sons, New York, Second Edition, 1966.

Data-over-Cable Security

Chet Birger, Ph.D
Consulting Engineer
Network Analysis and Engineering
Nortel Networks

Abstract

Data-over-cable's use of the hybrid fiber/coax (HFC) network for two-way data communications carries with it security risks inherent to communications over a shared medium. The cable industry is addressing these security issues by incorporating cryptographic mechanisms for encryption and authentication into link-layer communications across the HFC network. By employing cryptographic tools such as symmetric and public-key encryption, signed one-way hash functions, and digital certificates and signatures in link layer security systems, data-over-cable systems can be built to support a range of security capabilities. Multimedia Cable Network System's DOCSIS baseline privacy is one particular data-over-cable security solution that provides privacy, message integrity, and a limited degree of service protection.

Introduction

Data communications over cable television networks provides efficient, high-bandwidth network access for computers in homes, schools, and small businesses. Cable television networks use a shared communications medium (a radio frequency [RF]) network built over a hybrid fiber/coax cable plant) to connect users; the aggregation of users' traffic that the architecture enables contributes significantly to data-over-cable's overall efficiency.

There are security risks, however, inherent in data communications over a cable network's shared medium. In the absence of any security measures, the following must be true:

- Downstream transmissions from a head end, which are broadcast throughout the RF network, can potentially be received by any device connected to that RF network, making the service susceptible to malicious eavesdropping and threatening the privacy of users' data communications.

- A head end receiving upstream transmissions cannot authenticate the source of these transmissions; they can originate from any location on the cable network's tree topology for which the head end serves as the root. Without the ability to authenticate the source of data transmissions, the network is susceptible to service piracy.

In contrast, wireline telephone networks provide a dedicated line (the "local loop") between each subscriber and the telephone central office. Eavesdropping on communications across a particular local loop would require physical access to that telephone line, representing a significant deterrent. Similarly, pirating wireline telephone service (as opposed to wireless) is difficult because telephone companies know which lines go to which homes; i.e., telephone companies use their wiring plant to authenticate the source of a transmission. Wireline security is a far greater issue for cable-based data communications, with its shared broadcast media, than it is for telephone-based network access services.

The cable industry is addressing its privacy and security issues by incorporating cryptography into data-over-cable communications architectures. In particular, the latest generations of cable modems (CMs) and head ends support link-level security, integrating cryptographic mechanisms for authentication and encryption into the RF media access control (MAC) protocol.

Note that similar privacy and security issues are present in wireless communication systems such as cellular telephone service, which, like cable systems, relies on a shared communications medium. While cellular service providers early on introduced technical methods to address theft of service, they have only recently begun to address communications privacy.

This paper discusses CM security issues and demonstrates how modern cryptographic technology is employed to address them. Following this introduction, we outline the security requirements of a data-over-cable communications system. The next section describes the cryptographic tools that are used to address data-over-cable security requirements and presents how these cryptographic tools are specifically employed to secure link-layer communications over a cable network. The paper's final section presents an overview of a particular data-over-cable security solution: DOCSIS baseline privacy, developed by the MCNS consortium.

Security Requirements

Network Architecture

The predominant architecture for data-over-cable communications, as reflected by both the MCNS DOCSIS and IEEE 802.14 standards efforts, relies on centralized control of the cable network. In particular, a head end controller at the root of the cable network's tree topology is the only network node permitted to transmit on the downstream channel and the only network node permitted to receive on the upstream channel. CM nodes are designed to receive only on the downstream channel and transmit only on the upstream channel. Thus, all link-layer communications on the cable network occur between a head-end controller and one or more CMs. Downstream unicast flows have a single intended recipient, while downstream multicast flows can have multiple intended recipients. All upstream flows may be viewed as unicast, having a single intended recipient, the head end controller. *Figure 1* depicts the upstream and downstream flows between a CM and head-end controller.

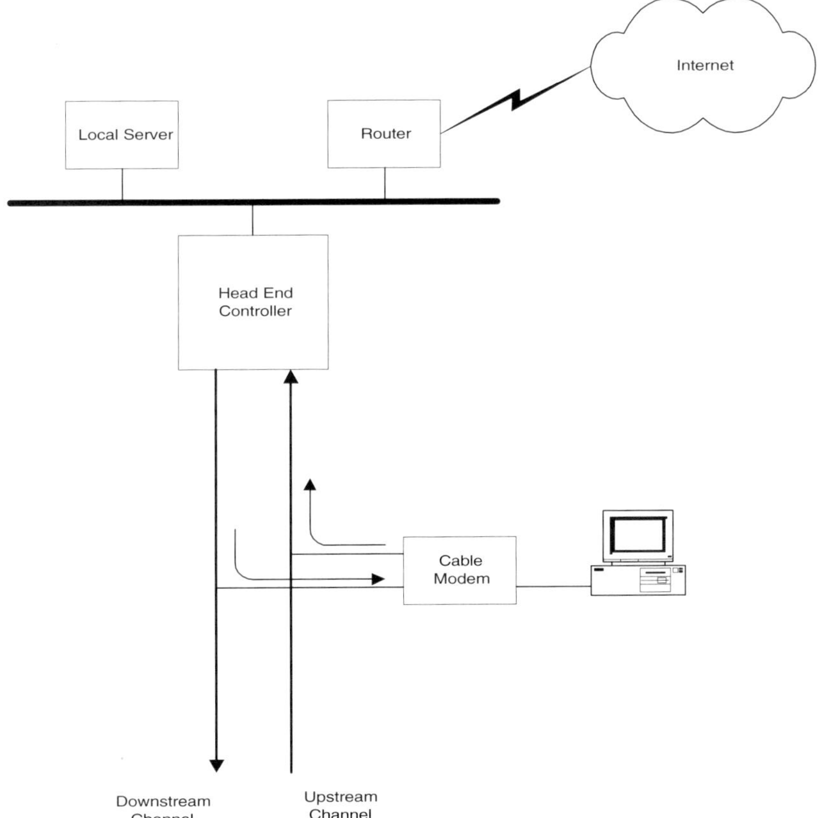

Figure 1: *All Communications on the Cable Network Occur Between a Head-End Controller and One or More Cable Modems*

The data service supported by this architecture allows bidirectional transfer of data traffic between a cable system head end and customer locations over the hybrid-fiber/coax (HFC) cable network. The head end will provide access to locally based data services and to the wider Internet.

Principal Security Goals

In the context of the link-layer transport of unicast and multicast traffic flows between head end and cable modems, the principal security goals of a data-over-cable system can be stated as follows:

Privacy: ensure the privacy of users' link-layer data communications across the shared medium cable network.

Integrity: ensure the integrity of users' link-layer data communications across the shared medium cable network.

Service protection: restrict to authorized users access to data services accessible through the head end.

Securing Link-Layer Communications over the Cable Network

Privacy

Link Layer Encryption Data-over-cable systems provide user data privacy by encrypting link-layer traffic flows between CMs and head-end controllers. CMs and head-end controllers encrypt the payload data of link-layer frames they transmit onto the cable network. The encrypted link-layer frames carry an encryption header identifying the security association (SA) under which the frame data is encrypted.

Security Associations An SA is the set of security parameters, including keying data, related to a given connection or set of connections. Connections, in this case, are data link-layer flows across the cable network. Link-layer frames carrying encrypted data must include an SA identifier to allow the decrypting entity (either a CM or head-end controller) to identify the keying material required to decrypt the frame data.

Prior to encrypting frames for transmission across the cable network, CM and head-end controllers need to identify the SA whose keying data will encrypt the frame data. SA selection is trivial for a CM: all of its upstream transmissions can flow across a single link-layer connection, terminating at the head end controller. SA selection is more complex for the head-end controller: keying data to be used for a particular downstream transmission must be associated with one of a large number of potential SAs, because each downstream CM has its own SA. Selecting the appropriate SA requires mapping the destination address within the forwarded user data packet to the target CM, or group of CMs in the case of multicast traffic.

Data Encryption Algorithms CMs and head-end controllers encrypt data traffic using symmetric algorithms, such as the data encryption standard (DES), triple DES, and IDEA. The alternative public-key algorithms are too slow, given today's processor speeds, to keep up with broadband cable data transmission rates, which are about 10 Mbps. In fact, even with the use of the speedier symmetric algorithms, CMs and head-end controllers are being built with specific hardware support for encryption.

CM security systems, however, do use public-key cryptography in the key management protocols they employ for securing and distributing symmetric traffic encryption keys. Their use will be discussed later in this paper.

Integrity

Integrity is a security service that guarantees modifications to transmitted frames are detectable. While link-layer protocols employ checksums to detect transmission bit errors, these checksums do not protect against intentional modification of messages by a malicious attacker. To address the integrity of the link-layer communications across a cable network, a keyed one-way hash function, also known as a message authentication code, is computed over the encrypted frame data and encryption header and transmitted with the encrypted frame data. A message authentication code is a key-dependent one-way hash function; the message data is combined with a secret key value, and the one-way hash is computed over the combination.

A malicious attacker attempting to forge or alter a transmission will be unable to generate a valid hash for the altered message without access to the secret key. The cryptographic strength of the one-way hash (e.g., MD5, SHA-1) will prevent the attacker from constructing a forged message with the same hash value, i.e., the digital "fingerprint," as the original.

A separate authentication header may accompany the message authentication code. The authentication header identifies the security association under which the message authentication code was calculated. Alternatively, one security header may identify a single SA shared by both encryption and integrity mechanisms.

Figure 2 depicts the cryptographic processing CMs and head-end controllers apply to message data when receiving or transmitting frames across their cable network interfaces. Prior to link-level forwarding across an RF network, packet data is first encrypted, and then a message authentication code is calculated across the encrypted packet data and encryption header and attached to the outgoing encrypted packet data. Note that encryption and message integrity may be employed separately or in tandem, depending on the security features a particular data-over-cable security system is addressing.

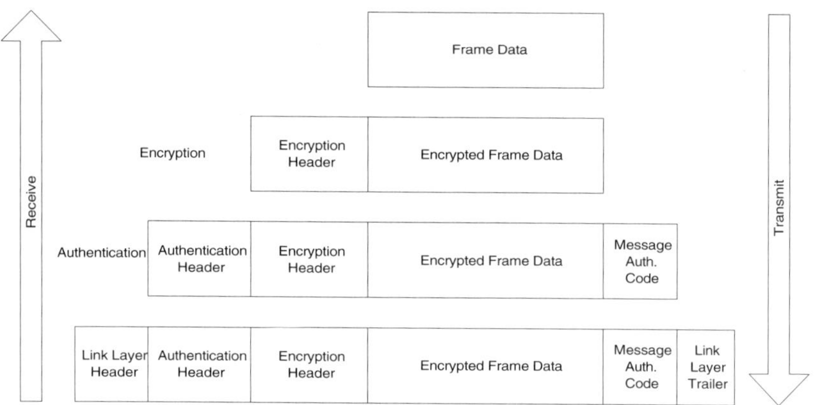

Figure 2: *Cryptographic Processing of Link Layer Frames*

Service Protection

Data-over-cable service providers use their cable network to deliver various data services to subscribers. Service protection allows providers to restrict access to these services to authorized users. This paper has described how encrypting link-layer payloads and employing message authentication codes can address the data privacy and integrity goals of a data-over-cable security system. Service protection is an outgrowth of link-layer encryption and message authentication, provided key distribution includes mutual authentication of the participating entities.

Symmetric encryption algorithms and keyed one-way hash functions both assume the participating entities (CMs and head-ends, in our case) share the appropriate sets of SA parameters. A data-over-cable security solution must include a key management protocol that ensures the secure distribution of SA parameters (i.e., encryption keys) to CMs and head ends.

If a head-end controller encrypts all downstream packet data, users will not have access to those transmissions unless their CMs have the keying material necessary to decrypt the packet data. Incorporating CM authentication into the key management protocol forces the verification of a CM's identity prior to sending the modem the specific keying material it is authorized to receive. A key management protocol supporting authentication uses public-key cryptography both to verify a CM's identity and to ensure that only the intended cable modem receives the keying material—a significant issue because keying material, like user data, is sent across the shared medium RF network. A service subscriber is paired with a uniquely identifiable CM, or with a removable security module the subscriber inserts into a CM (e.g., a smart card). A subscriber's access to the data that a head-end controller broadcasts across the RF network is restricted to the particular encrypted downstream traffic flows for which the subscriber's CM is authorized to retrieve.

Authenticated key distribution also permits the head-end controller to restrict its forwarding of the upstream transmissions it receives from CMs. If link-layer transmissions carry message authentication codes, and if distribution of the corresponding authentication keys is restricted to authenticated modems, the head end can authenticate the source of each upstream link-layer frame it receives across its RF network interface and discard the frames that fail authentication.

If link-layer transmissions are encrypted, but do not carry message authentication codes, forwarding beyond the RF network is still controlled. Authenticated key distribution ensures only authenticated CMs receive encryption keys. While an unauthenticated CM could transmit frames with encryption headers containing valid SA identifiers, the CM would not be able to obtain the keying material for those SAs. Without a message authentication code, the head-end controller cannot authenticate the transmission's origin; the head end's decryption of the frame would result in garbage data, which would subsequently be dropped within the switching logic due to unrecognizable packet addressing. The advantage in employing message authentication codes is that it catches the unauthorized transmissions at the earliest possible stage, and it prevents an attacker from injecting garbage data into the network beyond the head-end controller.

As a final comment, note that service providers have no control over what is attached to the RF network downstream from the head-end controller. Since the hybrid fiber/coax network is passive (other than the presence of amplifiers), the head-end controller is the first point in the upstream path at which to exercise access control. In particular, the controller's link-layer interface, running a protocol incorporating security mechanisms, is the first opportunity the network has to exercise access control.

Key Management

As is the case with any cryptography-based security system, key distribution is fundamental to the integrity of cable network security systems. Key distribution must not have weaknesses that would undermine the security of link-layer encryption and message integrity checks. If strong service protection is required (e.g., defense against the use of "cloned" or pirated modems, descramblers, etc.), the key management protocol must support cryptography–based CM authentication.

If data-over-cable security requirements call for privacy and integrity, but not strong service protection, the key management protocol need not support CM authentication. DOCSIS baseline privacy, developed by the MCNS consortium, is a cable network security solution designed to provide privacy, some degree of message integrity, and weak service protection based on the uniqueness of the CM MAC addresses. Its key management protocol does not employ cryptography–based CM authentication. We will discuss DOCSIS baseline privacy and its key management protocol in greater detail in the final section of this paper.

Key management is generally recognized as the most challenging component in the design of any communications security system. In this brief introduction to the topic, the paper will consider some of the principal issues influencing the design of key distribution systems for data-over-cable services.

Decoupling of the Key Management Protocol from Link Layer Security Mechanisms In the security architecture outlined in this paper, the key management protocol is coupled to the link-layer security mechanisms (i.e., encryption of frame data, attachment of message authentication codes) only through the SA identifier and the association's constituent parameters. The key management protocol establishes the SA and makes the SA's parameters available to the link-layer security protocol, which includes both the cryptographic algorithms and the protocol defining the specific application of these algorithms.

This separation of key management mechanisms from the link-layer security mechanisms allows replacement or upgrading of key management protocols without having to modify implementations of link-layer mechanisms. Note that within CM systems, link-layer security mechanisms are often implemented in hardware, while key management protocols are likely to be implemented in software. The flexibility that this decoupling supports allows a single CM hardware platform to operate with a variety of key management protocols and, thus, provide varying degrees of communications security, depending upon the particular key management protocol in use.

Reducing Key Exposure and Two Tiered Key Distribution The longer a particular traffic encryption key is employed, the greater the amount of data encrypted by that key, and the more attractive the key becomes as a target for cryptanalytic attack. Reducing a key's "exposure" reduces its value and defends against cryptanalytic attack. Two-tier key distribution schemes are frequently employed to reduce the exposure of any particular traffic encryption key. Rather than employ a long-term service key to encrypt data, the long-term key is a key encryption key that encrypts traffic keys. Traffic keys are regularly updated with new keys, and the key updates are sent encrypted with the long-term key encryption key. Since the traffic keys themselves are short compared to the frame data, and are pseudo-random numbers, the long term key receives little exposure, and it would not be subject to a known-plaintext or a chosen-plaintext attack.

Public Key Algorithms for Key Exchange and Authentication The key management protocols that CMs and head-end controllers employ to synchronize keying material rely on public-key algorithms, such as RSA or Diffie-Hellman, for securing key distribution. Public key algorithms permit the secure distribution of keys without establishing a previously arranged, shared secret. In two-tiered key distribution systems, a public key algorithm would secure distribution of the long-term key encryption key, which in turn would work with a symmetric encryption algorithm for securing periodic distribution of shorter-lived traffic encryption keys.

Public-key algorithms are also employed in digital signature mechanisms. A digital signature associates signed data with the owner of a specific private key, and allows an authenticator to verify that signature with the signer's public key. Digital signatures can authenticate the sender and recipient of the key management protocol exchanges that distribute a long-term key encryption key. Authenticating distribution of a long-term key encryption key ensures that subsequent exchanges of the shorter-term traffic encryption keys will only be made between authenticated entities.

MCNS DOCSIS Baseline Privacy

DOCSIS baseline privacy is one of two data-over-cable security standards developed by the MCNS consortium. The MCNS designed baseline privacy as a low-cost security solution providing privacy, some degree of message integrity, and weak service protection based on the uniqueness of CM MAC addresses. Baseline privacy does not provide service providers strong service protection from a sophisticated attacker employing a cable modem clone programmed to operate with a legitimate cable modem MAC address. The alternative, higher-end security solution provides strong protection through the use of digital certificates and signatures for authentication. The higher-end solution also specifies the use of tamper-resistant hardware to prevent malicious subscribers from extracting keying material from a CM and redistributing the pirated keys.

Baseline privacy consists of a protocol for encrypting link-layer frames across an RF network and a key management protocol for securing the distribution of encryption keys. The link-layer protocol defines a mechanism for encrypting link-layer frame data and a packet format for carrying encrypted data within RF MAC frames. CMs and cable modem termination systems (CMTS), MCNS's terminology for a head end controller, use baseline privacy's key management protocol to share and synchronize encryption keys in a manner that ensures a user's data privacy.

Below is a quick overview of baseline privacy's key management and link encryption protocols. While the overview skips over many significant details, it does outline the protocols' principal design components and illustrates a real-world application of the security concepts discussed in the previous sections.

Key Management Protocol Exchanges The baseline privacy key management (BPKM) protocol runs between CMs and CMTSs, and uses the MCNS MAC protocol's management messaging. The protocol adheres to a client/server model in which the CM is a BPKM client, requesting authorization and traffic keys, and the CMTS is a BPKM server, authorizing CMs and responding to CMs' requests for keys.

BPKM partitions key management into two distinct activities:

- CM authorization, in which a CM requests and obtains from the CMTS an authorization key and a list of SA identifiers the CM is authorized for, and
- Traffic encryption key retrieval, in which a CM requests and obtains current traffic encryption key data associated with a particular authorized SA.

The authorization key is the long-term key of a two-tiered key distribution system. The CMTS uses a long-term authorization key to secure its distribution of shorter-term traffic encryption keys to a CM.

BPKM exchanges begin with an individual CM sending an authorization request message to its head-end controller, the CMTS. The authorization request message includes the CM's MAC address and an RSA public key. The CM's MAC address serves as an identifier, linking the CM to a known service subscriber. The RSA public key is paired with a secret RSA private key known only to the requesting CM.

Upon receiving the authorization request, the CMTS checks the CM's authorization. If the advertised CM MAC address is linked to an authorized subscriber, the CMTS responds with an authorization reply containing a list of SA identifiers for which the CM is authorized to have keying material. The authorization reply also contains a long-term authorization key encrypted with the requesting CM's RSA public key. By encrypting the authorization key with the CM's public key, the CMTS ensures that only the requesting CM will be capable of decrypting and retrieving the long-term key. Note that the authorization exchange is not authenticated; the CMTS cannot assume the requesting modem is authentic, i.e., is not a MAC clone operated by a service pirate. Nevertheless, the CMTS knows that the authorization key it is distributing in its authorization reply is accessible to only a single CM: the CM owning the RSA private key paired with the RSA public key reported in the authorization request.

Upon receiving the authorization reply, the requesting CM sends individual key request messages to the CMTS, each message requesting a traffic encryption key associated with one of the SAs in which the CM is authorized to participate. Key request messages include the requesting CM's MAC address, uniquely identifying the requesting CM. The key request message also carries a message authentication code (HMAC SHA-1), which the CMTS uses to authenticate the source of the message. The key that calculates the message authentication code is derived from the authorization key obtained in the earlier authorization exchange. Validation of the key request's message authentication code verifies that the requesting modem is authorized, i.e., has a valid authorization key.

Upon validating a CM's key request, the CMTS responds with a key reply containing the traffic encryption key (TEK) associated with the SA identifier included in the validated key request. The CMTS DES encrypts the TEK with a long-term key encryption key, derived from the same authorization key that validates the key request. The CMTS also attaches a message authentication code to its key reply, allowing the requesting CM to validate the response.

Upon receiving the key reply, the CM decrypts the traffic encryption key and may begin using the TEK to encrypt and decrypt traffic data covered by the SA associated with that key. *Figure 3* illustrates these BPKM protocol exchanges.

The traffic encryption key that a CMTS provides to a CM has a limited lifetime. A CMTS's key reply messages include a remaining lifetime for the keying data delivered in the key reply. The CMTS manages its active traffic keys by continually generating new generations of keying data for each SA and flushing older generations of keying data at the end of their lifetimes. In baseline privacy, it is a CM's responsibility to ensure it has current keys. The CM does this by sending new key requests to a CMTS for the next generation of the SA's keying data prior to the expiration of the current generation. CMTSs and CMs distinguish between successive key generations by attaching a key sequence number to each key generation.

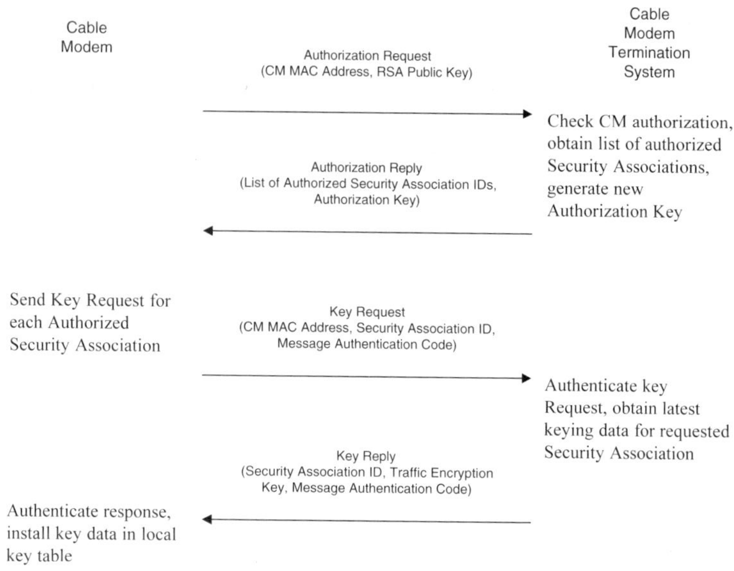

Figure 3: *Cable Modem Authorization and Traffic Encryption Key Retrieval*

A CM's authorization key, and the authorization status it implies, also has a limited lifetime. The CMTS includes in its authorization reply messages both the lifetime of an authorization and the authorization key. It is the responsibility of the CMs to re-authorize before their current authorization lapses. The CM reauthorizes by sending the CMTS a new authorization request prior to the expiration of its current authorization. The CMTS responds with a new authorization key, and the current list of SAs the re-authorizing CM is permitted to access. If a CM allows its authorization to expire, the CMTS will flush the associated authorization key, and reject any requests by that CM for traffic keys.

Link-Layer Encryption Protocol: After completing the initial BPKM authorization and key exchanges, CM and CMTS have synchronized keys and may begin securing (providing privacy for) user data communications. The CMTS encrypts downstream traffic, and the CM encrypts upstream traffic. Prior to encrypting user data packets, CMTS and CM must map the packets' addressing information to an SA and ensure the subsequent RF MAC frame carries a security header identifying the SA.

Baseline privacy data encryption services are defined as an extended service within the MCNS MAC sublayer. Packet header information specific to baseline privacy is located within an MCNS MAC extended header. The baseline privacy security header includes an SA identifier (the service ID or SID) and a 4-bit key sequence number to distinguish between an SA's successive generations of traffic keys. Together they uniquely identify a set of keying material for encrypting a particular link-layer traffic flow.

Figure 4 depicts the encryption and encapsulation of user packet data by CMs and head-end controllers under baseline privacy. The user packet data consists of an Ethernet frame. In the absence of any security processing, this frame would be encapsulated within an MCNS RF MAC frame for transparent transport across the RF network. When operating with baseline privacy, the Ethernet frame's type/length field, user data field, and CRC are encrypted; the frame's destination and source Ethernet addresses are left in the clear.

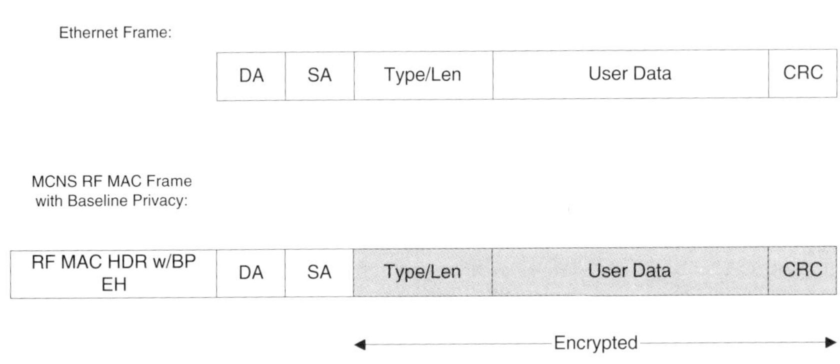

Figure 4: *Encapsulation of BP-Encrypted User Packet in MCNS RF MAC Frame*

Encrypting the Ethernet CRC provides some of the functionality of a message authentication code. After decrypting a received RF MAC frame, a CM or CMTS calculates the Ethernet user data's CRC and compares it with the decrypted CRC. If the encrypting device used the incorrect key, the CRCs would not match and the decrypted frame would be discarded. While encrypting the CRC guards against message forgery (any alterations of ciphertext will be detected) without requiring per-frame calculations of message authentication codes, it does introduce known dependencies within the cleartext that may be used in a cryptanalysis of the ciphertext. MCNS determined this tradeoff was justifiable, given baseline privacy's limited security objectives.

Baseline privacy employs the DES block cipher, in the cipher block chaining mode, for encryption of user frame data. Baseline privacy allows encryption hardware to be integrated directly within the RF MAC hardware interface. This is less costly, but less secure, than MCNS's higher-end security solution, which requires encryption operations to be done within a tamper-resistant security module.

Summary and Final Comments

Data-over-cable's use of a shared communications medium introduces security risks that could undermine the privacy and integrity of user's communications, and make the network susceptible to service piracy. The cable industry is addressing these security issues by incorporating cryptographic mechanisms for authentication and encryption into link-layer communications across the RF network. This paper identifies the principal security requirements associated with link-layer security for data communications across an HFC RF network, and discusses the particular ways in which cryptographic mechanisms can address these requirements.

The final section of the paper presents MCNS DOCSIS baseline privacy as an example of how security is being designed into cable-based data communications systems. The protocol implements many of the security mechanisms described previously. As is the case with any engineering solution, baseline privacy's design reflects tradeoffs of cost and functionality. Baseline privacy principally provides users with data privacy. It also provides cable service providers with service protection based on the uniqueness of a CM's MAC address, which, during provisioning, is registered with a particular subscriber. While this level of service protection prevents an unsophisticated hacker from gaining unauthorized access to data services, it is not intended to protect against a sophisticated hacker employing a MAC clone.

If baseline privacy were to support a stronger form of service protection, the key management protocol would have to employ strong, cryptography-based authentication mechanisms, which the protocol currently does not use. Baseline privacy represents a compromise: it foregoes the cost of incorporating digital certificates and digital signature mechanisms into the protocol's key exchanges, assuming that the use of MAC clones will not pose a significant threat (in terms of lost revenues) in the near future.

References

Gingold, D., "Integrated Digital Services for Cable Networks," Master's Thesis, Massachusetts Institute of Technology, 1996.

MCNS Baseline Privacy Interface Specification, SP-BPI-I01-970922, September 22, 1997.

Schneier, B., *Applied Cryptography* (2nd ed.), New York: John Wiley & Sons, 1996.

Acronym Guide

AAL ATM adaptation layer
ABR Available bit rate
ACI Adjacent channel interference
ADAP Adaptive digital access protocol
ADC Analog to digital converter
ADSL Asymmetric digital subscriber line
AFC Automatic frequency control
AFT Advanced file transfer
AGC Automatic gain control
ANSI American National Standards Institute
API Application programming interface
ARP Address resolution protocol
ARQ Automatic repeat request
ASIC Application specific integrated circuit
ATM Asynchronous transfer mode
AVT Audio/video transport
BBS Bulletin board system
BER Bit error rate
BOOTP Boot protocol
BOP Bit oriented protocol
BPI Baseline privacy interface
BPKM Baseline pprivacy key management
BRF Bandwidth request field
BSI Block sync interval
CATV Cable television
CBC Cipher block chaining
CBR Constant bit rate
CDF Cutoff decrease factor
CDMA Code division multiple access
CDN Cable data network
CDV Cell delay variation
CIDR Classless interdomain routing
CIR Committed information rate
CLP Cell loss priority
CM Cable modem
CMCI CM to customer premises equipment interface
CMOS Complementary metal oxide superconductor
CMTRI CM telephony return interface
CMTS Cable modem termination system
CMTS NSI CM termination system network side interface
CNR Carrier to noise ratio
CODEC Coder decoder or compression decompression
CP Connection point
CPE Customer premises equipment
CRA Collision resolution algorithm
CS Contention slot
CSMA/CD Carrier sense multiple access with collision detection
DAC Digital to analog converter
DAVIC Digital audio video council
DCO Digitally controlled oscillator
DDS Direct digital synthesis
DES Data encryption standard
DHCP Dynamic host configuration protocol
DMT Discrete multitone
DNS Domain name server
DOCSIS Data over cable service inter face specifications
DOCSIS TRI Data over cable service interface specification with telephony return interface
DQPSK Differentially coherent QPSK
DRF DOCSIS RFI specification
DS Data slots
DTR DOCSIS TRI specification
ECB Electronic codebook
ECN Engineering change notice
ECO Engineering change order
ECR Engineering change request
EOP End of procedure

ETSI European Telecommunications Standards Institute
F/TDMA Frequency and time division multiple access
FAMM Frequency Agile Multimode
FB Feedback
FCC Federal Communications Commission
FCFS First come first served
FDM Frequency division multiplexing
FDMA Frequency division multiple access
FEC Forward error correction
FEP Front end processor
FEQ Frequency domain equalizer
FF Feedforward
FFT Fast Fourier transform
FIFO First in first out
FIR Finite impulse response
FTP File transfer protocol
FTR First transmission rule
GDV Group delay variation
HC Head end controller
HDLC High level data link control
HEC Header error control
HFC Hybrid fiber/coax cable
IA Information appliance
IANA Internet Assigned Numbers Authority
IDEA International data encryptions algorithm
IDFT Inverse discrete Fourier Transform
IDOC Internet data over cable
IEEE Institute of Electrical and Electronic Engineers
IETF Internet Engineering Task Force
IF Intermediate frequency
IFDAC Intermediate frequency digital to analog converter
IGMP Internet group management protocol
IM Intermodulation
IP Internet protocol
IPCP Internet protocol control protocol
IPSEC IP security protocol
IPX/SPX Internet packet exchange/sequenced packet exchange
ISA Integrated services architecture or Interactive Services Association
ISDN Integrated services digital network
ISI Intersymbol interference
ISM Industrial scientific medical
ISP Internet service provider
ITFS Instructional television fixed service
ITU International Telecommunications Union
IUC Interval usage code
KSG Key stream generators
L2TP Layer 2 tunneling protocol
LAN Local area network
LIFO Last in first out
LPTV Low power television
LTBC Local timebase counter
MAC Media access control
MAN Metropolitan area network
MAP Manufacturing automation protocol
MAX_DATA Maximum number of DS in a frame
MCM Multicarrier modulation
MCNS Multimedia cable network sys tem
MDS Multipoint distribution service
MIB Management information base
MIC Message integrity check
MMDS Microwave multipoint distribution system

MP Multilink PPP
MPEG Motion Pictures Experts Group
MPU Minimum policed unit
MSO Multiple system operator
MSS Maximum segment size
MTU Maximum transmission unit
NAC Network access control
NAPT Network address port translator
NAS Network access server
NAT Network address translation
NCO Numerically controlled oscillator
NCTA National Cable Television Association
NIC Network interface card
NIST National Institute of Standards and Technology
NIU Network interface unit
NTSC National Television System Committee
OAM&P Operations, administration, maintenance, and provisioning
OSS Operation support system
OSSI Operation support system interface
PCS Personal communication service
PDU Protocol data units
PF Pointer field
PG Processing gain
PHY Physical
PLL Phase locked loop
PON Passive optical network
POP Point of presence
POTS Plain old telephone service
PPP Point to point
PPTP Point to point tunneling protocol
PSTN Public switched telephone network
PTI Payload type indicator
PVC Provisioned virtual circuit
QAM Quadrature amplitude modulation
QoS Quality of service
QPSK Quaternary phase shift keying
RAS Remote access server
RES Residential enhanced service
RF Radio frequency
RFC Request for comment
RFI Radio frequency interface
RFP Request for proposal
RMS Root mean square
RNG Random number generator
RS Reed Solomon Codes
RSMI Removable security module interface
RSVP Resource reservation protocol
RTCP Real time control protocol
RTP Real time protocol
SA Security association
SCAN Split channel asymmetric network
SCM Subscriber cable modem
SCTE Society of Cable Telecommunications Engineers
SDMT Synchronized discrete multitone
SID System identification number
SLIP Serial line interface protocol
SNMP Simple network management protocol
SNR Same number redial
SNR Signal to noise radio
SONET Synchronous optical network
SSI Security system interface
SSL Secure socket layer
STI Studio transmitter link
STL Studio transmitter link
STM Synchronous transfer mode
TB_CLOCK Timebase clock resolution

TB_COUNT Timebase count
TB_MARKER Timing tick in the form of a timing marker
TBL_MARKER Local time marker
TCD Telephony channel descriptor
TCM Trellis coded modulation
TCP/IP Transmission control protocol/Internet protocol
TDM Time division multiplexing
TDMA Time division multiple access
TEK Traffic encryption key
TFS Television fixed service
TFTP Trivial file transfer protocol
TH Tomlinson Harashima
ToS Type of service
TRAC Telephony remote access con centrator
TRO Temporary restraining order
TRTS Telephone return termination system
UCD Upstream channel descriptor
UDP User datagram protocol
UHF Ultra high frequency
UW Unique word
VBR Variable bit rate
VCI Virtual channel identifier
VHF Very high frequency
VL Variable length
VPN Virtual private network
VSB Vestigial sideband
WAN Wide area networks
WCS Wireless communication service
WIN Maximum window size

Index